Molecules in Electromagnetic Fields

Molecules in Electromagnetic Fields

From Ultracold Physics to Controlled Chemistry

Roman V. Krems

WILEY

Registered Office
John Wiley & Sons, Inc., 111 River Street, Hoboken, NJ 07030, USA

Editorial Office
111 River Street, Hoboken, NJ 07030, USA

For details of our global editorial offices, customer services, and more information about Wiley products visit us at www.wiley.com.

Wiley also publishes its books in a variety of electronic formats and by print-on-demand. Some content that appears in standard print versions of this book may not be available in other formats.

Library of Congress Cataloging-in-Publication Data

Names: Krems, Roman V., 1977- author.
Title: Molecules in electromagnetic fields : from ultracold physics to
controlled chemistry / by Roman V. Krems.
Description: 1st edition. | Hoboken, NJ : John Wiley & Sons, 2019. | Includes
index. |
Identifiers: LCCN 2017053373 (print) | LCCN 2017060297 (ebook) | ISBN
9781119387350 (pdf) | ISBN 9781119387398 (epub) | ISBN 9781118173619
(cloth)
Subjects: LCSH: Molecules. | Electromagnetic fields.
Classification: LCC QC173.3 (ebook) | LCC QC173.3 .K74 2019 (print) | DDC
539/.6–dc23
LC record available at https://lccn.loc.gov/2017053373

Cover design by Wiley
Cover image: © Roman Krems

Set in 10/12pt Warnock by SPi Global, Chennai, India

Printed in the United States of America

V097253_052318

Contents in Brief

Contents

List of Figures

List of Tables

Preface

Much of our knowledge about molecules comes from observing their response to electromagnetic fields. Molecule–field interactions provide a "lense" into the microscopic structure and dynamics of molecules. Molecule–field interactions also provide a knob for controlling molecules.

The focus of much recent research has been on controlling the *translational* and *rotational* motions of molecules by tunable fields. This effort has transformed molecular physics. New experimental techniques – unimaginable 10 to 20 years ago – have been introduced. We have learnt to interrogate molecules and molecular interactions with extremely high precision. Most importantly, this work has allowed – and stimulated! – us to ask new questions and has built bridges between molecular physics and other areas of physics.

For example, the interrogation of translationally controlled molecules is now considered to be the most viable route to determining the magnitude of the electric dipole moment of the electron. The outcome of such experiments may restrict, and maybe even resolve, the debate about the extensions of the Standard Model of particle physics. Interfacing with a completely different field, molecules trapped in optical lattices can be used as quantum simulators of a large variety of lattice spin models. Probing the structure and dynamics of such molecules is expected to identify the phases of many lattice spin models that are currently either unknown or under debate. Spectroscopy measurements of molecules in external field traps approach the fundamental accuracy limit of the nonrelativistic quantum mechanics, prompting quantum chemists to reconsider their toolbox for molecular structure calculations.

The effort aimed at controlling the three-dimensional motion of molecules has resulted in many unique experiments. Molecules can be spun by cleverly crafted laser fields all the way until the centrifugal force pulls the nuclei apart, breaking chemical bonds. This provides potentially new sources of radiation and singular quantum objects – "superrotors" – for the study of collision physics and kinetics of chain reactions. Slow molecular beams, which can be grabbed and guided by external fields, provide new unique opportunities to study chemical encounters with extremely high control over the collision energy. The isolation of molecules from ambient environments into samples maintained at an ultracold temperature opens the pathway to studying chemistry near absolute zero, controlled chemistry and, as argued later in this book, a conceptually new platform for assembling complex molecules.

At the core of all this exciting research is the manipulation of molecules by external fields. This book is an attempt to describe basic quantum theory needed to understand and compute molecule–field interactions of relevance to the abovementioned work. The formal theory is accompanied with examples from the recent literature, predictions about what may be happening next, and the discussion of current and future problems that need to be overcome for new major applications of molecules in electromagnetic fields.

Goals of This Book

This book is intended to serve as a tutorial for students beginning research, experimental or theoretical, in an area related to molecular physics. The focus of the book is on theoretical approaches of relevance to the recent exciting developments in molecular physics briefly mentioned above.

There are two kinds of chapters in this monograph:

- Chapters 1–4 and 8–10 are written to provide a detailed summary of rigorous theory of molecule–field interactions, quantum scattering theory, and the theory of scattering resonances. The goal of these chapters is practical – to provide enough details that will allow the reader to write computer codes for calculating the rotational energy levels of molecules, the AC and DC Stark shifts of molecules, the Zeeman shifts, molecular collision observables, and the features of field-induced scattering resonances. These chapters also gradually introduce angular momentum and spherical tensor algebra with the application to problems involving molecules in electromagnetic fields. Each of the chapters is largely self-contained. I attempt to present the complete derivations of all important equations so the material in these chapters requires little more than basic knowledge of quantum theory.
- Chapters 5–7 and 11–12 illustrate the applications of field-induced interactions for controlling the motion of molecules in three-dimensional space, trapping molecules, controlling molecular collisions, and ultracold controlled chemistry. The purpose of these chapters is to illustrate the extent and power of field control of microscopic behavior of molecules. The discussion in these chapters is more descriptive, with references to relevant sources in the recent literature. However, this discussion relies in many places on equations derived in Chapters 1–4 and 8–10.

There are also five appendices at the end of the monograph. The main purpose of the appendices is to provide the theoretical background for understanding the details in Chapters 1–4 and 8–10. I hope that these appendices can serve as mini-tutorials on angular momentum algebra, coordinate rotations, and spherical tensors.

What This Book Is Not

I have recently heard a famous scientist and author saying that "one never finishes a book; one abandons the writing." I have now experienced this myself. There are many things I would have liked to do for this book, which I must abandon. To ensure that I do not mislead the reader, let me point out the following.

- This book is not a comprehensive review. It would be impossible to cite all relevant papers. The references presented are isolated, sample articles, which should direct the reader either to the pioneering work or some of the most widely cited work. If your paper has not been referenced, and you think it should have, please forgive me.
- This book is not a complete account of theories used by the practitioners in the field. There are many topics related to molecule–field interactions and the corresponding theoretical approaches that have been left out. For example, there is no discussion of beautiful work on atto-second spectroscopy or coherent control of intramolecular dynamics. Other topics had to be left out due to space and time constraints.
- This is not a standard textbook. Some of the approaches taken in this monograph are unconventional, reflecting the author's preferences and views. The hope is that the reader will benefit as much from the discussion of the approaches and the derivations of the intermediate steps as from the final equations.

How to Read This Book

Most of the chapters in this monograph can be read independently. Those interested in specific calculations of specific field effects should be able to simply peruse the corresponding isolated chapter.

If the algebra in Chapters 1–4 and 8–10 appears unfamiliar, it may be useful first to study the appendices. The appendices provide the background on angular momentum algebra required for understanding the details of these chapters. Some of the derivation details are left to the reader as exercises that appear at the end of the corresponding chapters. The exercises are usually accompanied with hints or references to other books.

Chapters 5–7 and 11–12 are largely independent and require little background knowledge. Some of the material in these chapters are based on our recent review article [1], which provides a more comprehensive list of references than the present monograph.

Most importantly, the reader should be critical of every statement and equation presented in this monograph.

Vancouver, British Columbia Roman V. Krems
August 2017

Acknowledgments

This book is a testament to the patience and support of my family. I am greatly indebted to my wife, Zhiying Li, who has not only allowed me to work on this manuscript, but also immensely helped by rederiving many of the equations, proofreading many of the pages, and offering constructive, albeit at times hurtful, critique. I am also very thankful to my students and colleagues, particularly, Fernando Luna and Rodrigo Vargas, who have read most of the chapters and provided useful feedback, and Chris Hemming for the notes that were used to prepare a part of Chapter 10. This project was initiated by a letter from a Wiley consulting editor Edmund Immergut. I would like to thank Ed sincerely for his trust in me and for his cheerful yearly reminders prompting me to continue the work. I would also like to acknowledge Wikipedia that I have had to consult on more than a few occasions to calibrate my understanding of the subject matter.

1

Introduction to Rotational, Fine, and Hyperfine Structure of Molecular Radicals

A diatomic molecule is a molecule with one atom too many.
—Arthur Shawlow, 1981 Physics Nobel Laureate

1.1 Why Molecules are Complex

A molecule is an ensemble of positive and negative charges. This is a blessing and a problem. It is a blessing because the interaction forces between positive and negative particles are all known so an accurate quantum theory of molecular structure can, in principle, be formulated. We can write the exact nonrelativistic Hamiltonian for any molecular system. In this sense, quantum chemistry is more fortunate than, for example, nuclear physics. The trouble is that there are a lot of particles in any given molecule – too many to allow the exact solutions of the Schrödinger equation. Precise numerical solutions of the Schrödinger equation can be obtained for a system of three (and even a whooping four!) particles. However, molecular physics is not just about H_2^+ (three particles) or H_2 (four particles). In order to compute the properties of molecules with more than two electrons, we must rely on tricks, such as the Born–Oppenheimer approximation, and approximate numerical techniques.

A diatomic molecule is the simplest kind of molecules. There are many books written about the structure of diatomic molecules. My favorites are "The Theory of Rotating Diatomic Molecules" by Mizushima [2], "Perturbations in the Spectra of Diatomic Molecules" by Lefebvre-Brion and Field [3], and "Rotational Spectroscopy of Diatomic Molecules" by Brown and Carrington [4]. A glance at the introductory pages of the Mizushima's book is sufficient to tell that the subject is far from simple. So, what makes the diatomic molecules complex?

First of all, a diatomic molecule is much more than just two atoms. An atom possesses spherical symmetry: the electrons are moving in a centrally symmetric potential of the nucleus. This symmetry is broken when two atoms are brought together. Symmetries can be used to reduce the dimension of

Molecules in Electromagnetic Fields: From Ultracold Physics to Controlled Chemistry, First Edition. Roman V. Krems.

the Hilbert space required to solve a quantum mechanical problem. That's one reason to like symmetries! Broken symmetry means fewer conserved quantum numbers and bigger Hilbert spaces, which leads to more complex computations or makes the computations impossible. Second, there are two heavy particles in a diatomic molecule (as opposed to one in an atom). More particles of similar mass usually lead to more coupled degrees of freedom. This results in many interactions that have similar magnitudes and that must be considered simultaneously.

A fundamental theory of molecules must ensure that the description of the electrons is consistent with the special theory of relativity. This can be achieved within the framework of the Dirac equation that describes a relativistic spin-1/2 particle [4]. The Dirac equation is, however, not without problems. First of all, it cannot be properly generalized to an ensemble of more than two particles. Second, it deals with both the electron and the positron and we, in molecular physics, are usually not interested in the latter. The workaround is to derive an effective Hamiltonian, which can be inserted into the Schrödinger equation, by a transformation of the Dirac equation that separates the electron and positron subspaces. For a simple molecule, such Hamiltonian may have to include the electron kinetic energy with relativistic corrections, the Coulomb energy, the Darwin correction to the Coulomb energy, the spin–orbit interaction, the spin–other–orbit interaction, the orbit–orbit interaction, and the spin–spin interaction [4]. That is a lot of interactions to deal with! And this does not include the interactions of the electrons with the nuclear spins or the interactions stimulated by the motion of the nuclei.

The rotational motion of a diatomic molecule is an extremely complex process because it affects the entire system of the electrons and the nuclei. It induces a plethora of interactions that perturb the molecular energy levels in all kinds of ways. These interactions can be cast in the language of angular momentum theory. The rotation of a rigid body is classically described by an angular momentum. The angular momentum for the rotational motion of a diatomic molecule is a vector sum of the spin and orbital angular momenta of the individual electrons, the spin angular momenta of the nuclei, and the orbital angular momentum of the nuclei. That is a lot of angular momenta to deal with!

The complexity increases when more than two atoms join together to form a polyatomic molecule. The diatomic molecules possess a cylindrical symmetry, which means that the rotation of the molecule about the axis joining the two atoms must leave the properties of the molecular system unchanged. When three atoms form a nonlinear triatomic molecule, this symmetry is broken. Nonlinear triatomic molecules may possess a C_{2v} symmetry. This symmetry is broken when four atoms form a nonplanar molecule. When the number of atoms in a molecule is large, it is often no longer possible to treat the molecule as a single closed quantum system, and the quantum description may involve having to treat a part of a molecule as a quantum bath. Fortunately, for the reader and the writer, most of this book is about diatomic molecules.

1.2 Separation of Scales

It is clear from the above that it is impossible to compute the energy levels of a diatomic molecule exactly. Given that a molecule is a many-body quantum system, is it feasible at all to obtain an accurate description of molecular properties? Fortunately, nature offers help. The theory of molecular structure is based on *the separation of scales*. We can think in terms of different *time* scales or different *energy* scales.

From the dynamical point of view, the electrons – the lightest particles in a molecule – move the fastest. The typical timescale for an electron to traverse a molecule is on the order of one atomic unit of time or $\sim 2.4 \times 10^{-17}$ s. When given a chance, the electrons can jump between molecules in as little as ~ 10 atomic units of time [5]. The massive nuclei are much more sluggish than the electrons, moving over a distance comparable with the size of a molecule in about $\sim 10^{-12}$ to 10^{-14} s. This is so much slower than the timescale of the electron motion that the electrons adjust instantaneously to any change in the nuclear configuration. Therefore, we can freeze the nuclei and compute the energy of the electrons and repeat this computation at different fixed positions of the nuclei, thus obtaining an electronic energy curve (for diatomic molecules) or surface (for polyatomic molecules).

It is important to keep in mind that the electronic energy curves are not observable. The separation of the electronic and nuclear degrees of freedom is just a maneuver that allows us to develop a hierarchical model of molecular energy levels and simplifies the computations. If we could, we would solve the Schrödinger equation in one step, treating nuclei and electrons on an equal footing. This would produce the eigenvalues of the full molecular Hamiltonian, which would represent the internal energies of the molecule. These energies are observable in a spectroscopic experiment. The purpose of the hierarchical theory of molecular structure discussed here is to solve the Schrödinger equation in steps, leading to energies that approximate the exact eigenvalues of the full Hamiltonian.

An electronic energy curve serves as a potential that determines the vibrational motion of the nuclei. The nuclei roll in the electronic potentials like a marble in a spoon…well, not exactly because the nuclei are quantum particles (while the marble is not), but this analogy is useful to appreciate the role of the electronic potential. The time of 10^{-12}–10^{-14} s is the typical timescale of a vibrational period as the nuclei move back and forth in the electronic potential.

The rotational motion of the nuclei is even slower than the vibrational motion. As a typical molecule undergoes one round of rotation, its nuclei complete multiple rounds of vibration. Therefore, if we are interested in the rotational motion only, instead of describing a molecule as a vibrating rotor, we can often treat it as a rigid rotor with a size representing an average over the vibrational motion. The typical timescale of a rotational period is $\sim 10^{-8} - 10^{-11}$ s.

Since time and frequency are conjugate variables, shorter-time periodic dynamics corresponds to larger frequencies (ω), and hence larger energy gaps ($\hbar\omega$) between quantum states. Therefore, it is often convenient to think about molecular structure in terms of the separation of energy scales. The energy required to excite electrons from their lowest-energy state to their lowest-excited state sets the largest energy scale in a molecule. The energy of the vibrational motion of the nuclei is a perturbation to the electronic energy. In other words, for each electronic state, there are multiple vibrational states (these are the quantized motional states of the quantum marble in the electronic potential), and the energy splittings between the vibrational states are much smaller than the energy gap between different electronic states. In turn, the vibrational energy is significantly greater than the energy of the rotational motion of the molecule. For each vibrational state, there are multiple rotational states, and the energy splittings between the rotational states are much smaller than the energy gaps between different vibrational states. Therefore, we can treat the rotational energy as a perturbation to the vibrational energy, which is a perturbation to the electronic energy. This hierarchical structure of the energy scales is illustrated schematically in Figure 1.1.

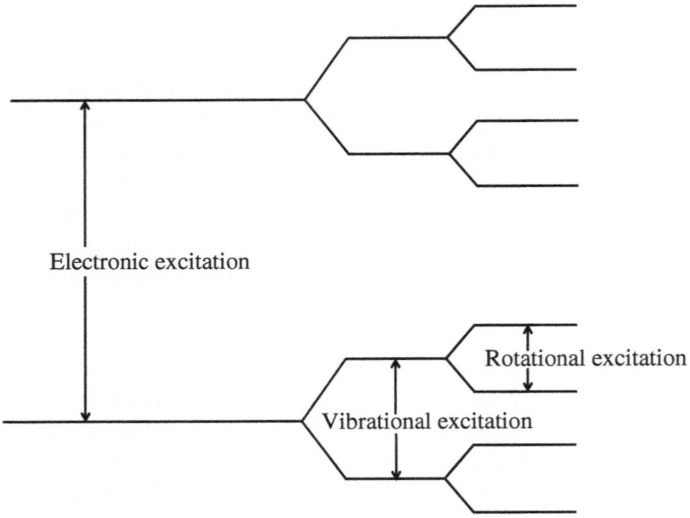

Figure 1.1 Schematic diagram (not to scale) of the hierarchical structure of the electronic, vibrational, and rotational energy levels for a typical molecule.

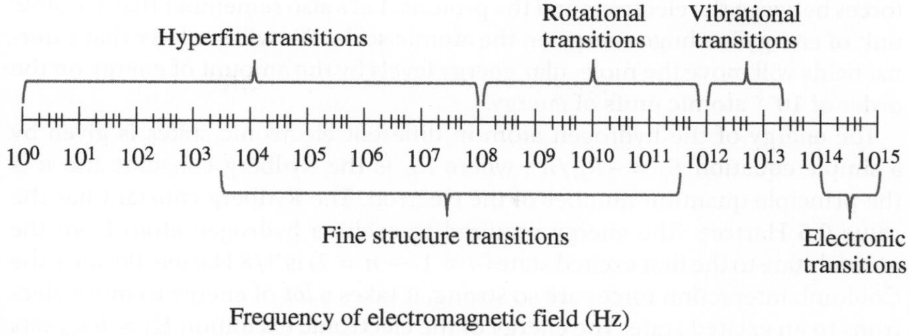

Figure 1.2 Typical frequencies of electromagnetic field (in Hz) required to excite hyperfine, fine structure, rotational, vibrational, and electronic transitions in diatomic molecules.

There are also fine structure interactions, which arise, for example, due to the magnetic interactions between unpaired electrons. And then, there are hyperfine interactions, which occur, for example, due to the magnetic field generated by the atomic nuclei. I wish the fine structure interactions could be treated as a perturbation to the rotational energy and the hyperfine interactions to the fine structure energy. For many molecules, this is indeed the case. However, for many other molecules, these interactions are of similar magnitude so they have to be treated at the same order of perturbation theory. Figure 1.2 gives an overview of the hierarchical structure of molecular energy levels, presented on the scale of electromagnetic field frequencies that excite the corresponding transitions. In the following few sections, we will zoom in on each part of this energy spectrum.

1.2.1 Electronic Energy

The energy of the Coulomb attraction between elementary particles in a molecule is huge! To see this, let's calculate the interaction energy between an electron and a proton separated by 1-Bohr radius – a distance about the size of the hydrogen atom in the ground state:

$$U = \frac{1}{4\pi\epsilon_0}\frac{q_1 q_2}{r} = -8.98755 \times 10^9 \times \frac{(1.60217 \times 10^{-19})^2}{0.529177 \times 10^{-10}} = 4.3597 \times 10^{-18}\ \text{J}.$$

Expressed in atomic units of energy (see Appendix A), this is equal to 1 Hartree. The energy of the hydrogen atom in the ground state is actually 0.5 Hartree. Imagine now, hypothetically, that we form 1 g of hydrogen atoms from a mixture of noninteracting protons and electrons. This will release energy equal to about 3×10^{23} Hartree, which is sufficient to lift an object weighing 100 tons by 1 m (or lift an economy-size car up by 100 m). We learn from this that there is a lot of energy stored in 1 g of hydrogen atoms due to the enormous attraction forces between the electrons and the protons. Let's also remember that 1 atomic unit of energy is a huge energy on the atomic scale. We will see later that external fields will move the molecular energy levels by the amount of energy on the order of 10^{-6} atomic units of energy.

The energy of the hydrogen atom in different electronic states is given by a simple equation $E_n = -R_H/n^2$, where R_H is the Rydberg constant and n is the principle quantum number of the electron. The Rydberg constant has the value 0.5 Hartree. The energy required to excite a hydrogen atom from the ground state to the first excited state ($n = 1 \rightarrow n = 2$) is 3/8 Hartee. Because the Coulomb interaction forces are so strong, it takes a *lot* of energy to move electrons to an excited state. The energy of the electronic excitation $E_{\rm el} = \hbar\omega_{\rm el}$ sets the largest energy scale for the description of molecular structure. It is customary to express this energy in electron volts (eV). For most atoms in the ground state, the excitation energy $\hbar\omega_{\rm el}$ to the lowest-energy excited electronic state is between 2 and 6 eV.

When two atoms (call them atoms A and B) are brought together, the electronic energy levels of each of the atoms are perturbed by the interactions of the electrons of atom A with the electrons and the nucleus of atom B and vice versa. It is useful first to consider the atoms separated by a large but finite distance R. Since the atoms are charge neutral, each electron of atom A interacts with a charge neutral ensemble of charges (atom B) and vice versa. At the same time, the electrons of atom A are much more distant from the charges in atom B than from the charges in atom A. Therefore, at large R, the interatomic interactions are necessarily much weaker than the Coulomb interactions within the atoms. At large R, the interatomic interactions simply shift the electronic energies of the atoms by a small amount.

As the atoms come closer together, the molecule must be treated as an ensemble of electrons moving in the combined Coulomb field of the two nuclei. Since the electrons are moving much faster than the nuclei, one can fix the nuclei, calculate the energy of the electrons by solving the Schrödinger equation, and repeat this procedure at different values of the internuclear separation R. If the nuclei are fixed, the Hamiltonian of the molecule is

$$\hat{H}^{\rm el}_{A-B} = -\sum_i \frac{\hbar^2}{2m_e}\nabla_i^2 - \frac{1}{4\pi\epsilon_0}\sum_i \frac{Z_A e}{r_{A,i}} - \frac{1}{4\pi\epsilon_0}\sum_i \frac{Z_B e}{r_{B,i}} + \frac{1}{4\pi\epsilon_0}\sum_i\sum_{j>i}\frac{{\rm e}^2}{r_{ij}} + \frac{1}{4\pi\epsilon_0}\frac{Z_A Z_B}{R}. \tag{1.1}$$

This Hamiltonian describes an ensemble of electrons each with charge $-e$ and mass m_e placed in the Coulomb field of two nuclei carrying the charges Z_A and Z_B separated by a *fixed* distance R. The variables $r_{A,i}$, $r_{B,i}$, and r_{ij} specify the distance between ith electron and nucleus A, ith electron and nucleus B, and between two electrons labeled i and j. The internuclear separation R enters the Hamiltonian (1.1) as a constant parameter.

Solving the Schrödinger equation amounts to solving the eigenvalue problem

$$\hat{H}^{\text{el}}_{A-B}|n\rangle = E_n|n\rangle, \tag{1.2}$$

where the eigenstates $|n\rangle$ represent the electronic states and the eigenvalues E_n – the electronic energies of the molecule AB. Since the Hamiltonian changes as R is changed, both the eigenvectors $|n\rangle$ and the eigenvalues E_n depend on R. If the electronic energies E_n of the molecule are plotted as functions of R, from small to large values of R, it is often (although not always) possible to correlate a particular electronic state of the molecule AB with a particular electronic state of the isolated atoms. See Figure 1.3 for an example.

Figure 1.3 shows the energies of the electronic states of the OH molecule formed by an oxygen atom in two lowest-energy electronic states (labeled ^3P and ^1D) and a hydrogen atom in the ground electronic state, computed at different values of R. The limit of $R = \infty$ describes two isolated atoms. Figure 1.3 reflects the characteristic shape of the R dependence of the molecular electronic energies and illustrates two important points. First, the

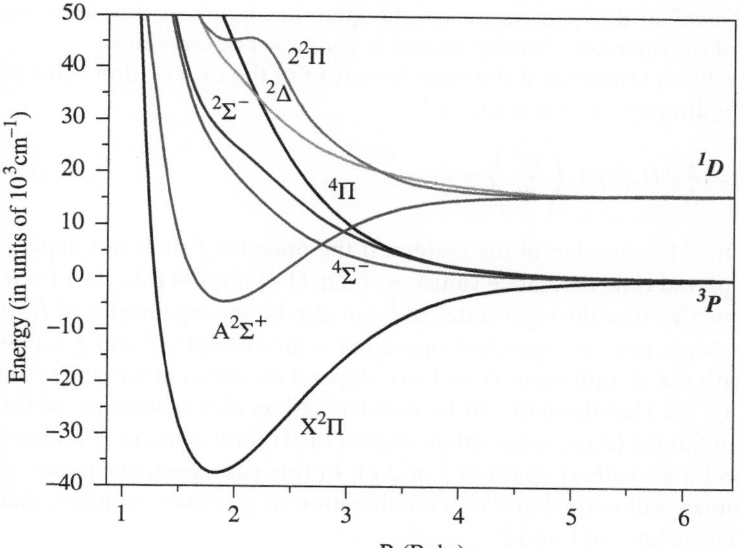

Figure 1.3 The electronic potentials of the molecule OH arising from the interaction of the oxygen atom in two lowest-energy electronic states labeled ^3P and ^1D with the hydrogen atom in the ground electronic state. The curves are labeled using standard spectroscopic notation [3]. In particular, the symbol X is the standard label for the lowest-energy (ground) electronic state of the molecule. These potential energies were calculated in Ref. [6]. *Source:* Adapted with permission from Krems et al. 2006. [7]. © 2006, American Astronomical Society.

interatomic interactions can lower the energy of the electronic states of the atoms by up to a few eV, thus producing stable molecules. This is the typical energy of a chemical bond. Second, the interaction of atoms in particular electronic states gives rise to multiple nondegenerate electronic states of the molecule. Exercise 1.7 will ask you to determine the number of nondegenerate electronic states that arises from the interaction of atoms A and B in particular electronic states.

It is important to remember that the Hamiltonian (1.1) describes a molecule with "clamped" nuclei. The *full* nonrelativistic Hamiltonian of the two-atom system is the sum of the Hamiltonian (1.1) and the kinetic energy of the nuclei:

$$\hat{H}_{A-B} = -\frac{\hbar^2}{2m_A}\nabla_A^2 - \frac{\hbar^2}{2m_B}\nabla_B^2 + \hat{H}^{\text{el}}_{A-B}, \tag{1.3}$$

where m_A and m_B denote the mass of the nuclei A and B.

The curves in Figure 1.3 are labeled by strange-looking symbols. What are they? Just as we did in Eq. (1.2), we could have labeled the eigenvalues (and the corresponding eigenvectors) of the Hamiltonian (1.1) for the OH molecule by some generic quantum number n. However, instead of using a generic label n, it is more meaningful to label each molecular state by a set of *good* quantum numbers. "*Good*" in this context means the quantum numbers that label the eigenstates of the operators corresponding to *conserved* observables.

An observable is *conserved* if the time derivative of the expectation value of the corresponding operator is zero, i. e.[1]

$$\frac{\mathrm{d}\langle\hat{f}\rangle}{\mathrm{d}t} = \frac{i}{\hbar}\langle[\hat{H},\hat{f}]\rangle + \left\langle\frac{\partial\hat{f}}{\partial t}\right\rangle = 0, \tag{1.4}$$

where $\hat{H}$ is the Hamiltonian of the system. If the operator $\hat{f}$ does not depend on time, its partial time derivative vanishes so Eq. (1.4) implies that $[\hat{H},\hat{f}] = 0$, and, consequently, that the eigenstates of $\hat{f}$ can also be the eigenstates of $\hat{H}$. If there are multiple time-independent operators – for example, $\hat{f}$ and $\hat{g}$ – that commute with the Hamiltonian $\hat{H}$ and that depend on different variables, the eigenstates of the Hamiltonian can be constructed as *direct products* of the eigenstates of $\hat{f}$ and $\hat{g}$ (times some other states if the Hilbert space of $\hat{H}$ is larger than the combined Hilbert space of $\hat{f}$ and $\hat{g}$). In this case, each eigenstate of the Hamiltonian will correspond to a combination of quantum numbers that specify the eigenstates of $\hat{f}$ and $\hat{g}$.

So, when solving a quantum mechanics problem, one of the first questions we should ask is, what is the maximum set of independent operators that commute with the Hamiltonian?

---

1 This well-known equality can be easily derived by noting that $\mathrm{d}\langle\hat{f}\rangle/\mathrm{d}t = \partial\langle\hat{f}\rangle/\partial t$, differentiating $\langle\hat{f}\rangle = \langle\psi|\hat{f}|\psi\rangle$ as a product of three terms and using the time-dependent Schrödinger equation and its complex conjugate for the time derivatives of $|\psi\rangle$ and $\langle\psi|$.

There are, at least, two reasons to do this. First, the operators $\hat{f}$ and $\hat{g}$ generally depend on fewer variables than the Hamiltonian; so it is generally easier to solve the eigenvalue problem with $\hat{f}$ or $\hat{g}$ than with $\hat{H}$. In many cases, especially if $\hat{f}$ or $\hat{g}$ is composed of angular momentum operators, the eigenstates of these operators can be found analytically. Second, if the operators $\hat{f}$ and $\hat{g}$ correspond to physical observables, the labels of their eigenstates contain information on the physical properties – such as symmetry with respect to the inversion of the coordinate system – of the eigenstates. Writing the eigenstates of the Hamiltonian as products of states with known symmetry properties will go a long way in our analysis of the matrix elements induced by external field perturbations.

Let us return to Figure 1.3, i.e. the problem with the Hamiltonian (1.1). Can we identify a set of independent operators that commute with this Hamiltonian? A good place to start is by examining symmetries and the consequences of these symmetries on conservation, or lack thereof, of angular momenta. In classical mechanics, the conservation of total angular momentum for an ensemble of bodies is a consequence of the isotropy of space. In other words, if a three-dimensional rotation of an ensemble of particles does not change their energy, the total angular momentum of the ensemble is conserved. Similarly, if a rotation of an ensemble of particles *about an axis* does not change their energy, the *projection* of the total angular momentum on this axis is conserved. In this case, we talk about an *axis of symmetry.*

We may associate an orbital angular momentum $\boldsymbol{l}_i$ and a spin angular momentum $\boldsymbol{s}_i$ with each electron in the molecule. The total electronic orbital and spin angular momenta can then be defined as $\boldsymbol{L} = \sum_i \boldsymbol{l}_i$ and $\boldsymbol{S} = \sum_i \boldsymbol{s}_i$, where the sums extend over all electrons in the molecule. The total orbital angular momentum of an ensemble of particles in a spherically symmetric potential is conserved.[2] However, there is no spherical symmetry in a molecule. The electrons are in the Coulomb field generated by two spatially separated nuclei. Therefore, the total orbital angular momentum of the electrons is *not* conserved. This means that the Coulomb interactions couple states corresponding to different values of L, so we cannot use L to label the molecular states. The story is different for the spin angular momentum. The nonrelativistic Hamiltonian given by Eq. (1.1) does not contain any spin variables. Therefore, $\boldsymbol{S}^2$ commutes with the electronic Hamiltonian (1.1). This means that the total spin is conserved, and each molecular state corresponds to a well-defined value of S.

The internuclear axis is obviously an axis of symmetry in a diatomic molecule. If the electrons are collectively moved around the internuclear axis, their energy must remain unchanged. The projection of $\boldsymbol{L}$ on the internuclear

---

2 That is why the energy levels of atoms can be labeled by the quantum number of the total orbital angular momentum, hence the notation ^{2}S and ^{3}P we used for the hydrogen and oxygen atoms. If this notation is unfamiliar, consult Ref. [8].

axis is commonly denoted by Λ and the projection of $\boldsymbol{S}$ by Σ. Both Λ and Σ are conserved. The conservation of Λ is a consequence of the axial symmetry of the diatomic molecule. Σ is conserved because the Hamiltonian (1.1) is spin independent.

We can thus define the electronic states of a diatomic molecule as

$$|n\rangle \equiv |\eta, S, \Lambda, \Sigma\rangle, \tag{1.5}$$

where η is introduced to label the states with the same values of S, Λ, and Σ but with different energies. The molecular states with $\pm\Lambda$ are degenerate, but the molecular states corresponding to different values of $|\Lambda|$ have different energies. The molecular states with different values of Σ are degenerate. Thus, for each S, there are $2S + 1$ degenerate nonrelativistic electronic states. In the spectroscopic literature, the molecular potentials are labeled using Greek letters[3] Σ, Π, Δ corresponding to the values of $|\Lambda| = 0, 1, 2$, etc. The standard notation also includes the superscript in front of the Greek letter that indicates the spin multiplicity $(2S + 1)$ of the molecular potentials. For example, the label $^2\Pi$ in Figure 1.3 denotes an electronic state with $|\Lambda| = 1$ and $S = 1/2$. The prefactor X is used to denote the ground electronic state.

It is important to keep in mind that the labels S, Λ, and Σ are good only when we treat a molecule as an ensemble of electrons moving around two nuclei *fixed in space*. This is not a complete description of a molecule. For the complete description, we need to consider the vibrational and rotational motion of the nuclei, i.e. we need to consider the problem with the Hamiltonian (1.3). The first two terms in Eq. (1.3) may induce couplings between the electronic states with different values of Λ or Σ. We will consider the vibrational and rotational motions of the molecule in the next sections.

1.2.2 Vibrational Energy

We froze the atomic nuclei in order to compute the eigenenergies of the electronic Hamiltonian presented in Figure 1.3. This is an approximation. Unfreezing the nuclei gives rise to perturbations that couple different electronic states of the molecule and, in principle, spoils the quantum labels in Eq. (1.5). These perturbations arise from the relative radial motion of the nuclei and the rotational motion of the internuclear axis. For most diatomic molecules, these perturbations are generally weak – often on the order of a few cm^{-1}, at best. It can be seen from Figure 1.3 that, when oxygen and hydrogen are brought together to the distance of a chemical bond length

---

3 Unfortunately, the convention is to use the same letter Σ for the term symbol of the electronic state with $\Lambda = 0$ and to label the eigenvalue of the z-component of $\boldsymbol{S}$.

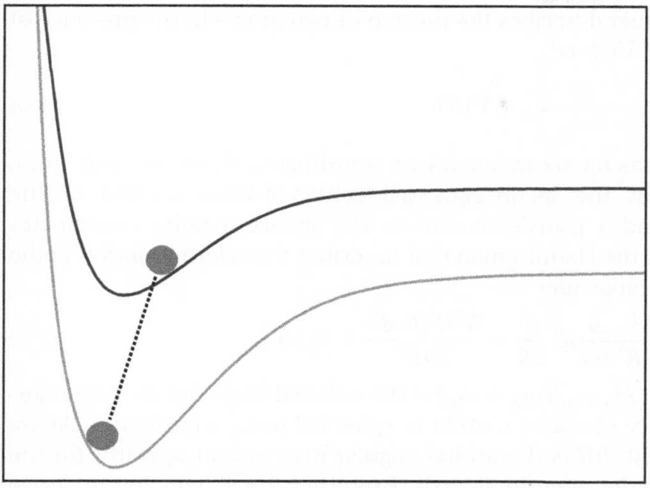

Figure 1.4 Couplings between different electronic states may affect the vibrational motion of molecules. These couplings become negligible when the electronic states are separated by a large amount of energy.

(~2-Bohr radii), the lowest-energy electronic state of OH is separated from other electronic states by >30 000 cm^{-1}. At these interatomic distances, the few cm^{-1} couplings between different electronic states will have no noticeable effect on the lowest-electronic energy of OH, and the molecule can be thought of as residing in a single electronic state $X^2\Pi$.

To appreciate this, one can use the analogy with two marbles rolling in different – vertically displaced – potential energy wells, as shown in Figure 1.4. Imagine that the marbles are magnetic so they can interact with each other as they vibrate. If the distance between the marbles is increased, the effect of the interaction between the marbles becomes weaker, until it is no longer noticeable. In the limit of a vanishingly small interaction or a very large separation between the potential curves, the marbles can be treated as independent. So, we may consider a molecule as described by a single electronic energy potential.

This is good news because if we want to understand the dynamics of vibrational motion of a chemically bound OH molecule, we are allowed to forget about any but the $X^2\Pi$ electronic state. In this book, we will discuss mostly molecules in the ground electronic state. Our starting point is thus a single electronic potential, usually of the form represented by the $X^2\Pi$ potential in Figure 1.3. We will call this potential $V(R)$, with R denoting the distance between the two atoms.

The Hamiltonian that describes the motion of two atoms in the presence of an external potential $V(R)$ is[4]

$$\hat{H} = -\frac{\hbar^2}{2m_A}\nabla_A^2 - \frac{\hbar^2}{2m_B}\nabla_B^2 + V(R). \tag{1.6}$$

This operator depends on six independent coordinates, three for each atom. After separating out the inconsequential center-of-mass motion of the two-atom system and a transformation to the spherical polar coordinates (R, θ, ϕ), we arrive at the Hamiltonian that describes the *relative* motion of the atoms in a diatomic molecule:

$$\hat{H}_{\text{ro-vib}} = -\frac{\hbar^2}{2\mu R^2}\frac{\partial}{\partial R}R^2\frac{\partial}{\partial R} + \frac{\hbar^2\boldsymbol{R}^2(\theta,\phi)}{2\mu R^2} + V(R). \tag{1.7}$$

In this equation, $\mu = m_A m_B/(m_A + m_B)$ is the reduced mass and $\boldsymbol{R}$ is the angular part of the Laplace operator written in spherical polar coordinates. As we shall see in a moment, $\hbar\boldsymbol{R}$ is the orbital angular momentum operator for the rotational motion of the interatomic axis. Exercise 1.2 will help you to derive Eq. (1.7) from Eq. (1.6).

The relative motion of two atoms can be separated into the vibrational motion and the orbital motion. Imagine for the moment that the orbital motion of the atoms is frozen so there is no second term in Eq. (1.7). The resulting Hamiltonian

$$\hat{H}_{\text{vib}} = -\frac{\hbar^2}{2\mu R^2}\frac{\partial}{\partial R}R^2\frac{\partial}{\partial R} + V(R) \tag{1.8}$$

depends only on one variable R. The eigenstates of this Hamiltonian – usually labeled by the quantum number υ – are the vibrational states of the molecule. The energy of the lowest vibrational excitation $E_{\upsilon=1} - E_{\upsilon=0} = \hbar\omega_{\text{vib}}$ sets the second largest scale in the theory of diatomic molecules.

To solve the Schrödinger equation describing the vibrational motion of a molecule, it is convenient to write the eigenfunctions of the Hamiltonian (1.8) as

$$\psi_\upsilon(R) = \chi_\upsilon(R)/R. \tag{1.9}$$

There are two reasons to do this. First, the volume element written in spherical polar coordinates is $\mathrm{d}X\mathrm{d}Y\mathrm{d}Z = R^2 \sin\theta \mathrm{d}R\mathrm{d}\Theta\mathrm{d}\phi$ so the integration over

4 I have swept a few "unimportant" details under the rug here. The nitpicky reader will notice that I use the same notation for the masses of the particles in Eq. (1.6) as in Eq. (1.3). Strictly speaking, this is not right. The first two terms in Eq. (1.3) describe the kinetic energy of the atomic nuclei, whereas the first two terms in Eq. (1.6) describe the motion of the atoms. So the masses in Eq. (1.6) represent the atomic masses (the sum of the nuclear mass and the mass of all electrons), and the derivative operators in Eq. (1.6) correspond to the kinetic energy of the atomic centers of mass. The Hamiltonian (1.3) is reduced to the Hamiltonian (1.6) by introducing the center-of-mass coordinates for each atom, integrating out the electronic degrees of freedom and assuming that the operators in Eq. (1.3) do not couple different eigenstates of the operator (1.1). From here on, we will assume that m_A and m_B represent the atomic masses.

R involves the factor R^2. Thus, the expressions for the radial matrix elements become simpler if the eigenfunctions are represented in the form (1.9), i.e.

$$\int_0^\infty \psi_v \hat{f}(R) \psi_{v'} R^2 \mathrm{d}R = \int_0^\infty \chi_v \hat{f}(R) \chi_{v'} \mathrm{d}R. \tag{1.10}$$

Second, the differential equation for χ_v is simpler than the one for ψ_v. In particular,

$$\begin{aligned} R^{-2} \frac{\partial}{\partial R} R^2 \frac{\partial}{\partial R} \frac{\chi(R)}{R} &= -R^{-2} \frac{\partial}{\partial R} \chi(R) + R^{-2} \frac{\partial}{\partial R} R \frac{\partial}{\partial R} \chi(R) \\ &= -R^{-2} \frac{\partial}{\partial R} \chi(R) + R^{-2} \frac{\partial}{\partial R} \chi(R) + R^{-1} \frac{\partial^2}{\partial R^2} \chi(R) \\ &= R^{-1} \frac{\partial^2}{\partial R^2} \chi(R). \end{aligned} \tag{1.11}$$

The transformation (1.9) thus removes the first derivatives from the differential equation.

Figure 1.5 shows the eigenvalues of the vibrational Hamiltonian (1.8) of the OH molecule in the ground electronic state $X^2\Pi$ and the wave functions $\chi_v(R)$ for three lowest-energy vibrational states. In general, the magnitude of $\hbar\omega_{\mathrm{vib}}$ depends on the reduced mass μ and the shape of the electronic potential $V(R)$. Typically, it ranges from 50 cm^{-1} for heavy molecules, such as RbCs, to

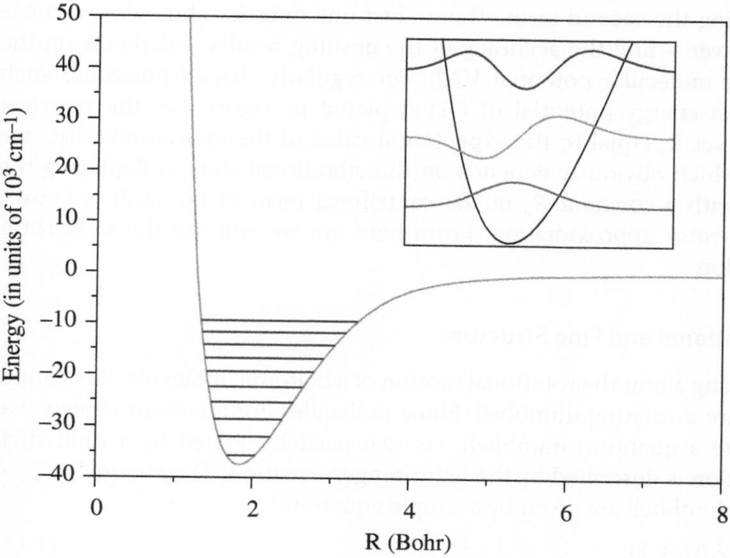

Figure 1.5 The vibrational energy levels of the OH molecule in the ground electronic state $X^2\Pi$. Only 10 lowest-energy levels are shown. The inset shows the vibrational wave functions $\chi_v(R)$ for $v = 0$, $v = 1$, and $v = 2$.

4401 cm^{-1} for H_2. In this book, we will discuss only molecules in the ground vibrational state ($v = 0$).

If the orbital motion of the atoms is not frozen – which is, of course, generally the case – the second term in Eq. (1.7) distorts the vibrational wave functions $\chi_v(R)$. Introducing the eigenfunctions of the operator $\boldsymbol{R}^2$ as

$$\boldsymbol{R}^2(\theta,\phi)Y_\lambda(\theta,\phi) = \epsilon_\lambda Y_\lambda(\theta,\phi), \tag{1.12}$$

we can write

$$\hat{H}_{\text{ro-vib}} Y_\lambda(\theta,\phi) = \left[-\frac{\hbar^2}{2\mu R^2}\frac{\partial}{\partial R}R^2\frac{\partial}{\partial R} + \frac{\hbar^2\epsilon_\lambda}{2\mu R^2} + V(R)\right] Y_\lambda(\theta,\phi), \tag{1.13}$$

and see that the expression in the square brackets is still a one-dimensional operator that depends on R only. The eigenfunctions of this operator are $\chi_v^\lambda(R)$. They are still one-dimensional functions of R, but they are now affected (distorted) by the presence of ϵ_λ in the square brackets so they must now be labeled by two quantum numbers, v and λ.

This can be avoided, if we replace the variable R in the denominator of the second term of Eq. (1.7) by a constant R_e. Doing so will separate the ro-vibrational Hamiltonian (1.7) into a part that depends on R only and the term $\hbar^2\boldsymbol{R}^2(\theta,\phi)/2\mu R_e^2$ that depends on the angles θ and ϕ only. The vibrational and rotational motions are thus uncoupled, and the vibrational wave functions are again $\chi_v(R)$. This simplifies the problem because we can obtain all vibrational wave functions $\chi_v(R)$ from a single diagonalization of the Hamiltonian (1.7) excluding the second term. There's just one question: how do we chose R_e? The answer – and the accuracy of the ensuing results – depends on the shape of the molecular potential $V(R)$. For regularly shaped potentials such as the lowest-energy potential of OH depicted in Figure 1.3, the smartest choice is to set R_e equal to the expectation value of the interatomic distance $\langle\chi_v|R|\chi_v\rangle$, which obviously depends on the vibrational state v. Replacing the variable R with a constant R_e in the centrifugal term of Eq. (1.7) is known as the rigid-rotor approximation. From here on, we will use the rigid-rotor approximation.

1.2.3 Rotational and Fine Structure

When thinking about the rotational motion of a diatomic molecule, it is tempting to picture a rotating dumbbell. Since molecules are quantum objects, we often picture a quantum dumbbell, i.e. two particles joined by a rigid stick whose rotation is described by the Schrödinger equation. The energy levels of a quantum dumbbell are given by a simple equation

$$E_J = BJ(J+1), \tag{1.14}$$

where B is a constant and $J = 0, 1, \ldots, \infty$ is an integer quantum number. For some molecules – such as H_2 in low vibrational levels – this picture may be adequate. However, we should not forget that molecules are made of electrons as well as nuclei, and the rotational motion of the interatomic axis drags the electrons around. The coupling of electrons to the rotational motion of the molecule often leads to the rotational energy level patterns, which look nothing like those of a quantum dumbbell. Picturing a molecule as a cloud of electrons revolving around the internuclear axis, it is reasonable to ask when and why a molecule is likely to behave as a quantum dumbbell. And when a molecule is not a quantum dumbbell, what is the pattern of its rotational energy levels?

The rotational motion of the internuclear axis is described by the operator $B_e\boldsymbol{R}^2(\theta,\phi)$ in Eq. (1.7). In that equation,

$$\boldsymbol{R}^2 = -\left[\frac{1}{\sin^2\theta}\frac{\partial^2}{\partial^2\phi} + \frac{1}{\sin\theta}\frac{\partial}{\partial\theta}\left(\sin\theta\frac{\partial}{\partial\theta}\right)\right]. \tag{1.15}$$

The right-hand side of Eq. (1.15) is a result of writing the Laplace operator in spherical polar coordinates. In classical mechanics, angular momentum of a particle with position $\boldsymbol{r}$ and linear momentum $\boldsymbol{p}$ is $\boldsymbol{l} = \boldsymbol{r}\times\boldsymbol{p}$. If we substitute in this equation the quantum operators for p_x, p_y, and p_z and transform the derivative operators to those written in terms of the spherical polar coordinates, we will find that $\boldsymbol{l}^2$ is given exactly by the right-hand side of Eq. (1.15) times $\hbar^2$. The operator $\hbar\boldsymbol{R}$ in Eq. (1.8) is thus the operator of angular momentum that describes the rotational motion of the internuclear axis.[5]

I took this opportunity to introduce the rotational constant

$$B_e = \hbar^2/2\mu R_e^2, \tag{1.16}$$

which sets the scale of the rotational energy. Because R_e, as defined above, depends on the vibrational states and the vibrational states depend on the shape of the electronic potential $V(R)$, the rotational constant B_e must be labeled by the vibrational and electronic state quantum numbers. I will omit these labels, whenever possible, in order to simplify the notation. The magnitude of the rotational constant B_e depends on the reduced mass (μ) and the size (R_e) of the molecule. For typical molecules, $R_e \sim 1-3$-Bohr radii and the reduced mass μ ranges from about 1 to about 100 proton masses. The rotational constant B_e can thus take values from 0.01 cm^{-1} for heavy diatomic molecules to 60.9 cm^{-1} for H_2.

And this is where things get really messy!

For starters, the spin–orbit interaction effects, coming from the relativistic Dirac equation and which we have neglected so far, are often comparable with the energy of the rotational motion. This means that in order to describe the rotational structure of diatomic molecules correctly, we must account for the relativistic interactions. What this means is that the Hamiltonian (1.7) is

5 From here on, we will define the operator $\boldsymbol{l}$ of the orbital angular momentum by the relation $\hbar\boldsymbol{l} = \boldsymbol{r}\times\boldsymbol{p}$. This will remove the $\hbar$ from most of the angular momentum algebra. See Appendix B for more details.

not good enough. We will make this Hamiltonian more accurate by adding an extra term $V_{\rm fs}$, which accounts for the spin-dependent, fine structure interactions – hence the subscript. This term, if written in all the detail, often looks complex. The actual form of $V_{\rm fs}$ depends on the nature of the molecule (i.e. the masses of the nuclei and the total number of electrons) as well as the electronic state of the molecule.

Second, we are down to such a fine energy scale that we may have to reexamine the approximations previously made. For example, those couplings to the electronically excited states that we were so happy to neglect may, in fact, affect the rotational motion of molecules. In addition, we have to worry about new interactions induced by the orbital motion of the nuclei – as we shall see in a moment, these interactions may also couple different electronic states. All these effects give rise to the fine structure of molecules – the splitting of rotational states into fine structure levels.

To allow for the possibility of including couplings between different electronic states, we can no longer consider the problem with a single electronic state potential $V(R)$, as in Eq. (1.7). Let us replace $V(R)$ with an operator that allows us to include multiple electronic states into consideration:

$$V(R) \Rightarrow \sum_{\eta S\Sigma\Lambda} |\eta S\Sigma\Lambda\rangle E_{\eta S\Sigma\Lambda}(R)\langle \eta S\Sigma\Lambda|. \tag{1.17}$$

In this equation, $E_{\eta S\Sigma\Lambda}(R)$ plays the role of $V(R)$ for each electronic state $|\eta S\Sigma\Lambda\rangle$.

This substitution leads to the following effective Hamiltonian that describes the relative motion of two atoms:

$$\hat{H}_{\rm ro-vib,fs} = \sum_{\eta S\Sigma\Lambda} |\eta S\Sigma\Lambda\rangle \left[-\frac{\hbar^2}{2\mu R^2}\frac{\partial}{\partial R}R^2\frac{\partial}{\partial R} + E_{\eta S\Sigma\Lambda}(R)\right] \langle \eta S\Sigma\Lambda| + \frac{\hbar^2 \boldsymbol{R}^2(\theta,\phi)}{2\mu R^2} + V_{\rm fs}. \tag{1.18}$$

In order to write this equation, we assumed that the derivative operators do not couple different electronic states.[6] This allowed us to put the derivative operators inside the square brackets in the above equation.

The matrix elements of the operator (1.18) in the subspace of the electronic states $|\eta S\Sigma\Lambda\rangle$ are

$$\begin{aligned}\langle \eta S\Sigma\Lambda|\hat{H}_{\rm ro-vib,fs}|\eta' S'\Sigma'\Lambda'\rangle &= \delta_{\eta\eta'}\delta_{SS'}\delta_{\Sigma\Sigma'}\delta_{\Lambda\Lambda'}\left[-\frac{\hbar^2}{2\mu R^2}\frac{\partial}{\partial R}R^2\frac{\partial}{\partial R} + E_{\eta S\Sigma\Lambda}(R)\right] \\ &+ \langle \eta S\Sigma\Lambda|\frac{\hbar^2 \boldsymbol{R}^2(\theta,\phi)}{2\mu R^2} + V_{\rm fs}|\eta' S'\Sigma'\Lambda'\rangle. \end{aligned} \tag{1.19}$$

---

6 This is not generally true because the electronic wave functions do depend, albeit weakly, on R. However, these derivative couplings are either forbidden by symmetry or quite small in magnitude so we rarely have to worry about them.

If the electronic states $|\eta S\Sigma\Lambda\rangle$ in Eq. (1.18) have very different energies $E_{\eta S\Sigma\Lambda}$, the off-diagonal matrix elements in Eq. (1.19) are of little significance. However, some nonrelativistic electronic states may happen to be close in energy, or even be degenerate, as is the case for states with the same values of η, S, and $|\Lambda|$, but different Σ. When this happens, the off-diagonal matrix elements in Eq. (1.19) mix different nonrelativistic states $|\eta S\Sigma\Lambda\rangle$ significantly so (some of) the quantum numbers S, Σ, or Λ are no longer good labels. This mixing of the electronic states often affects the rotational energy of the molecule to a great extent. We will see how in the next section.

1.3 Rotation of a Molecule

The kets $|\eta S\Sigma\Lambda\rangle$ describe the states of the electrons in the molecule-fixed frame with the z-axis directed along the internuclear axis. In order to understand the rotational structure of molecules, we need to consider what happens to these states as we let the interatomic axis rotate in a space-fixed coordinate frame. It is important to pay attention to the coordinate systems in this section. The space-fixed quantization axis is here and hereafter denoted by uppercase Z, to distinguish it from the molecule-fixed (interatomic) z-axis. The Z-axis defines an external frame of reference, in which the molecule rotates. In the absence of an external field, the Z-axis can be directed arbitrarily.

There are three operators that always commute with the Hamiltonian of a closed quantum system (a molecule in the absence of external fields): the square of the total angular momentum $\boldsymbol{J}^2$, the Z-component of the total angular momentum J_Z in the *space-fixed* coordinate system, and the parity operator $\hat{P}$. This is another way to say that the total angular momentum, the Z-component of the angular momentum, and the parity are conserved. The conservation of the total angular momentum and its Z-component is a consequence of the isotropy of space.

As a result of these commutations, we can define the states $|\gamma JM\varepsilon\rangle$ that are simultaneously the eigenstates of the Hamiltonian, $\boldsymbol{J}^2$, J_Z, and $\hat{P}$:

$$\hat{H}|\gamma JM\varepsilon\rangle = E_\gamma|\gamma JM\varepsilon\rangle, \tag{1.20}$$

$$\boldsymbol{J}^2|\gamma JM\varepsilon\rangle = J(J+1)|\gamma JM\varepsilon\rangle, \tag{1.21}$$

$$J_Z|\gamma JM\varepsilon\rangle = M|\gamma JM\varepsilon\rangle, \tag{1.22}$$

$$\hat{P}|\gamma JM\varepsilon\rangle = \varepsilon|\gamma JM\varepsilon\rangle. \tag{1.23}$$

Every molecular state can thus be labeled by a set of four quantum numbers: $\gamma, J, M, \varepsilon$. These quantum numbers are strictly conserved, which means that all matrix elements of the Hamiltonian (of a closed quantum system) are

$$\langle\gamma JM\varepsilon|\hat{H}|\gamma'J'M'\varepsilon'\rangle \propto \delta_{\gamma,\gamma'}\delta_{J,J'}\delta_{M,M'}\delta_{\varepsilon,\varepsilon'}. \tag{1.24}$$

The label γ is some generic quantum number that enumerates different energy states. The question now is: Can we replace the generic quantum number γ with a set of physical quantum numbers (such as S, Σ, Λ, or their combination) that are *nearly* conserved and that provide some physical insight into the rotational motion of the molecule? This will be the goal of this section.

First, let us identify a complete basis for a molecular Hamiltonian. We have used the labels $|\eta S\Sigma\Lambda\rangle$ for the electronic states of a molecule. The basis $|\eta S\Sigma\Lambda\rangle$ covers the electronic part of the Hilbert space of a molecular Hamiltonian but does not describe the nuclear degrees of freedom. We need to "dress" this basis with the states that describe the vibrational and rotational motions of the molecule.

For a quantum dumbbell, the total angular momentum $\boldsymbol{J}$ would be simply the rotational angular momentum $\boldsymbol{R}$. In the case of molecules, the total angular momentum $\boldsymbol{J}$ is the vector sum of the total orbital angular momentum of the electrons $\boldsymbol{L}$, the total spin angular momentum of the electrons $\boldsymbol{S}$, and the orbital angular momentum of the nuclei $\boldsymbol{R}$:

$$\boldsymbol{J} = \boldsymbol{L} + \boldsymbol{S} + \boldsymbol{R}. \tag{1.25}$$

It is also convenient to introduce the angular momentum $\boldsymbol{N}$ as

$$\boldsymbol{N} = \boldsymbol{L} + \boldsymbol{R}. \tag{1.26}$$

Note that, because orbital angular momentum is a cross product of coordinate and linear momentum, $\boldsymbol{R}$ is necessarily perpendicular to the interatomic axis so the projection of $\boldsymbol{N}$ on the interatomic axis z is the same as that of $\boldsymbol{L}$. The projection of $\boldsymbol{J}$ on the interatomic axis z is $\Omega = \Lambda + \Sigma$.

Note that M in Eq. (1.22) is the projection of $\boldsymbol{J}$ on a *space-fixed* quantization axis. As discussed in Appendix B, we can generally expand any angular momentum state with specific values of J and M as

$$\begin{aligned} |JM\rangle &= \sum_N \sum_{M_N} C^{JM}_{NM_N SM_S} |NM_N\rangle |SM_S\rangle \\ &= \sum_N \sum_{M_N} \sum_L \sum_{M_L} \sum_R C^{JM}_{NM_N SM_S} C^{NM_N}_{LM_L RM_R} |LM_L\rangle |SM_S\rangle |RM_R\rangle, \end{aligned} \tag{1.27}$$

where M_N, M_L, M_S, and M_R denote the projections of the corresponding angular momenta on the space-fixed quantization axis and $C^{j_3m_3}_{j_1m_1j_2m_2}$ are the Clebsch–Gordan coefficients (see Appendix B).

At the same time, the states $|JM\rangle$ can be written in terms of the total angular momentum states in the molecule-fixed coordinate frame as (see Appendix D)

$$|JM\rangle = \sum_\Omega D^{J*}_{M\Omega} |J\Omega\rangle, \tag{1.28}$$

where the expansion coefficients – known as the Wigner D-functions [9] – are the probability amplitudes for the state $|J\Omega\rangle$ to have a space-fixed projection M.

The D-functions depend on three Euler angles that describe the rotation of the molecule-fixed coordinate system in the space-fixed frame. We can think of the D-functions simply as the coefficients of the unitary transformation (1.28) that depend on three quantum numbers: J, M, and Ω.

Equation (1.28) is general and can also be written for the eigenstates of $\boldsymbol{L}^2$ and L_Z, $\boldsymbol{S}^2$ and S_Z, and $\boldsymbol{R}^2$ and R_Z:

$$|LM_L\rangle = \sum_{\Lambda} D^{L*}_{M_L\Lambda}|L\Lambda\rangle, \tag{1.29}$$

$$|SM_S\rangle = \sum_{\Sigma} D^{S*}_{M_S\Sigma}|S\Sigma\rangle, \tag{1.30}$$

$$|RM_R\rangle = \left(\frac{2R+1}{4\pi}\right)^{1/2} D^{R*}_{M_R0}. \tag{1.31}$$

Remember that $\boldsymbol{R}$ is always perpendicular to the interatomic axis, which simplifies the last equation.

If we now substitute Eqs. (1.29)–(1.31) into Eq. (1.27), the latter becomes

$$|JM\rangle = \sum_L \sum_\Lambda \sum_\Sigma |L\Lambda\rangle|S\Sigma\rangle \sum_N \sum_{M_N} \sum_{M_L} \sum_R C^{JM}_{NM_NSM_S} C^{NM_N}_{LM_LRM_R} \times \left(\frac{2R+1}{4\pi}\right)^{1/2} D^{R*}_{M_R0} D^{S*}_{M_S\Sigma} D^{L*}_{M_L\Lambda}. \tag{1.32}$$

Instead of using the D-functions labeled by the superscripts S and L and R, it would be nice to use just a single set of D-functions labeled by the total angular momentum quantum number, as in Eq. (1.28). To do this, we can exploit the following useful property of the D-functions [10] discussed in Appendix D:

$$\sum_{M_1} \sum_{M_2} D^{J_1*}_{M_1N_1} D^{J_2*}_{M_2N_2} C^{JM}_{J_1M_1J_2M_2} = C^{JN}_{J_1N_1J_2N_2} D^{J*}_{MN}. \tag{1.33}$$

Applying Eq. (1.33) twice transforms Eq. (1.32) into

$$|JM\rangle = \sum_L \sum_\Lambda \sum_\Sigma D^{J*}_{M\Omega}|L\Lambda\rangle|S\Sigma\rangle \sum_{N=|J-S|}^{J+S} \sum_{R=|N-L|}^{N+L} C^{J\Omega}_{N\Lambda S\Sigma} C^{N\Lambda}_{L\Lambda R0} \left(\frac{2R+1}{4\pi}\right)^{1/2}. \tag{1.34}$$

We have just related the space-fixed molecular states $|JM\rangle$ with the states describing the electrons in the molecule-fixed frame. However, we are not done yet, because $\boldsymbol{L}$ is not a proper angular momentum for electrons in a molecule (because there is no spherical symmetry for electrons in a molecule).

The quantum number L is meaningful only at infinite interatomic separations, where $\boldsymbol{L}$ can be defined as the vector sum of the total electronic orbital angular momenta $\boldsymbol{L}_A$ and $\boldsymbol{L}_B$ of the atoms A and B. At finite interatomic distances, the atom–atom interaction couples states with different L. The states

$|L\Lambda\rangle$ can be generally expressed as linear combinations of the eigenstates $|\eta\Lambda\rangle$,

$$|L\Lambda\rangle = \sum_{\eta} C_{\eta}^{L,\Lambda} |\eta\Lambda\rangle, \tag{1.35}$$

where $C_{\eta}^{L,\Lambda}$ are the coefficients of a unitary transformation. The coefficients of this unitary transformation generally have to be found numerically by diagonalizing the matrix of the electronic Hamiltonian in the basis of $|L\Lambda\rangle$ states. The quantum number η is used to distinguish the electronic states with the same value of Λ but different energies. Keep in mind that, because of the cylindrical symmetry of the diatomic molecule, Λ remains a good quantum number.

The angular part of the molecular states can thus be written as

$$|JM\rangle = \sum_{\Lambda} \sum_{\Sigma} \sum_{\eta} |JM\Omega\rangle |\eta\Lambda S\Sigma\rangle f(J, \Lambda, S, \Sigma, \eta), \tag{1.36}$$

where, following Appendix D, we have made the substitution

$$|JM\Omega\rangle \equiv \left(\frac{2J+1}{8\pi^2}\right)^{1/2} D_{M\Omega}^{J*}, \tag{1.37}$$

and absorbed everything else in Eq. (1.34) into the coefficients $f(J, \Lambda, S, \Sigma, \eta)$.

The states $|JM\Omega\rangle|\eta\Lambda S\Sigma\rangle$ form a complete angular basis set for the expansion of molecular states. The full Hamiltonian matrix in this basis is generally nondiagonal in all quantum numbers, except J and M. However, some diagonal elements of the Hamiltonian matrix may be well separated from and/or weakly coupled to the rest of the matrix. When this is the case, we can ignore the off-diagonal elements and identify the molecular states as

$$|\gamma JM\rangle \approx |v\rangle |JM\Omega\rangle |\eta\Lambda S\Sigma\rangle \tag{1.38}$$

so the molecular states can be labeled by the quantum numbers Λ, Σ, Ω in addition to J, M, and the parity eigenvalue.

But what if the different electronic states *are* strongly mixed? For example, what if the fine structure interaction induces large couplings between states with the same Λ and S but different values of Σ? Then, we could search for another basis related to the basis $|JM\Omega\rangle|\eta\Lambda S\Sigma\rangle$ by a unitary transformation, hoping that the matrix of the Hamiltonian in the new basis will be nearly diagonal. For the case just hypothesized, the alternative basis set can be defined as

$$|JM\rangle|\eta N\Lambda S\rangle = \sum_{\Sigma} C_{N\Lambda S\Sigma}^{J\Omega} |JM\Omega\rangle |\eta\Lambda S\Sigma\rangle, \tag{1.39}$$

where $C_{N\Lambda S\Sigma}^{J\Omega}$ are the Clebsch–Gordon coefficients of the transformation between the coupled and uncoupled representations in the molecule-fixed frame:

$$|J\Omega\rangle = \sum_{\Lambda} \sum_{\Sigma} C_{N\Lambda S\Sigma}^{J\Omega} |N\Lambda\rangle |S\Sigma\rangle. \tag{1.40}$$

Since $\Omega = \Lambda + \Sigma$, the summation over Σ, while keeping Λ fixed, is equivalent to a summation over Ω. In essence, Eq. (1.39) couples the total angular momentum to the spin angular momentum to produce the conserved states $|N\Lambda\rangle$.

If the Hamiltonian matrix in the basis (1.39) contains diagonal matrix elements that are well separated from and/or weakly coupled to the rest of the matrix, we can label the molecular states as

$$|\gamma JM\rangle \approx |v\rangle |JM\rangle |\eta N\Lambda S\rangle. \tag{1.41}$$

Whether the molecular states are well described by Eq. (1.38) or by Eq. (1.39) matters a lot for the rotational structure of molecules and for how molecules interact with electromagnetic fields. Therefore, the molecular states are classified into categories that correspond to different basis sets. These categories are called Hund's coupling cases (a), (b), (c), (d), etc. The molecular states well described by Eq. (1.38) are Hund's case (a). Those described by Eq. (1.39) are Hund's case (b). The patterns of the rotational energy levels (i.e. whether they look like those of the quantum dumbbell or – for example – come in doublets or have a more complex structure) depend on the Hund's case.

1.4 Hund's Cases

Let me reiterate that neither of the basis sets – (1.38) or (1.39) – diagonalizes fully the Hamiltonian. Beyond γ, J, M, and ε, none of the quantum numbers are perfectly good. We have used the term "nearly good" and talked about a basis set "representing well" the molecular states. What does it really mean? If we were to evaluate the matrix of the full molecular Hamiltonian, for example, in Hund's case (b) basis (1.39) and diagonalize it, the eigenvectors for a *Hund's case (b) molecule* would still be represented as linear combinations of states (1.39) with *different* values of N. However, the relative contribution of one specific state with one specific value of N in this linear combination would be 99, 99 ... %. In other words, for all intents and purposes, this specific value of N can be used to characterize a particular eigenstate of the full molecular Hamiltonian. That's what we typically refer to as "good," even in cases when the quantum numbers are not truly good.

So, ultimately, different Hund's cases tell us in which of the possible basis sets the matrix of the full Hamiltonian is *nearly* diagonal. Let's consider some of the most typical Hund's cases, case by case.

1.4.1 Hund's Coupling Case (a)

Imagine that $V_{\rm fs} = 0$. If this is the case, what are the (nearly) conserved observables? To answer this question, we need to rewrite the rotational Hamiltonian in terms of the angular momenta $\boldsymbol{J}$, $\boldsymbol{L}$, and $\boldsymbol{S}$

$$B_{\rm e}\boldsymbol{R}^2 = B_{\rm e}[\boldsymbol{J} - \boldsymbol{L} - \boldsymbol{S}]^2. \tag{1.42}$$

Expanding the brackets, we obtain

$$[\boldsymbol{J} - \boldsymbol{L} - \boldsymbol{S}]^2 = \boldsymbol{J}^2 + \boldsymbol{L}^2 + \boldsymbol{S}^2 - 2J_zS_z - 2J_zL_z + 2L_zS_z$$
$$- J_+S_- - J_-S_+ - J_+L_- - J_-L_+ + S_-L_+ + S_+L_-, \qquad (1.43)$$

where I use standard notation for the raising and lowering angular momentum operators (see Appendix B).

It is easy to show (Exercise 1.3) that $\boldsymbol{J}^2, \boldsymbol{L}^2$, and $\boldsymbol{S}^2$ commute with the operator (1.42). Therefore, J and S are good quantum numbers! As we already know, L cannot be a good quantum number because $\boldsymbol{L}^2$ does not commute with the full Hamiltonian of the molecular system.

The raising and lowering angular momentum operators $J_\pm$, $L_\pm$, and $S_\pm$ raise or lower the z-component of the corresponding angular momentum by one. For example, $L_+|\Lambda\rangle = const|\Lambda + 1\rangle$. We can see that the terms $S_\pm L_\mp$ couple states with different values of Λ and Σ. For example, the term S_-L_+ couples the state $|\Lambda, \Sigma\rangle$ to the state $|\Lambda + 1, \Sigma - 1\rangle$. Similarly, the terms $J_\pm L_\mp$ couple states with different Λ and the terms $J_\pm S_\mp$ couple states with different Σ. So the rotational operator (1.42) itself induces couplings between different electronic states. These couplings are called the Coriolis couplings.

Are the Coriolis couplings important? To answer this question, recall that the typical value of $B_e < 10$ cm^{-1} and the typical energy separation between electronic states corresponding to different values of $|\Lambda|$ is $|E_{v,\eta S\Sigma\Lambda} - E_{v,\eta S\Sigma\Lambda'}| >$ 10 000 cm^{-1}. The mixing of different Λ states due to the Coriolis couplings is therefore expected to be small. This suggests that Λ may still be treated – very often – as a good quantum number. The story is different for Σ. In the absence of spin-dependent interactions and the Coriolis couplings, the electronic states with different Σ are degenerate. The Coriolis couplings must therefore mix different Σ states significantly. And this is where $V_{\rm fs}$ often helps. It is, of course, nonzero for most molecules. Imagine now that $V_{\rm fs}$ contains a term AL_zS_z with the constant $A \gg B_e$. This term, coming from the spin–orbit interaction, must then split the states of different Σ and thereby suppress[7] the role of the $L_\pm$ and $S_\pm$ operators in Eq. (1.43). If this happens, Λ and Σ can be treated as good quantum numbers. Because $\Omega = \Lambda + \Sigma$, it is also a good quantum number. If this is the case, we're dealing with Hund's case (a) molecules.

Hund's case (a) molecules are molecules whose energy levels can be labeled by J, M, Ω, S, Σ, and Λ.

1.4.2 Hund's Coupling Case (b)

It is to deal with this case that we introduced the angular momentum $\boldsymbol{N} = \boldsymbol{J} - \boldsymbol{S}$. In the absence of spin-dependent interactions, $\boldsymbol{N}$ must be conserved. Why? Because both $\boldsymbol{J}$ and $\boldsymbol{S}$ are conserved, as we established in Section 1.43.

---

7 An interested reader may notice that the matrix elements of the operators $J_\pm S_\mp$ in Eq. (1.43) depend on the value of J so the relative importance of these couplings changes as J increases. Therefore, some Hund's case (a) molecules may change their orientation – and become Hund's case (b), for example – when they are promoted to rotational levels with high J.

Imagine now that V_{fs} does *not* contain large AL_zS_z terms. Or that the diagonal matrix elements of this operator are zero, as is the case for molecules in states with $\Lambda = 0$. If this is the case, Λ can still be a good quantum number (see the discussion above). However, Σ is less fortunate. If Σ is not defined, $\Omega = \Lambda + \Sigma$ suffers, too. We are dealing with Hund's case (b) molecules.

Hund's case (b) molecules are molecules whose energy levels can be labeled by N, S, Λ, J and M.

1.4.3 Hund's Coupling Case (c)

Imagine now the case when the molecular potentials corresponding to different Λ happen to be close in energy and V_{fs} contains the spin–orbit interaction terms that couple $\boldsymbol{L}$ and $\boldsymbol{S}$. Λ is no longer a good quantum number. Neither is Σ. When both Λ and Σ are bad, they conspire to preserve Ω. We are dealing with Hund's case (c) molecules.

Hund's case (c) molecules are molecules whose energy levels can be labeled by J and M, which are always conserved, and Ω, and no other angular momentum quantum numbers.

We will refer back to this section as we discuss the rotational structure of molecules in external fields in the next chapters. We will then see how the different Hund's cases correspond to different patterns of the rotational energy levels of molecules and how they determine the interaction of molecules with external electric and magnetic fields.

1.5 Parity of Molecular States

In addition to $\boldsymbol{J}^2$ and J_Z, there is another operator that always commutes with the Hamiltonian of a closed quantum system, namely, the parity operator $\hat{P}$. The molecular states are therefore the eigenstates of the parity operator with the eigenvalues ε. The goal of this section is to discuss how to attach the parity label ε to the molecular states. This is particularly important for our discussions in the subsequent chapters because interactions induced by electric fields couple states of different parity. The energy gap between states of different parity therefore determines the response of a molecule to an external electric field.

The parity operator inverts the *space-fixed* coordinate system

$$(X, Y, Z) \rightarrow (-X, -Y, -Z). \tag{1.44}$$

Since reversing the coordinate axes twice leaves them unchanged, $\hat{P}^2$ is the identity operator. For any state $|\varphi\rangle$ satisfying the eigenvalue equation

$$\hat{P}|\varphi\rangle = \varepsilon|\varphi\rangle, \tag{1.45}$$

we can write

$$\hat{P}^2|\varphi\rangle = \varepsilon^2|\varphi\rangle = |\varphi\rangle, \tag{1.46}$$

which shows that the eigenvalues of $\hat{P}$ must be $\varepsilon = \pm 1$.

In this section, we will consider the effect of $\hat{P}$ on molecular states. For example, what happens when $\hat{P}$ is applied to Hund's case (a) states $|v\rangle|JM\Omega\rangle|\eta\Lambda S\Sigma\rangle$?

The vibrational part is easy. The wave functions of the vibrational states $|v\rangle$ depend on a single *scalar* variable – the distance between the atoms. This distance is the same in the original or inverted coordinate system so the vibrational states remain unchanged upon inversion of the coordinate frame

$$\hat{P}|v\rangle = |v\rangle. \tag{1.47}$$

The electronic part is tricky. The problem is the parity operator inverts the *space-fixed* coordinate axes, but the electronic states $|\eta\Lambda S\Sigma\rangle$ are defined in the *molecule-fixed* frame whose z-axis is directed along the internuclear axis. In order to determine the effect of $\hat{P}$ on the electronic states, it is helpful to establish that [11]

> The effect of inverting the *space-fixed* coordinate frame is the same as the result of a reflection in the *molecule-fixed* xz-plane.

To see this, we shall use the transformation between the space-fixed coordinates (X, Y, Z) and the molecule-fixed coordinates (x, y, z) given in Appendix C. The coordinates (x, y, z) and (X, Y, Z) can be those of an electron or a nucleus.

The operator for the reflection in the molecule-fixed xz-plane is commonly denoted as $\sigma_v(xz)$. The symmetry operation $\sigma_v(x, z)$ leaves x and z unchanged, while transforming y to $-y$. It must also somehow change the matrix of the direction cosines – the matrix that can be used to relate a vector in the molecule-fixed frame to a vector in a space-fixed frame – in Appendix C. If it didn't, we would have a problem because the transformation $(x, y, z) \to (x, -y, z)$ changes a right-handed coordinate system into a left-handed one. The direction cosine matrix is an operator that rotates vectors. It must preserve the handedness. It is easy to show that

> If a vector (x, y, z) in a right-handed coordinate system is related to a vector (X, Y, Z) by the transformation (C.1), a vector (x, y, z) in a left-handed coordinate system is related to a vector (X, Y, Z) by the transformation obtained from Eq. (C.1) by the substitutions $\theta = \pi - \theta$ and $\phi = \pi + \phi$.

Equation (C.1) shows that the transformation $(x, y, z) \to (x, -y, z)$ accompanied with $\theta \to \pi - \theta$ and $\phi \to \pi + \phi$ is equivalent to the transformation $(X, Y, Z) \to (-X, -Y, -Z)$.

Consider now an isolated atom with one electron whose coordinates are

$$\begin{aligned} x_e &= r_e \sin\theta_e \cos\phi_e \\ y_e &= r_e \sin\theta_e \sin\phi_e \\ z_e &= r_e \cos\theta_e. \end{aligned} \tag{1.48}$$

The symmetry operation $\sigma_v(x_e z_e)$ changing y_e to $-y_e$ corresponds to the transformation $\phi_e \to -\phi_e$. The wave function of an electron in an angular momentum state $|l\lambda\rangle$ is a spherical harmonic $Y_{l,\lambda}(\theta_e, \phi_e)$. The spherical harmonics all depend on ϕ_e through the factor $e^{i\lambda\phi_e}$ and the $\pm\lambda$ components of the spherical harmonics are related as $Y_{l,-\lambda} = (-1)^\lambda Y_{l,\lambda}$ so the transformation $\phi_e \to -\phi_e$ changes

$$|l\lambda\rangle \to (-1)^\lambda |l - \lambda\rangle. \tag{1.49}$$

For two electrons in the total angular momentum state $|L(l_1 l_2)\Lambda\rangle$, we have

$$\sigma_v(xz)|L(l_1 l_2)\Lambda\rangle = \sigma_v(xz)\sum_{\lambda_1}\sum_{\lambda_2} C^{L\Lambda}_{l_1\lambda_1,l_2\lambda_2}|l_1\lambda_1\rangle|l_2\lambda_2\rangle \tag{1.50}$$

$$= \sum_{\lambda_1}\sum_{\lambda_2}(-1)^{\lambda_1+\lambda_2} C^{L\Lambda}_{l_1\lambda_1,l_2\lambda_2}|l_1 - \lambda_1\rangle|l_2 - \lambda_2\rangle \tag{1.51}$$

$$= (-1)^{L+l_1+l_2-\Lambda}\sum_{\lambda_1}\sum_{\lambda_2} C^{L-\Lambda}_{l_1-\lambda_1,l_2-\lambda_2}|l_1 - \lambda_1\rangle|l_2 - \lambda_2\rangle \tag{1.52}$$

$$= (-1)^{L+l_1+l_2-\Lambda}|L(l_1 l_2) - \Lambda\rangle, \tag{1.53}$$

where I used the symmetry properties of the Clebsch–Gordan coefficients [10] and the equality $\Lambda = \lambda_1 + \lambda_2$. The phase factor in Eq. (1.53) may seem inconvenient because l_1 and l_2 are not well-defined quantum numbers, even in an atom. There are many states with different l_1 and l_2 that can contribute to the same state $|L\Lambda\rangle$. The proper way of writing the state $|L\Lambda\rangle$ should include a sum over l_1 and l_2, i.e.

$$|L\Lambda\rangle = \sum_{l_1}\sum_{l_2}\sum_{\lambda_1}\sum_{\lambda_2} C^{L\Lambda}_{l_1\lambda_1,l_2\lambda_2}|l_1\lambda_1\rangle|l_2\lambda_2\rangle. \tag{1.54}$$

However, it turns out that the sum $l_1 + l_2$ determines the parity of the atomic state. This can be readily checked by acting with the parity operator on the right-hand side of Eq. (1.54). Exercise 2.2 will help us to establish that a spherical harmonic $Y_{l,m}$ acquires a factor $(-1)^l$ when the coordinate frame is inverted. Thus, we note that[8]

$$\hat{P}|l_1\lambda_1\rangle|l_2\lambda_2\rangle = (-1)^{l_1+l_2}|l_1\lambda_1\rangle|l_2\lambda_2\rangle \tag{1.55}$$

so the parity of the product states $|l_1\lambda_1\rangle|l_2\lambda_2\rangle$ is determined by the factor $(-1)^{l_1+l_2}$. The atomic states $|L\Lambda\rangle$ are the eigenstates of $\hat{P}$ so

$$\hat{P}|L\Lambda\rangle = \sum_{l_1}\sum_{l_2}(-1)^{l_1+l_2}\sum_{\lambda_1}\sum_{\lambda_2} C^{L\Lambda}_{l_1\lambda_1,l_2\lambda_2}|l_1\lambda_1\rangle|l_2\lambda_2\rangle = \varepsilon|L\Lambda\rangle. \tag{1.56}$$

8 Keep in mind that at this point we are discussing an isolated atom, so the projections λ_1, λ_2, and Λ are with respect to the z-axis of an arbitrary coordinate frame, which is inverted in Eq. (1.55). The z-axis of this coordinate frame will later be fixed to point in the direction of the interatomic axis.

The second equality can only hold if the sum in Eq. (1.54) is allowed to include the terms with $l_1 + l_2$ either even or odd, but not both.

This can, of course, be generalized to multielectron systems, reducing Eq. (1.53) to a general result

$$\sigma_v(xz)|L\Lambda\rangle = \varepsilon(-1)^{L-\Lambda}|L-\Lambda\rangle, \tag{1.57}$$

where $\varepsilon = \pm 1$ is the parity of the atomic state.

We now bring two atoms in states $|L_A\Lambda_A\rangle$ and $|L_B\Lambda_B\rangle$ together and fix the z-axis to point in the direction of the interatomic axis. We can form the total angular momentum states for the electrons in the molecule as follows:

$$|L\Lambda\rangle = \sum_{L_A}\sum_{L_B}\sum_{\Lambda_A}\sum_{\Lambda_B} C^{L\Lambda}_{L_A\Lambda_A,L_B\Lambda_B}|L_A\Lambda_A\rangle|L_B\Lambda_B\rangle. \tag{1.58}$$

Using Eqs. (1.53) and (1.57), we obtain

$$\sigma_v(xz)|L\Lambda\rangle = \varepsilon_A\varepsilon_B(-1)^{L-\Lambda}|L-\Lambda\rangle, \tag{1.59}$$

where ε_A and ε_B are the parities of the two atoms.

Combining two atoms in a molecule spoils the quantum number L, leading to molecular states

$$|\eta\Lambda\rangle = \sum_L C^{\eta}_L|L\Lambda\rangle, \tag{1.60}$$

where the coefficients C^{η}_L form the inverse of the transformation in Eq. (1.35). This creates a bit of a problem. If we were to act with σ_v on the molecular states $|\eta\Lambda\rangle$, the right-hand side of Eq. (1.60) would give us a sum over terms with the phase factor $(-1)^L$. As in Eq. (1.56), these terms conspire to produce a phase factor in addition to $(-1)^{\Lambda}$. However, the states $|\eta\Lambda\rangle$ are the eigenstates of the electronic Hamiltonian so their overall phase is undefined. We will choose the phase of the electronic molecular states so that

$$\sigma_v|\eta\Lambda\rangle = (-1)^{\Lambda}|\eta-\Lambda\rangle, \tag{1.61}$$

for molecular states with $|\Lambda| \neq 0$.

The states with $\Lambda = 0$ deserve a special mention. For such states,

$$\sigma_v|\eta\Lambda = 0\rangle = \pm|\eta\Lambda = 0\rangle, \tag{1.62}$$

the $\pm$ sign determines the intrinsic symmetry, so that this sign is carried over as a subscript in the label of the electronic state. All Σ states are therefore classified as Σ^+ and Σ^-.

In order to determine the effect of σ_v on $|JM\Omega\rangle$, we must recall that the latter represent the D-functions, which depend on three Euler angles α, β, and γ

defined in Appendix D. The symmetry operation σ_v amounts to $\beta \to \pi - \beta$ and $\gamma \to \pi + \gamma$ [11], which, given the properties of the D-functions [10], produces

$$\sigma_v |JM\Omega\rangle = (-1)^{J-\Omega} |JM - \Omega\rangle. \tag{1.63}$$

The spin functions $|S\Sigma\rangle$ are the real tricky ones. They do not have any explicit dependence on the spatial coordinates so how do they transform under the inversion of the coordinate system? Well, we know that $\Omega = \Sigma + \Lambda$, and that both $\Lambda \to -\Lambda$ and $\Omega \to -\Omega$ under σ_v, so we must conclude that

$$\sigma_v |S\Sigma\rangle = (-1)^{S-\Sigma} |S - \Sigma\rangle. \tag{1.64}$$

We can now determine the effect of parity inversion on Hund's case (a) states. For states with nonzero Λ,

$$\hat{P}|JM\Omega\rangle|\eta\Lambda S\Sigma\rangle = (-1)^{J-\Omega-\Lambda+S-\Sigma}|JM-\Omega\rangle|\eta - \Lambda S - \Sigma\rangle. \tag{1.65}$$

Thus, the states $|JM\Omega\rangle|\eta\Lambda S\Sigma\rangle$ are clearly not the eigenstates of the parity inversion operator. However, Eq. (1.65) can be used to construct the states of well-defined parity.

Given a state $|\varphi\rangle$ that changes in some way under parity inversion, the eigenstates of the parity operator can always be constructed as $|\tilde{\varphi}\rangle = |\varphi\rangle \pm \hat{P}|\varphi\rangle$, which can be verified by acting with $\hat{P}$ on $|\tilde{\varphi}\rangle$

$$\hat{P}|\tilde{\varphi}\rangle = \hat{P}|\varphi\rangle \pm \hat{P}^2|\varphi\rangle = \pm(|\varphi\rangle \pm \hat{P}|\varphi\rangle) = \pm|\tilde{\varphi}\rangle. \tag{1.66}$$

The resulting state should, of course, be properly normalized. This suggests that the eigenstates of the parity operator can be constructed from Hund's case (a) states by combining the left-hand side of Eq. (1.65) with its right-hand side. This yields the so-called parity-adapted states. A similar procedure can be used to construct the parity-adapted states for molecules in rotational states described by other Hund's cases. We will return to this as we discuss specific examples later.

1.6 General Notation for Molecular States

Much of the discussion in the following chapters is general and applies to molecules in any electronic states. When this is the case, I will assume that the molecule is prepared in a particular vibrational state $|v\rangle$ of a particular electronic state $|n\rangle$, without specifying the angular momentum quantum numbers of this state. The rotational states of the molecules are described by the general notation $|JMK\varepsilon\rangle$, where the quantum number K is introduced to distinguish the molecular states with the same values of J, M, and ε but with different energies. In the case of Hund's case (a) states, K is $|\Omega|$. For Hund's case (b) states, K is N. The molecular states will thus be denoted by

$$|v\rangle|n\rangle|JMK\varepsilon\rangle. \tag{1.67}$$

1.7 Hyperfine Structure of Molecules

The proton – just like the electron – has a spin. So does the neutron. The vector sum of the spins of all protons and neutrons in an atomic nucleus gives rise to the nuclear spin, commonly denoted by $\boldsymbol{I}$. The ground state of some atomic nuclei happens to have $I = 0$, but, in general, $I \neq 0$. A nonzero nuclear spin generates a magnetic field that perturbs the electrons in the molecule. This effect is hyperly small, and the interactions between the nuclear spins and the electrons are called hyperfine interactions.

Most often, we will be dealing with three types of hyperfine interactions in diatomic molecules [4]: the Fermi contact interaction, the long-range magnetic dipole–dipole interaction, and the interaction between the electric quadrupole moments of the nuclei and the gradient of the electric field generated by the charged particles within the molecule. The first two is a result of the magnetic interactions between the nuclei and the electrons. There are other sources of hyperfine interactions. Here, we will consider, very briefly, these three, just to introduce the reader to hyperfine interactions, explain why they are usually small, and introduce the spherical tensor description discussed in more detail in Appendix E and used later in the book.

1.7.1 Magnetic Interactions with Nuclei

The electron is a spin-1/2 particle and, as such, it possesses a spin magnetic moment

$$\boldsymbol{\mu}_S = -g_S \mu_B \boldsymbol{s}, \tag{1.68}$$

where $\mu_B = e\hbar/2m_e$ is the Bohr magneton and $g_S = 2.00231930419922\ldots$ is the electron spin g value. As discussed in Section 3.1.1, this follows from the Dirac equation. A nucleus α with a nonzero spin $\boldsymbol{I}$ likewise has a magnetic moment

$$\boldsymbol{\mu}_I = g_\alpha \mu_N \boldsymbol{I}, \tag{1.69}$$

where g_α is the nuclear g-factor and μ_N is the nuclear magneton equal to $e\hbar/2m_p$ with m_p being the proton mass. Note that because $m_p \gg m_e$, the nuclear magneton $\mu_N \ll \mu_B$. Therefore, the magnetic moment of a nucleus is much smaller than the magnetic moment of the electron. This alone tells us that we should expect the magnetic hyperfine interactions to be much weaker than, for example, the interactions between two electron spins. The electron spin–spin interactions contribute to the fine structure of molecules with more than one unpaired electrons (see Section 2.4.3).

From classical electrodynamics, we know that the potential energy of a magnetic moment $\boldsymbol{\mu}$ placed in a magnetic field $\boldsymbol{B}$ is

$$U = -\boldsymbol{\mu} \cdot \boldsymbol{B}. \tag{1.70}$$

We also know that the magnetic field $\boldsymbol{B}(\boldsymbol{\mu}, \boldsymbol{r})$ generated by a magnetic dipole $\boldsymbol{\mu}$ a distance $\boldsymbol{r}$ away is

$$\boldsymbol{B}(\boldsymbol{\mu}, \boldsymbol{r}) = \frac{2\mu_0}{3}\boldsymbol{\mu}\delta^3(\boldsymbol{r}) + \frac{\mu_0}{4\pi r^3}\{3(\boldsymbol{\mu}\cdot\hat{\boldsymbol{r}})\hat{\boldsymbol{r}} - \boldsymbol{\mu}\}, \tag{1.71}$$

where μ_0 is the permeability of free space and $\hat{\boldsymbol{r}}$ is the unit vector directed along $\boldsymbol{r}$. Thus, we can think of the magnetic hyperfine interaction as the potential energy of the electron placed in a magnetic field generated by the magnetic moment of the nucleus. As can be seen from Eq. (1.71), this potential energy can be written as a sum of two contributions. The one arising from the first term in Eq. (1.71) is the Fermi contact interactions. The second term in Eq. (1.71) gives rise to a long-range magnetic dipole–dipole interaction.

1.7.2 Fermi Contact Interaction

For an ensemble of electrons and two nuclei, the Fermi contact interaction is simply the sum over the contact dipole–dipole interactions:

$$\hat{H}^{\mathrm{F}}_{\mathrm{hf}} = \frac{2\mu_0}{3}\sum_i\sum_\alpha g_S\mu_B g_\alpha \mu_N \delta^3(\boldsymbol{r}_{\alpha i})\boldsymbol{s}_i\cdot\boldsymbol{I}_\alpha. \tag{1.72}$$

This equation is not very convenient because it depends explicitly on the spins of the individual electrons.

Our ultimate goal is to calculate the perturbation of the molecular energy levels due to the hyperfine interactions. In first order, the perturbation is determined by the expectation value of the hyperfine interaction evaluated for a particular electronic state of the molecule: $\langle \eta, S|\hat{H}^{\mathrm{F}}_{\mathrm{hf}}|\eta, S\rangle$. The molecular states $|\eta, S\rangle$ are determined by the total spin of the molecule and do not depend explicitly on the spins of the individual electrons.

It is clear now why Eq. (1.72) is inconvenient: in order to evaluate the matrix elements $\langle \eta, S|\hat{H}^{\mathrm{F}}_{\mathrm{hf}}|\eta, S\rangle$, we must find the integrals $\langle \eta, S|\boldsymbol{s}_i\delta^3(\boldsymbol{r}_{\alpha i})|\eta, S\rangle$ of single-electron operators over multielectron states. To deal with these integrals, one can introduce the projection operators Φ^S_i defined as follows [4]:

$$\boldsymbol{s}_i = \Phi^S_i \boldsymbol{S}. \tag{1.73}$$

By taking the scalar product of each side of Eq. (1.73) with $\boldsymbol{S}$ and rearranging the terms [4], we can write

$$\Phi^S_i = \boldsymbol{s}_i \cdot \boldsymbol{S}/[S(S+1)]. \tag{1.74}$$

Given these operators, one can reintroduce the Fermi contact interaction in a more convenient *effective* operator form

$$\hat{H}_{\rm hf}^{\rm F,eff} = \langle \eta, S|\hat{H}_{\rm hf}^{\rm F}|\eta, S\rangle$$
$$= \frac{2\mu_0}{3}\sum_i\sum_\alpha g_S\mu_B g_\alpha \mu_N \langle \eta, S|\Phi_i^S\delta^3(\boldsymbol{r}_{\alpha i})|\eta, S\rangle \boldsymbol{S}\cdot\boldsymbol{I}_\alpha \equiv \sum_\alpha b_{\rm F}^\alpha \boldsymbol{S}\cdot\boldsymbol{I}_\alpha. \tag{1.75}$$

Equation (1.75) is nice because it allows us to evaluate the matrix elements of the Fermi contact interaction in the *molecular* basis of $|S\Sigma\rangle|I_A M_{I_A}\rangle|I_B M_{I_B}\rangle$ vectors using angular momentum algebra, as discussed in Appendix B. We have absorbed the integrals over the single-electron operators into the constants

$$b_{\rm F}^\alpha = \sum_i g_S\mu_B g_\alpha \mu_N \langle \eta, S|\Phi_i^S\delta^3(\boldsymbol{r}_{\alpha i})|\eta, S\rangle.$$

These constants can be computed by evaluating the integrals or determined by fitting the eigenvalues of Eq. (1.75) to observed spectroscopic lines. These constants are tabulated in reference books for specific molecules in specific electronic states.

1.7.3 Long-Range Magnetic Dipole Interaction

The second term in Eq. (1.71) gives rise to long-range interactions of the electron spin magnetic moment with the nuclear spin magnetic moment. We see that it can be written in the following form:

$$\hat{H}_{\rm hf}^{\rm dd} = -\sum_\alpha\sum_i g_s\mu_B g_\alpha \mu_N \frac{\mu_0}{4\pi}\left\{\frac{\boldsymbol{s}_i\cdot\boldsymbol{I}_\alpha}{r_{\alpha i}^3} - \frac{3(\boldsymbol{s}_i\cdot\boldsymbol{r}_{\alpha i})(\boldsymbol{I}_\alpha\cdot\boldsymbol{r}_{\alpha i})}{r_{\alpha i}^5}\right\}, \tag{1.76}$$

where *I* replaced the unit vectors $\hat{\boldsymbol{r}}_{\alpha i}$ with $\boldsymbol{r}_{\alpha i}/r_{\alpha i}$. The expression in the curly brackets of Eq. (1.76) can be written much more nicely as scalar product of two rank-2 spherical tensors as follows (see Appendix E and Exercise 1.8 for more details):

$$\frac{\boldsymbol{s}_i\cdot\boldsymbol{I}_\alpha}{r^3} - \frac{3(\boldsymbol{s}_i\cdot\boldsymbol{r})(\boldsymbol{I}_\alpha\cdot\boldsymbol{r})}{r^5} = \left(\frac{4\pi}{5}\right)^{1/2}\frac{\sqrt{6}}{r^3}\sum_q(-1)^q Y_{2-q}(\hat{\boldsymbol{r}})[\boldsymbol{s}_i\otimes\boldsymbol{I}_\alpha]_q^{(2)}, \tag{1.77}$$

where $Y_{2q}(\hat{\boldsymbol{r}})$ is a spherical harmonic of rank 2 that depends on the angles describing the orientation of the unit vector $\hat{\boldsymbol{r}}$ and

$$[\boldsymbol{s}\otimes\boldsymbol{I}]_q^{(2)} = \sum_{q_1}\sum_{q_2} C_{1q_1 1q_2}^{2q}\hat{T}_{q_1}^1(\boldsymbol{s})\hat{T}_{q_2}^1(\boldsymbol{I}) \tag{1.78}$$

is a product of two rank-1 spherical tensors (see Appendix E and Exercise 1.8). Equation (1.76) can thus be rewritten as

$$\hat{H}_{\rm hf}^{\rm dd} = -\sum_{\alpha}\sum_{i} g_s\mu_B g_\alpha \mu_N \frac{\mu_0}{4\pi}\left(\frac{4\pi}{5}\right)^{1/2}\frac{\sqrt{6}}{r_{\alpha i}^3} \times \sum_{q}\sum_{q_1}\sum_{q_2}(-1)^q Y_{2-q}(\hat{\boldsymbol{r}}_{\alpha i}) C_{1q_1 1q_2}^{2q}\,\hat{T}_{q_1}^1(\boldsymbol{s}_i)\hat{T}_{q_2}^1(\boldsymbol{I}_\alpha). \tag{1.79}$$

Once again, one can introduce the spin projection operators Φ_i^S and write the effective operator [4]

$$\hat{H}_{\rm hf}^{\rm dd,eff} = -\sum_{\alpha}\frac{\sqrt{6}}{3}\sum_{q} d_\alpha^q[\boldsymbol{S}\otimes\boldsymbol{I}_\alpha]_q^{(2)}, \tag{1.80}$$

where the coefficients d_α^q are defined as follows:

$$d_\alpha^q = \sum_i 3 g_s\mu_B g_\alpha\mu_N\frac{\mu_0}{4\pi}\left(\frac{4\pi}{5}\right)^{1/2}(-1)^q\langle\eta, S|\frac{\Phi_i^S}{r_{\alpha i}^3}Y_{2-q}(\hat{\boldsymbol{r}}_{\alpha i})|\eta, S\rangle. \tag{1.81}$$

1.7.4 Electric Quadrupole Hyperfine Interaction

Atomic nuclei have an effective shape so some nuclei (namely, those with $I \geq 2$) have an electric quadrupole moment. The quadrupole moments of the nuclei may interact with the electric field gradient ∇E_α generated by the positive and negative charges within the molecule. This interaction can be written in a compact form of a scalar product of two spherical tensors of rank 2

$$\hat{H}_{\rm hf}^{\rm Q} = -e\sum_\alpha T^2(\nabla\boldsymbol{E}_\alpha)\cdot T^2(\boldsymbol{Q}_\alpha), \tag{1.82}$$

and rewritten in the effective form [4]

$$\hat{H}_{\rm hf}^{\rm Q,eff} = \sum_\alpha\sum_k\frac{eQ_\alpha q_k^\alpha}{I_\alpha(2I_\alpha - 1)}\sqrt{6}T_k^2(\boldsymbol{I}_\alpha, \boldsymbol{I}_\alpha), \tag{1.83}$$

where the magnitudes of Q_α and the integrals

$$q_k^\alpha = -2\langle\eta, \Lambda|T_k^2(\nabla\boldsymbol{E}_\alpha)|\eta, \Lambda'\rangle \tag{1.84}$$

are molecule specific and can be found in reference books.

Exercises

1.1 Obtain Eq. (1.6) from Eq. (1.3).
Equation (1.3) is the full nonrelativistic Hamiltonian of a diatomic molecule, while Eq. (1.6) describes the motion of the nuclei in the electronic potential of a single electronic state. This equation is obtained

using the Born–Oppenheimer approximation. To obtain Eq. (1.6), we can represent an eigenstate of the full molecular Hamiltonian by a basis set expansion in terms of the eigenstates of the electronic Hamiltonian (1.1); act with the full Hamiltonian on this basis set expansion and project the result on a particular electronic state (say, the ground state). While doing this, we should neglect the matrix elements of the kinetic energy of the nuclei coupling different electronic states.

1.2 Obtain Eq. (1.7) from Eq. (1.6).
Equation (1.6) describes the motion of two particles with masses m_A and m_B, while Eq. (1.7) describes the motion of a single particle with the reduced mass μ in a coordinate system with the origin placed at the center of mass of the two particles. To obtain Eq. (1.7), it is best to write the kinetic energies of the two particles as $\boldsymbol{p}_A^2/2m_A$ and $\boldsymbol{p}_B^2/2m_B$, introduce the relative coordinate

$$\boldsymbol{R} = \boldsymbol{r}_A - \boldsymbol{r}_B$$

and the center-of-mass coordinate

$$\boldsymbol{R}_{\mathrm{CM}} = (m_A\boldsymbol{r}_A + m_B\boldsymbol{r}_B)/(m_a + m_B).$$

Assuming $\boldsymbol{R}_{\mathrm{CM}} = 0$, we can write the kinetic energy operator in terms of the time derivatives of $\boldsymbol{R}$. The final step is to replace the classical kinetic energy with the quantum expression and write the Laplace operator in the spherical polar coordinates. This leads to Eq. (1.7).

1.3 Use Eq. (1.42) to show that $\boldsymbol{J}^2$, $\boldsymbol{S}^2$, and $\boldsymbol{L}^2$ commute with $\boldsymbol{R}^2$.
To prove this, we just need to expand the right-hand side of Eq. (1.42) and write out the different commutators.

1.4 Find the commutator of $\boldsymbol{L}^2$ with the Hamiltonian given by Eq. (1.1).
Based on symmetry considerations, we know that this commutator must be nonzero. Which of the terms in Eq. (1.1) make this commutator nonzero?

1.5 Find the commutator of $\boldsymbol{L}_z$ with the Hamiltonian given by Eq. (1.1).
Based on symmetry considerations, we know that this commutator must vanish. To show this, it might be helpful to note that $\boldsymbol{L}_z = \sum_i \boldsymbol{l}_{z_i}$, where l_{z_i} is the z-component of the orbital angular momentum of ith electron, the sum is over all the electrons in the molecule, and z is the interatomic axis.

1.6 The ground electronic state of hydrogen is H:1s ^2S and that of oxygen is O:$(1s)^2(2s)^2(2p)^4$ ^{3}P. Write the molecular term for all the electronic states of H_2, OH, and O_2 that can be built up by bringing together their respective ground-state atoms.

1.7 The electronic states of diatomic molecules are characterized by the projection Λ of the total electronic orbital angular momentum and the spin multiplicity $(2S + 1)$. Consider the interaction of two atoms in states of the electronic orbital angular momenta L_A and L_B and spin angular momenta S_A and S_B. Find the maximum number of molecular states that arise from the interaction of such atoms. The result was obtained by Wigner and Witmer [12] and is known as one of the Wigner–Witmer rules. To find the maximum number of states, it might be useful first to find the number of states with $\Lambda = 0$ and consider the maximum value of $|\Lambda|$ possible. This exercise is based on a similar problem in Ref. [9].

1.8 Prove the equality

$$\boldsymbol{s}\cdot\boldsymbol{I} - 3(\boldsymbol{s}\cdot\hat{\boldsymbol{r}})(\boldsymbol{I}\cdot\hat{\boldsymbol{r}}) = \left(\frac{4\pi}{5}\right)^{1/2}\sqrt{6}\sum_q(-1)^q Y_{2-q}(\hat{\boldsymbol{r}})[\boldsymbol{s}\otimes\boldsymbol{I}]_q^{(2)},$$

used in Eq. (1.77).

Note that the spin angular momentum is a vector operator so it is a rank-1 spherical tensor, whose components are given by

$$\hat{T}_1^1(\boldsymbol{s}) = -\frac{1}{\sqrt{2}}\hat{s}_+$$

$$\hat{T}_{-1}^1(\boldsymbol{s}) = \frac{1}{\sqrt{2}}\hat{s}_-$$

$$\hat{T}_0^1(\boldsymbol{s}) = \hat{s}_z,$$

where $\hat{s}_\pm$ are the raising/lowering operators and $\hat{s}_z$ is the Z-component of angular momentum $\boldsymbol{s}$. The same equations hold for the components of $\hat{T}_q^1(\boldsymbol{I})$.

With the above equations, all that is needed to prove the equality is the transformation between the Cartesian basis and the basis of spherical harmonics. Given the relations between the Cartesian coordinates and the spherical polar coordinates

$$z = r\cos\theta$$

$$x = r\sin\theta\cos\phi$$

$$y = r\sin\theta\sin\phi$$

and the mathematical expressions for the spherical harmonics of rank 1

$$Y_{1,0} = \frac{1}{2}\sqrt{\frac{3}{\pi}}\cos\theta$$

$$Y_{1,-1} = \frac{1}{2}\sqrt{\frac{3}{2\pi}} \sin\theta e^{-i\phi}$$

$$Y_{1,+1} = -\frac{1}{2}\sqrt{\frac{3}{2\pi}} \sin\theta e^{i\phi},$$

it is easy to establish the following relations:

$$Y_{1,0} = \frac{1}{2}\sqrt{\frac{3}{\pi}}\frac{z}{r}$$

$$Y_{1,-1} = \frac{1}{2}\sqrt{\frac{3}{2\pi}}\frac{x - iy}{r}$$

$$Y_{1,+1} = -\frac{1}{2}\sqrt{\frac{3}{2\pi}}\frac{x + iy}{r}.$$

Note also that a spherical harmonic $Y_{k,q}$ is a spherical tensor $\hat{T}^k_q$ so a rank-2 spherical harmonic can be written as

$$Y_{2,q} = \sum_{q_1}\sum_{q_2} C^{2q}_{1q_1 1q_2} Y_{1,q_1} Y_{1,q_2},$$

where $C^{2q}_{1q_1 1q_2}$ is a Clebsch–Gordan coefficient. See appendices B and E for more details.

2

DC Stark Effect

2.1 Electric Field Perturbations

A molecule is an ensemble of negative and positive charges. When placed in an external electric field, these charges must be perturbed, leading to changes in the molecular energy levels. To understand the effect of this perturbation, it is helpful to compare the electric field *inside* a molecule with the external electric field that can be generated by two charged metal plates.

From page 5, we know that the Coulomb force $\boldsymbol{F}$ acting on an electron placed at 1-Bohr radius away from a proton is equal to one atomic unit. The electric field generated by the proton is given by $\boldsymbol{E} = \boldsymbol{F}/e$, where e is the charge of the electron. In atomic units, $e = 1$ so the magnitude of $\boldsymbol{E}$ is equal to one atomic unit of the electric field strength. The electric field magnitude is often measured in volts per cm. From Appendix A, 1 atomic unit of electric field is equal to 5.14×10^9 V cm^{-1}. The electric field experienced by electrons inside a molecule is thus $\gtrsim 10^{10}$ V cm^{-1}.

In the laboratory, the electric field is produced by applying voltage to spatially separated electrodes. Using high-quality electrodes, it is relatively easy to generate electric fields $\sim$1–10 kV cm^{-1}. The state-of-the-art techniques can be used to produce fields of up to 100–200 kV cm^{-1}. The magnitude of about 500 kV cm^{-1} = 5×10^5 V cm^{-1} appears to be the largest strength of a static electric field that can be generated in the laboratory.

We see that an external electric field is much weaker than the electric field generated by the elementary particles inside a molecule! This means that the electrons are unlikely to be perturbed by the external field. The nuclear motion, on the other hand, must be affected – or else we wouldn't be discussing this.

In order to understand how an external electric field affects a molecule, we need to start by writing the Hamiltonian of a molecule in an electric field. In this chapter, we assume that the electric field is static and uniform.

In general, an electromagnetic field is defined by a four-potential consisting of a three-dimensional vector potential $\boldsymbol{A}$ and the scalar potential φ. The magnetic

Molecules in Electromagnetic Fields: From Ultracold Physics to Controlled Chemistry, First Edition. Roman V. Krems.

and electric fields are given in terms of $\boldsymbol{A}$ and φ by the following equations:[1]

$$\boldsymbol{B} = \nabla \times \boldsymbol{A}, \tag{2.2}$$

$$\boldsymbol{E} = -\nabla\varphi - \frac{\partial \boldsymbol{A}}{\partial t}. \tag{2.3}$$

In the case of a static field, the electric field is given simply by the negative gradient of the scalar potential

$$\boldsymbol{E} = -\nabla\varphi. \tag{2.4}$$

If the field is uniform, we can write

$$\varphi = -\boldsymbol{E} \cdot \boldsymbol{r}. \tag{2.5}$$

The potential energy of a particle with charge q placed in a uniform electricostatic field is given by

$$V = q\varphi \tag{2.6}$$

Therefore, using Eq. (2.5), we can write

$$V = -q\boldsymbol{E} \cdot \boldsymbol{r}. \tag{2.7}$$

where $\boldsymbol{r}$ is the position vector of the particle.

For an ensemble of particles, we simply sum over the potential energy of each particle

$$V = -\sum_i q_i \boldsymbol{E} \cdot \boldsymbol{r}_i = -\boldsymbol{E} \cdot \left(\sum_i q_i \boldsymbol{r}_i\right) \equiv -\boldsymbol{E} \cdot \boldsymbol{d}, \tag{2.8}$$

where we have defined the dipole moment vector

$$\boldsymbol{d} = \sum_i q_i \boldsymbol{r}_i. \tag{2.9}$$

It is useful to know that, if the ensemble of particles is charge neutral, i.e. if $\sum_i q_i = 0$, the dipole moment vector is independent of the choice of the

---

1 Where do these definitions of the electric and magnetic field come from? As described in Ref. [13], most fundamentally, a field is defined by a four-vector, with three components of this vector being the x, y, and z components and the fourth being the time component. For an electromagnetic field, we will denote the spatial part of the four-vector by $\boldsymbol{A}$ and the time component by φ. Given the four-potential, one can write the Lagrangian for a charged particle in an electromagnetic field. Given the Lagrangian, one can obtain the equation of motion for the charged particle that has the following form [13]

$$\frac{\mathrm{d}\boldsymbol{p}}{\mathrm{d}t} = -q\frac{\partial \boldsymbol{A}}{\partial t} - q\nabla\varphi + q\boldsymbol{v} \times (\nabla \times \boldsymbol{A}). \tag{2.1}$$

The first two terms on the right-hand side are independent of the velocity and represent the force acting on a motionless charge q. This force is due to the electric field. The last term is velocity dependent. It is due to the magnetic field.

coordinate system. To prove this, displace the origin of the coordinate system by a constant vector $\boldsymbol{a}$. The dipole moment vector in the new coordinate system is

$$\boldsymbol{d}' = \sum_i q_i(\boldsymbol{r}_i + \boldsymbol{a}) = \sum_i q_i\boldsymbol{r}_i + \boldsymbol{a}\sum_i q_i = \boldsymbol{d}. \tag{2.10}$$

This is, of course, not true if the charges do not add up to zero.[2]

In order to obtain the Hamiltonian of a diatomic molecule in a uniform electric field, all we have to do is add Eq. (2.8) to Eq. (1.3), assuming that the sum in Eq. (2.8) runs over all electrons and nuclei. We will treat this term as a perturbation and examine its effect on the molecular energy levels in first and second orders of perturbation theory. To do this, we need to evaluate the matrix elements of the operator (2.8) with the electronic–vibrational–rotational states of the diatomic molecule in the brakets. We will take the constant electric field vector outside the matrix elements

$$\langle\psi|\boldsymbol{E}\cdot\boldsymbol{d}|\psi\rangle = E_x\langle\psi|d_x|\psi\rangle + E_y\langle\psi|d_y|\psi\rangle + E_z\langle\psi|d_z|\psi\rangle = \boldsymbol{E}\cdot\langle\psi|\boldsymbol{d}|\psi\rangle \tag{2.11}$$

to express the perturbations in terms of the molecule-specific matrix elements of the dipole moment operator $\langle\psi|\boldsymbol{d}|\psi\rangle$. Note that Eq. (2.8) does not involve electron or nuclear spins so the electric dipole moment operator must be diagonal in spin degrees of freedom. The shift of the molecular energy levels due to the perturbation (2.8) is the dc Stark effect.

2.2 Electric Dipole Moment

What is the dipole moment of a molecule in a particular electronic–vibrational–rotational state $|\psi\rangle = |vnJMK\varepsilon\rangle$? The answer is zero! The expectation value

$$\langle\psi|\boldsymbol{d}|\psi\rangle = 0 \tag{2.12}$$

must vanish because $\boldsymbol{d}$ is a vector operator and the molecular states $|\psi\rangle$ have well-defined parity ε. The inversion of the coordinate axes changes the sign of vectors

$$\hat{P}\boldsymbol{d} = -\boldsymbol{d} \tag{2.13}$$

and introduces the factor $-\varepsilon^2 = -1$ to Eq. (2.12), where ε is the parity of state $|\psi\rangle$. However, the matrix elements of $\boldsymbol{d}$ must be the same in the original and inverted coordinate systems. This can be satisfied only if the matrix elements in Eq. (2.12) are zero.

2 Interestingly, this rule extends to higher-order moments. For example, if the overall charge *and* the dipole moment of a system are zero, the quadrupole moment is independent of the choice of the coordinate frame. Otherwise, the quadrupole moment is frame dependent.

This can be understood more intuitively by considering the integral

$$\int_{-\infty}^{\infty} f(x)xf(x)\mathrm{d}x, \tag{2.14}$$

which must be zero because x changes sign at $x = 0$ and $f(x)^2$ is an even function of x.

On the other hand, the integrals

$$\int_{-\infty}^{\infty} f(x_1, x_2)(x_1 + x_2)f(x_1, x_2)\mathrm{d}x_1 = x_2 \int_{-\infty}^{\infty} f(x_2, x_1)^2 \mathrm{d}x_1 = x_2 g(x_2) \tag{2.15}$$

do not have to be zero because the integration is performed only over a part of the variable space.

That is why the expectation value

$$\boldsymbol{d}_n(R) = \langle n|\boldsymbol{d}|n\rangle = d_n(R)\hat{R} \tag{2.16}$$

produces a nonvanishing function $d_n(R)$, which is the dipole moment function of the molecule in a particular electronic state n. In Eq. (2.16), we have integrated over the electronic, but not the nuclear coordinates, so we have integrated over some, but not all, the coordinates. The result is a vector $\boldsymbol{d}_n(R)$ directed along the internuclear axis $\hat{R} = \boldsymbol{R}/R$. The typical shape of the dipole moment function $d_n(R)$ for a heteronuclear diatomic molecule (which separates in the limit $R \to \infty$ into neutral atoms) is shown in Figure 2.1.

The expectation value

$$\boldsymbol{d}_{n,v} = \langle v|\boldsymbol{d}_n(R)|v\rangle = \langle v|d_n(R)|v\rangle\hat{R} \tag{2.17}$$

is the permanent electric dipole moment of a diatomic molecule in a particular vibrational state v of a particular electronic state n. The magnitude of the permanent dipole moment is conventionally measured in units of Debye. For typical polar diatomic molecules, $d_{n,v} = |\boldsymbol{d}_{n,v}|$ ranges from 0.01 to 9 Debye.

We can now rewrite Eq. (2.12) as $\langle JMK\varepsilon|\boldsymbol{d}_{n,v}(\hat{R})|JMK\varepsilon\rangle = 0$. This equation has profound consequences. In particular, it means that the first-order correction to the energy of the molecular states due to the perturbation (2.8) vanishes. This means that the Stark shifts of molecular energy levels E_{nvJMK} are determined by the second-order correction

$$\Delta E_{nvJMK\varepsilon} = \sum_{n'v'J'M'K'\varepsilon'} \frac{\langle nvJMK\varepsilon|\boldsymbol{E}\cdot\boldsymbol{d}|n'v'J'M'K'\varepsilon'\rangle\langle n'v'J'M'K'\varepsilon'|\boldsymbol{E}\cdot\boldsymbol{d}|nvJMK\varepsilon\rangle}{E_{nvJMK\varepsilon} - E_{n'v'J'M'K'\varepsilon'}} \tag{2.18}$$

and must be quadratic functions of the electric field strength. Note that perturbation theory and Eq. (2.18) may be inadequate at high electric fields.

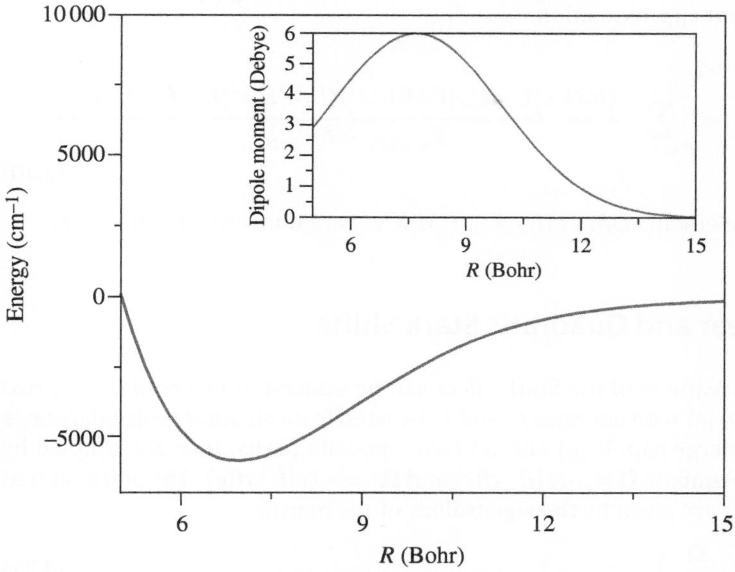

Figure 2.1 The molecular potential and the dipole moment function of the molecule LiCs in the ground electronic state $^1\Sigma$.

To get a feeling for the electric-field-induced perturbations, it is helpful to know the magnitude of the matrix elements in Eq. (2.18). While these matrix elements depend on the details of the electronic, vibrational, and rotational wave functions, it is safe to assume that

$$|\langle n v J M K \varepsilon | \boldsymbol{E} \cdot \boldsymbol{d} | n' v' J' M' K' \varepsilon' \rangle| < E d_{n,v}, \tag{2.19}$$

where $n \neq n'$ and/or $v \neq v'$. How big can the right-hand side be? Assuming $E =$ 514 kV cm^{-1} (= 10^{-4} atomic units) and $d_{n,v} = 9$ Debye (= 3.54 atomic units), the value $Ed_{n,v} = 3.54 \times 10^{-4}$ atomic units = 77.7 cm^{-1}. This is an extreme case. For less crazy fields and "normal" molecules, the typical value of these matrix elements is a few cm^{-1}, at best. If we compare this to the energy separation between electronic states, often >10 000 cm^{-1}, we need to conclude that the perturbation (2.8) is not likely to mix different electronic states. But we knew this already! What we can now also conclude is that the perturbation (2.8) does not appreciably mix the vibrational states, either.[3] This means that Eq. (2.18) is,

3 This is true, at least, for molecules in low vibrational states and electric fields <100 kV cm^{-1}. I should note that the off-diagonal matrix elements (2.19) are usually *much* smaller than $Ed_{n,v}$ because the integration is performed over a product of two functions whose overlap integral is zero.

to an excellent approximation,

$$\Delta E_{nvJMK\varepsilon} = \sum_{J'M'K'\varepsilon'} \frac{\langle JMK\varepsilon|\boldsymbol{E}\cdot\boldsymbol{d}_{n,v}|J'M'K'\varepsilon'\rangle\langle J'M'K'\varepsilon'|\boldsymbol{E}\cdot\boldsymbol{d}_{n,v}|JMK\varepsilon\rangle}{E_{nvJMK\varepsilon} - E_{nvJ'M'K'\varepsilon'}}. \tag{2.20}$$

The matrix elements $\langle JMK\varepsilon|\boldsymbol{E}\cdot\boldsymbol{d}_{n,v}|J'M'K'\varepsilon'\rangle$ are nonzero, only if $\varepsilon' \neq \varepsilon$.

2.3 Linear and Quadratic Stark Shifts

The general features of the Stark effect can be understood by considering two states $|g\rangle$ and $|e\rangle$ with energies E_g and E_e isolated from all other molecular states by a large energy gap. If $|g\rangle$ and $|e\rangle$ have opposite parity, they are coupled by the matrix elements $\Omega = -\langle g|\boldsymbol{E}\cdot\boldsymbol{d}|e\rangle$ and $\Omega^* = -\langle e|\boldsymbol{E}\cdot\boldsymbol{d}|g\rangle$. The Stark-shifted energy levels are given by the eigenvalues of the matrix

$$\begin{pmatrix} E_g & \Omega \\ \Omega^* & E_e \end{pmatrix}, \tag{2.21}$$

which are

$$E_{1,2} = \frac{(E_g + E_e)}{2} \pm \frac{\sqrt{4|\Omega|^2 + (E_g - E_e)^2}}{2}. \tag{2.22}$$

Equation (2.22) tells us all there is to know about the dc Stark effect. In particular, I want to point out two general features:

- If $|\Omega| \ll |E_g - E_e|$, the Taylor series expansion of the square root gives

  $$E_{1,2} = \frac{(E_g + E_e)}{2} \pm \left[\frac{(E_g - E_e)}{2} + \frac{|\Omega|^2}{(E_g - E_e)}\right], \tag{2.23}$$

  which is the same as Eq. (2.18). Explicitly,

  $$E_1 = E_g + \frac{|\Omega|^2}{(E_g - E_e)} \tag{2.24}$$

  and

  $$E_2 = E_e - \frac{|\Omega|^2}{(E_g - E_e)}. \tag{2.25}$$

  Because Ω is very small, $E_1 \sim E_g$ and $E_2 \sim E_e$. If $E_g < E_e$, then $E_g - E_e < 0$ and, consequently,

  $$E_1 < E_g$$

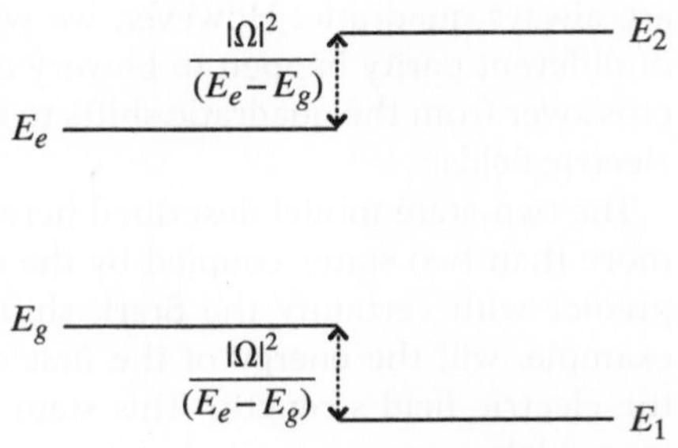

Figure 2.2 Shifts of the energy levels E_g and E_e in the presence of a perturbation Ω that couples the states $|g\rangle$ and $|e\rangle$.

and

$$E_2 > E_e.$$

We see that the lower energy level shifts *down* and the upper energy level shifts *up*, i.e. the states *repel* each other. This result is so important that I represent it graphically in Figure 2.2.

- If $|\Omega| \gg |E_g - E_e|$, then

$$E_{1,2} = \frac{(E_g + E_e)}{2} \pm |\Omega|, \tag{2.26}$$

Because $\Omega \propto E$, the energy separation between E_1 and E_2 increases *linearly* with the electric field strength. We can no longer tell which of the two states is $|g\rangle$ and which is $|e\rangle$ because the $|g\rangle$ and $|e\rangle$ states are strongly mixed and the g and e labels lose their meaning. We are dealing with the linear Stark effect.

We can now make a few predictions about how a static electric field changes the molecular energy levels:

- The energy of the lowest energy level *always* decreases in the presence of an electric field.
- At low electric fields, the energy shifts of the molecular levels are quadratic functions of the electric field strength.
- At high electric fields, the energy shifts of the molecular levels are linear functions of the electric field strength.
- The crossover from the quadratic to linear shifts depends on the energy separation between the molecular states of opposite parity. In particular, if $|g\rangle$ and $|e\rangle$ were degenerate, there would be no quadratic Stark shifts.

There are no diatomic molecules with truly degenerate states of different parity.[4] The Stark shifts of molecular energy levels in the limit of low electric fields

4 This statement must be made with a caveat. In Chapter 7, we will discuss molecules whose rotational energy levels of different parity may become degenerate in the presence of a magnetic field.

are always quadratic. However, we will find that, in some molecules, states of different parity happen to be very close in energy. For such molecules, the crossover from the quadratic shifts to the linear Stark effect occurs at very low electric fields.

The two-state model described here is an idealization. In general, there are more than two states coupled by the dipole operator. That is why we cannot predict with certainty the Stark shifts of the excited molecular levels. For example, will the energy of the first excited state increase or decrease with the electric field strength? This state must repel from the ground state *and* from higher energy excited states so the result is determined by the tug of war (or, more accurately, push of war) between two (or more) couplings. In order to compute the Stark shifts of molecular levels accurately, it is generally necessary to diagonalize the matrix of the operator (2.8) in the basis of electronic–vibrational–rotational states that describe the molecule at zero electric field.

2.4 Stark Shifts of Rotational Levels

The rotational energy levels of a diatomic molecule in an electric field can be computed by diagonalizing the matrix of the effective Hamiltonian given by the sum of Eq. (1.18) and the electric-field-induced interaction (2.8). It is usually most convenient to perform the calculations in the basis of the corresponding Hund's case, in which the field-free Hamiltonian is diagonal. For an accurate calculation, we need to include all relevant states in the basis set. The number of relevant states depends on the type of the molecule and the magnitude of the electric field. At low electric fields, the energy shifts are well described by Eq. (2.20). The departure from this equation occurs at electric fields strong enough to mix different parity states appreciably.

2.4.1 Molecules in a $^1\Sigma$ Electronic State

$$^1\Sigma : S = 0, \Lambda = 0, \Sigma = 0, \Omega = 0, \boldsymbol{J} = \boldsymbol{N}.$$

The rotational structure of molecules in the $^1\Sigma$ electronic state is particularly simple because $S = 0$ and $\Lambda = 0$. This electronic state is nondegenerate and there are no fine structure interactions to perturb the rotational energy levels. The rotational operator (1.42) is

$$B_e \boldsymbol{R}^2 = B_e[\boldsymbol{N} - \boldsymbol{L}]^2 = B_e[\boldsymbol{N}^2 + \boldsymbol{L}^2 - 2N_z L_z - N_+ L_- - N_- L_+]. \quad (2.27)$$

The operators including $L_\pm$ induce couplings to other electronic states that are too far away in energy to be important (see Section 1.4.1). Therefore, we can safely ignore these terms. The third term on the right-hand side gives no contribution because $\langle \eta S\Lambda\Sigma | L_z | \eta S\Lambda\Sigma \rangle = \Lambda = 0$ for a Σ electronic state. The expectation value $\langle \eta S\Lambda\Sigma | \boldsymbol{L}^2 | \eta S\Lambda\Sigma \rangle$ can be incorporated as a small correction into the molecular potential. The effective rotational Hamiltonian for $^1\Sigma$ molecules is then simply

$$B_e \boldsymbol{R}^2 \Rightarrow B_e \boldsymbol{N}^2. \tag{2.28}$$

The rotational states $|JMK\rangle \Rightarrow |NM_N\rangle$ can be labeled by the quantum numbers $N = J$ and $M_N = M$. The additional quantum number K is redundant because the states with different energy are uniquely labeled by a different value of J. The rotational states $|NM_N\rangle$ are the spherical harmonics $Y_{NM_N}(\theta, \phi)$. They are the eigenstates of $\boldsymbol{N}^2$ and N_Z:

$$\boldsymbol{N}^2 |NM_N\rangle = N(N+1)|NM_N\rangle \tag{2.29}$$

$$N_Z |NM_N\rangle = M_N |NM_N\rangle. \tag{2.30}$$

At zero electric field, the energy of the rotational states is

$$E_N = B_e N(N+1), \tag{2.31}$$

and the rotational states $|NM_N\rangle$ with different values of M_N are degenerate.

The energy levels of a $^1\Sigma$ molecule in an electric field are given by the eigenvalues of the following effective Hamiltonian:

$$\begin{aligned} \hat{H}_{\text{Stark}} &= \sum_{NM_N} |NM_N\rangle B_e N(N+1) \langle NM_N| \\ &\quad + \sum_{NM_N} \sum_{N'M'_N} |NM_N\rangle V^{\text{ef}}_{NM_N;N'M'_N} \langle N'M'_N|, \end{aligned} \tag{2.32}$$

where

$$V^{\text{ef}}_{NM_N;N'M'_N} = -\langle NM_N | \boldsymbol{E} \cdot \boldsymbol{d}_0 | N'M'_N \rangle. \tag{2.33}$$

Since we are discussing only one particular electronic state and only one vibrational state, I use a simplified notation $\boldsymbol{d}_0 \equiv \boldsymbol{d}_{n,v}$ for the dipole moment defined in Eq. (2.17).

It is most convenient[5] to direct the laboratory-fixed Z-axis parallel to the electric field vector $\boldsymbol{E}$. The operator $\boldsymbol{E} \cdot \boldsymbol{d}_0$ can then be written as

$$\boldsymbol{E} \cdot \boldsymbol{d}_0 = E d_0 \hat{Z} \cdot \hat{R} = E d_0 \cos\theta, \tag{2.34}$$

where E is the strength of the electric field and d_0 is the magnitude of the permanent dipole moment of the molecule. From Exercise 1.8 and Appendix E, we know that

$$\cos\theta = \sqrt{\frac{4\pi}{3}} Y_{10}(\theta, 0), \tag{2.35}$$

5 In principle, the choice of the space-fixed Z-axis is arbitrary. However, if the Z-axis is not pointing along the field direction, we lose the convenience of M – the Z-projection of $\boldsymbol{J}$ – being a good quantum number.

where Y_{10} is a spherical harmonic of rank 1. Thus, the matrix elements (2.33) reduce to the integrals over products of three spherical harmonics

$$V^{\text{ef}}_{NM_N;N'M'_N} = -Ed_0\sqrt{\frac{4\pi}{3}}\langle NM_N|Y_{10}|N'M'_N\rangle, \tag{2.36}$$

which can be evaluated using the Wigner–Eckart theorem (see Section 9.1 and Appendix D) to yield

$$\langle NM_N|Y_{10}|N'M'_N\rangle = (-1)^{M_N}\sqrt{\frac{3}{4\pi}}\sqrt{(2N+1)(2N'+1)} \times \begin{pmatrix} N & 1 & N' \\ -M_N & 0 & M'_N \end{pmatrix}\begin{pmatrix} N & 1 & N' \\ 0 & 0 & 0 \end{pmatrix}. \tag{2.37}$$

The 3*j* symbols – the symbols in the large brackets – are not just numbers. They have a name attached to them, which means that they have peculiar properties. Of particular relevance to our discussion are the following four:

- A 3*j* symbol $\begin{pmatrix} j_1 & j_2 & j_3 \\ m_1 & m_2 & m_3 \end{pmatrix} = 0$, unless $m_1 + m_2 + m_3 = 0$.
- A 3*j* symbol $\begin{pmatrix} j_1 & j_2 & j_3 \\ 0 & 0 & 0 \end{pmatrix} = 0$, unless $j_1 + j_2 + j_3$ is an even integer.
- A 3*j* symbol $\begin{pmatrix} j_1 & j_2 & j_3 \\ m_1 & m_2 & m_3 \end{pmatrix} = 0$, unless $|j_1 - j_3| \le j_2 \le j_1 + j_3$.
- Flipping the signs of all projections introduces a simple phase factor

$$\begin{pmatrix} j_1 & j_2 & j_3 \\ m_1 & m_2 & m_3 \end{pmatrix} = (-1)^{j_1+j_2+j_3}\begin{pmatrix} j_1 & j_2 & j_3 \\ -m_1 & -m_2 & -m_3 \end{pmatrix}$$

Given these properties of the 3*j*-symbols, Eq. (2.37) reveals a lot of insightful information. In particular, we see that

- The first of the 3*j* symbols in Eq. (2.37) vanishes unless $M_N = M'_N$. This means that the electric field directed along the Z-axis couples only states with the same projection of $\boldsymbol{N}$. This means that M_N is a good quantum number and the energy levels of a $^1\Sigma$ molecule in an electric field can be labeled by M_N. The conservation of the angular momentum projection on the electric field axis also follows from the symmetry of the problem, namely, the fact that the electric field potential is invariant under rotation about the electric field axis (provided, of course, the Z-axis is chosen to be along the field direction, see footnote on page 43).
- The second of the 3*j* symbols in Eq. (2.37) vanishes unless $N + N'$ is an odd integer. This means that the diagonal matrix elements, for which $N = N'$, of the dipole moment operator are zero. But we already knew that! This also means that the parity of the rotational states $|NM_N\rangle$ can be associated with the factor $\varepsilon = (-1)^N$. Exercise 2.2 will ask you to prove this independently.

- The second of the 3j symbols in Eq. (2.37) vanishes unless $|N - N'| = 1$. This proves the familiar selection rule $\Delta N = \pm 1$ for the matrix elements of the dipole moment operator.
- Because $(-1)^{N+N'+1} = 1$ (or else the second of the 3j symbols is zero), flipping the sign of M_N and M'_N leaves Eq. (2.37) unchanged. This means that the molecular states $|N, M_N\rangle$ and $|N, -M_N\rangle$ remain degenerate in the presence of an electric field.

Figure 2.3 illustrates the dc Stark effect for a molecule in the $^1\Sigma$ electronic state. At zero electric field, the energy of the rotational states is given by Eq. (2.31). The energy separation between the rotational ground state and the first excited rotational state is equal to $2B_e$. I express the energy of the molecular states in units of $2B_e$ and plot the energy levels as functions of the dimensionless parameter d_0E/B_e, which makes Figure 2.3 a universal illustration of the dc Stark effect for $^1\Sigma$ molecules. Note that the states with $N > 0$ are $(2N + 1)$-degenerate at zero electric field. This degeneracy is lifted in the presence of an electric field, splitting the rotational levels into the Stark levels with different values of $|M_N|$. The Stark shifts are quadratic at low electric fields and linear at high electric fields, as predicted. The Stark levels with the same value of M_N repel each other, also as predicted. Note, for example, that the energy shift of

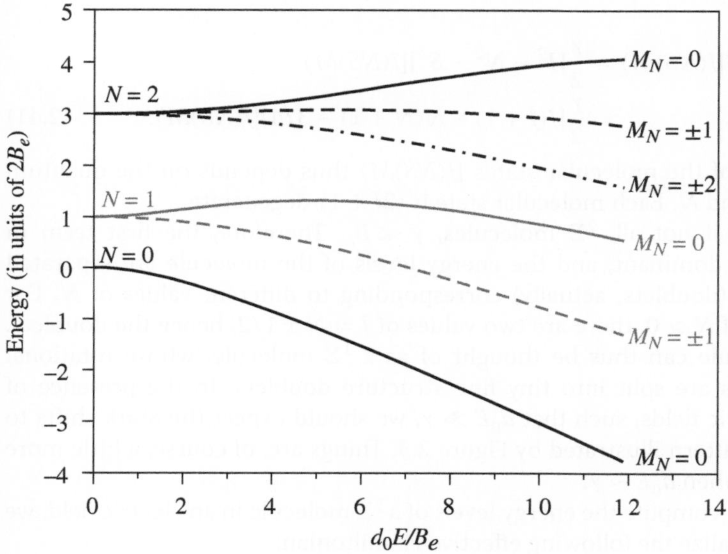

Figure 2.3 The Stark shifts of the rotational energy levels of a diatomic molecule in a $^1\Sigma$ electronic state with the permanent dipole moment d_0 and the rotational constant B_e as functions of the electric field strength E.

the states $|M_N = \pm 2\rangle$ is necessarily negative because this is the lowest-energy state with $M_N = \pm 2$.

2.4.2 Molecules in a $^2\Sigma$ Electronic State

$$^2\Sigma : S = 1/2, \Lambda = 0, \Sigma = 1/2, \Omega = 0, \boldsymbol{J} = \boldsymbol{N} + \boldsymbol{S}$$

Because $\Lambda = 0$, the effective rotational Hamiltonian for the diatomic molecule in this electronic state is also given by Eq. (2.28). However, the rotational structure of $^2\Sigma$ molecules is perturbed by the spin–rotation interaction, which can be written in an effective operator form $V_{\text{fs}} = \gamma \boldsymbol{N} \cdot \boldsymbol{S}$, where γ is the spin–rotation interaction constant. The molecular states are the eigenstates of the operator

$$\hat{H}_{\text{rot, fine}} = B_e \boldsymbol{N}^2 + \gamma \boldsymbol{N} \cdot \boldsymbol{S}. \tag{2.38}$$

They can be represented as

$$|J(NS)M\rangle = \sum_{M_N} \sum_{M_S} C^{JM}_{NM_N SM_S} |NM_N\rangle |SM_S\rangle, \tag{2.39}$$

which leads to the eigenvalue equations

$$B_e \boldsymbol{N}^2 |J(NS)M\rangle = B_e N(N+1) |J(NS)M\rangle \tag{2.40}$$

and

$$\begin{aligned} \gamma \boldsymbol{N} \cdot \boldsymbol{S} |J(NS)M\rangle &= \frac{\gamma}{2} [\boldsymbol{J}^2 - \boldsymbol{N}^2 - \boldsymbol{S}^2] |J(NS)M\rangle \\ &= \frac{\gamma}{2} [J(J+1) - N(N+1) - 3/4] |J(NS)M\rangle. \end{aligned} \tag{2.41}$$

The energy of the molecular states $|J(NS)M\rangle$ thus depends on the quantum numbers J and N. Each molecular state is $(2J+1)$-degenerate.

For most, if not all, $^2\Sigma$ molecules, $\gamma \ll B_e$. Therefore, the first term in Eq. (2.38) is dominant, and the energy levels of the molecule are separated into groups (doublets, actually) corresponding to different values of N. For each value of $N > 0$, there are two values of $J = N \pm 1/2$, hence the doublets. A $^2\Sigma$ molecule can thus be thought of as a $^1\Sigma$ molecule, whose rotational energy levels are split into tiny fine structure doublets. In the presence of strong electric fields, such that $d_0 E \gg \gamma$, we should expect the Stark shifts to follow the pattern illustrated by Figure 2.3. Things are, of course, a little more interesting when $d_0 E \sim \gamma$.

In order to compute the energy levels of a $^2\Sigma$ molecule in an electric field, we must diagonalize the following effective Hamiltonian:

$$\begin{aligned} \hat{H}_{\text{Stark}} &= \sum_{JNM} |J(NS)M\rangle \left\{ B_e N(N+1) + \frac{\gamma}{2} [\boldsymbol{J}^2 - \boldsymbol{N}^2 - \boldsymbol{S}^2] \right\} \langle J(NS)M| \\ &+ \sum_{JNM} \sum_{J'N'M'} |J(NS)M\rangle V^{\text{ef}}_{JNM;J'N'M'} \langle J'(N'S)M'|. \end{aligned} \tag{2.42}$$

To evaluate the matrix elements of the field-induced interaction

$$V^{\mathrm{ef}}_{JNM;J'N'M'} = -Ed_0\sqrt{\frac{4\pi}{3}}\langle J(NS)M|Y_{10}|J'(N'S)M'\rangle, \tag{2.43}$$

we can use again the Wigner-Eckart theorem, yielding (see Section 9.1)

$$\langle J(NS)M|Y_{10}|J'(N'S)M'\rangle$$
$$= (-1)^{J-M}\begin{pmatrix} J & 1 & J' \\ -M & 0 & M' \end{pmatrix}\langle J(NS)||\hat{T}^1||J'(N'S)\rangle. \tag{2.44}$$

Just like Eq. (2.33), this equation nicely illustrates the conservation of the total angular momentum projection M. The double-bar matrix element can be evaluated using Eq. (5.72) of Ref. [9] or as discussed in Section 9.3.2. It is important to keep in mind that the spherical tensor $\hat{T}^1$ acts on the subspace $|NM_N\rangle$. Copying from Ref. [9],

$$\langle J(NS)||\hat{T}^1||J'(N'S)\rangle = (-1)^{J'+3/2}[(2J+1)(2J'+1)]^{1/2}\begin{Bmatrix} N & J & 1/2 \\ J' & N' & 1 \end{Bmatrix}$$
$$\times [(2N+1)(2N'+1)]^{1/2}\sqrt{\frac{3}{4\pi}}\begin{pmatrix} N & 1 & N' \\ 0 & 0 & 0 \end{pmatrix}. \tag{2.45}$$

The symbol in the curly braces is a 6*j* symbol, which can be evaluated numerically with subroutines readily available in most computing environments such as Mathematica. The last 3*j* symbol ensures that the electric-field-induced interaction couples only states with $|N - N'| = 1$.

There is an easier way to solve the eigenvalue problem – one that bypasses the need to derive Eq. (2.45) and evaluate the 6*j* symbols. The molecular states $|J(NS)M\rangle$ are related to the product states $|NM_N\rangle|SM_S\rangle$ by a unitary transformation (2.39). Therefore, instead of using the basis of molecular states $|J(NS)M\rangle$, it is perfectly legitimate to use the basis of direct products $|NM_N\rangle|SM_S\rangle$, leading to the effective Hamiltonian

$$\hat{H}_{\mathrm{Stark}} = \sum_{M_S}|SM_S\rangle\langle SM_S|\left\{\sum_{NM_N}|NM_N\rangle B_e N(N+1)\langle NM_N| + \sum_{NM_N}\sum_{N'M'_N}|NM_N\rangle V^{\mathrm{ef}}_{NM_N;N'M'_N}\langle N'M'_N|\right\}$$
$$+\sum_{M_S}\sum_{M'_S}\sum_{M_N}\sum_{M'_N}|SM_S\rangle|NM_N\rangle V^{\mathrm{SR}}_{NM_NM_S,NM'_NM'_S}\langle SM'_S|\langle NM'_N|. \tag{2.46}$$

The terms in the big curly braces give the Stark Hamiltonian of a $^1\Sigma$ molecule. The projection operator $|SM_S\rangle\langle SM_S|$ ensures that this term is diagonal in the spin quantum number M_S. The last term requires a little work. It couples states with different M_S and M_N but preserves N. The matrix elements

$$V^{\mathrm{SR}}_{NM_NM_S,NM'_NM'_S} = \gamma \langle NM_N | \langle SM_S | \boldsymbol{N} \cdot \boldsymbol{S} | N'M'_N \rangle | SM'_S \rangle \tag{2.47}$$

can be evaluated by writing

$$\boldsymbol{N} \cdot \boldsymbol{S} = N_Z S_Z + \frac{1}{2}(N_+ S_- + N_- S_+) \tag{2.48}$$

and noting that the $N_\pm$ operators act only on the $|NM_N\rangle$ states, while the $S_\pm$ operators act only on the spin states. The ladder operators $N_\pm$ change the rotational states as follows:

$$N_\pm |NM_N\rangle = [N(N+1) - M_N(M_N \pm 1)]^{1/2} |NM_N \pm 1\rangle, \tag{2.49}$$

which leads to the following expression for the matrix elements of $N_\pm$:

$$\langle NM_N | N_\pm | N'M'_N \rangle = \delta_{N,N'} \delta_{M_N,M'_N \pm 1} [N(N+1) - M'_N(M'_N \pm 1)]^{1/2}. \tag{2.50}$$

The analogous equation holds for the matrix elements $\langle SM_S | S_\pm | SM'_S \rangle$.

Figure 2.4 illustrates what happens to a $^2\Sigma$ molecule in an electric field. As predicted, the Stark energy levels at high electric fields form the same pattern as the Stark levels of a $^1\Sigma$ molecule shown in Figure 2.3, except that some of the levels are split into doublets by the spin–rotation interaction. Note that the spin–rotation splittings become larger as the rotational energy increases. This is simply the result of the spin–rotation interaction depending on the rotational angular momentum as $\propto \boldsymbol{S} \cdot \boldsymbol{N}$.

In the limit of zero electric field, the energy levels can be labeled by the quantum numbers J and N. This can be done by inspection, noting that the ground rotational state $N = 0$ has no fine structure doublet and the excited states form pairs of J states that lie very close in energy. We know that the higher energy level of each fine structure doublets corresponds to a larger value of J. We know this because the higher energy level splits into more Stark levels. This happens because the spin–rotation interaction constant of the molecule chosen for this example is positive ($\gamma > 0$). Had the spin–rotation interaction constant γ been negative, the order of the fine structure levels would have been reversed, i.e. the higher J states would have been lower in energy. Using a similar logic, we can assign the Stark levels the values of the quantum number M. I have done this for the lowest four states. Exercise 2.4 will ask you to do the rest of the job.

2.4.3 Molecules in a $^3\Sigma$ Electronic State

$$^3\Sigma : S = 1, \Lambda = 0, \Sigma = 0, \pm 1, \Omega = \Sigma, \boldsymbol{J} = \boldsymbol{N} + \boldsymbol{S}.$$

Once again, $\Lambda = 0$, and the rotational energy is determined by $B_e \boldsymbol{N}^2$. The rotational levels are now perturbed by two fine structure interactions, the

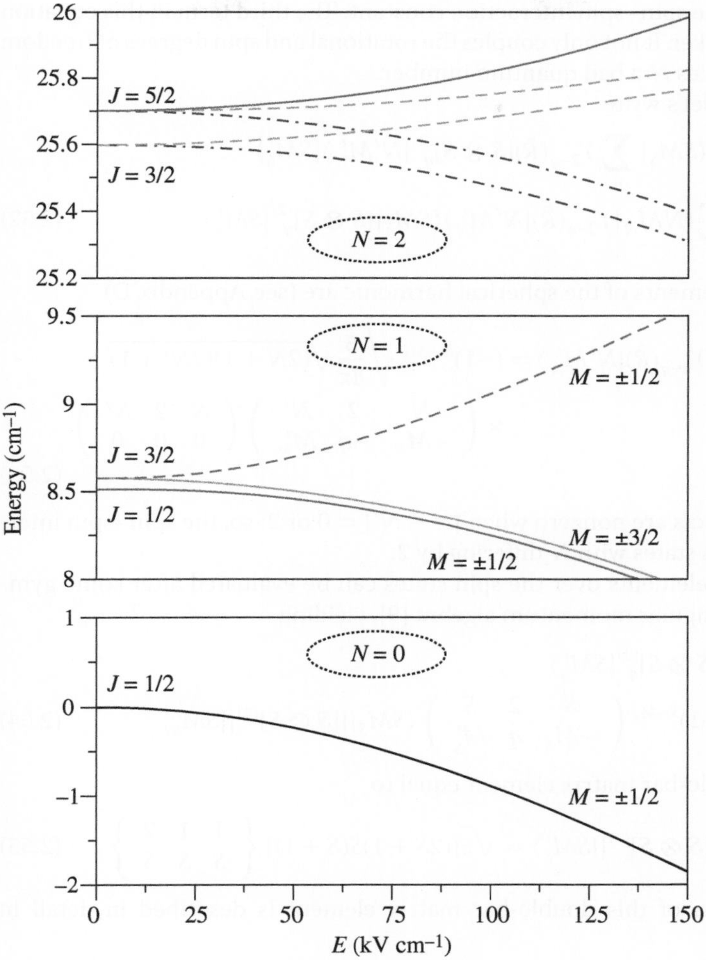

Figure 2.4 The Stark shifts of the rotational energy levels of a diatomic molecule in a $^2\Sigma$ electronic state. The energy levels presented are for the molecule CaH, which has $B_e = 4.277$ cm^{-1}, $\gamma = 0.0415$ cm^{-1}, and the dipole moment 2.94 Debye. The different panels correspond to different values of N in the limit of zero electric field. Bear in mind that N is not a good quantum number because the electric-field-induced interaction (2.36) couples states with different N.

spin–rotation interaction and the spin–spin interaction, leading to the effective Hamiltonian [2]

$$\hat{H}_{\text{rot,fine}} = B_e \boldsymbol{N}^2 + \gamma \boldsymbol{N} \cdot \boldsymbol{S} + \frac{2}{3}\lambda_{SS}\sqrt{\frac{4\pi}{5}}\sqrt{6}\sum_q Y_{2-q}(\hat{R})[\boldsymbol{S} \otimes \boldsymbol{S}]_q^{(2)}, \tag{2.51}$$

where λ_{SS} is the spin–spin interaction constant. The third term in this equation is a trouble maker. It not only couples the rotational and spin degrees of freedom but it also makes N a bad quantum number.

To see this, let's write

$$\begin{aligned}&\langle NM_N|\langle SM_S|\sum_q Y_{2-q}(\hat{R})[\boldsymbol{S}\otimes\boldsymbol{S}]^{(2)}_q|N'M'_N\rangle|SM'_S\rangle\\&\quad=\sum_q\langle NM_N|Y_{2-q}(\hat{R})|N'M'_N\rangle\langle SM_S|[\boldsymbol{S}\otimes\boldsymbol{S}]^{(2)}_q|SM'_S\rangle.\end{aligned}\tag{2.52}$$

The matrix elements of the spherical harmonic are (see Appendix D)

$$\begin{aligned}\langle NM_N|Y_{2-q}(\hat{R})|N'M'_N\rangle&=(-1)^{N-M_N}\sqrt{\frac{5}{4\pi}}\sqrt{(2N+1)(2N'+1)}\\&\quad\times\begin{pmatrix}N&2&N'\\-M_N&-q&M'_N\end{pmatrix}\begin{pmatrix}N&2&N'\\0&0&0\end{pmatrix}.\end{aligned}\tag{2.53}$$

These $3j$ symbols are nonzero when $|N-N'|=0$ or 2; so, the spin–spin interaction couples states with N differing by 2.

The matrix elements over the spin states can be evaluated after some gymnastics with angular momentum algebra [9], yielding

$$\begin{aligned}&\langle SM_S|[\boldsymbol{S}\otimes\boldsymbol{S}]^{(2)}_q|SM'_S\rangle\\&\quad=(-1)^{S-M_S}\begin{pmatrix}S&2&S\\-M_S&q&M'_S\end{pmatrix}\langle SM_S||[\boldsymbol{S}\otimes\boldsymbol{S}]^{(2)}||SM'_S\rangle\end{aligned}\tag{2.54}$$

with the double-bar matrix element equal to

$$\langle SM_S||[\boldsymbol{S}\otimes\boldsymbol{S}]^{(2)}||SM'_S\rangle=\sqrt{5}[(2S+1)S(S+1)]\begin{Bmatrix}1&1&2\\S&S&S\end{Bmatrix}.\tag{2.55}$$

The evaluation of this double-bar matrix element is described in detail in Section 9.3.2.

Because the spin–spin interaction couples states with different N, the total angular momentum states as defined in Eq. (2.39) are *not* the eigenstates of the Hamiltonian (2.51). The eigenstates can generally be written as

$$|JMK\rangle=\sum_N a^K_N|J(NS)M\rangle,\tag{2.56}$$

where for each state K there are, at most, two terms on the right-hand side. Two, because Eq. (2.53) couples, at most, two N states. Why not more? Simply because $\boldsymbol{J}=\boldsymbol{N}+\boldsymbol{S}$ and we know that J is conserved, meaning that there cannot be any couplings between states of different J. If $|N-N'|>2$ and $S=1$, there is no way $J=N\pm S$ and $J'=N'\pm S$ can be equal. Note that the second $3j$ symbol in Eq. (2.53) forbids all couplings between states with $|N-N'|=1$. Remember that a 3j symbol with vanishing projections is zero unless the sum of the three angular momentum quantum numbers is even.

For most $^3\Sigma$ molecules, $\gamma \ll \lambda_{SS} \ll B_e$. This means that the first term in Eq. (2.51) sets the scale of the rotational energy and the other two terms lead to fine structure of the rotational levels. For molecules with $\lambda_{SS} \ll B_e$, the matrix of the coefficients a_N^K in Eq. (2.56) is very close to the identity matrix, making N a nearly good quantum number. For such molecules, the rotational energy levels form fine structure triplets, corresponding to $J = N - 1$, $J = N$, and $J = N + 1$, separated by large energy gaps equal to $B_e N(N + 1)$.

Given the matrix elements (2.52), we know all we need to know to convert a $^2\Sigma$ molecule into a $^3\Sigma$ molecule. In particular, the Stark energy levels can be computed by diagonalizing the sum of the spin–spin interaction matrix and the matrix of the Hamiltonian (2.46), directly in the basis of the product states $|NM_N\rangle|SM_S\rangle$.

2.4.4 Molecules in a $^1\Pi$ Electronic State – Λ-Doubling

$$^1\Pi : S = 0, |\Lambda| = 1, |\Sigma| = 0, |\Omega| = 1.$$

Here's a different case. The rotational Hamiltonian (1.42) is

$$B_e \boldsymbol{R}^2 = B_e[\boldsymbol{J} - \boldsymbol{L}]^2. \tag{2.57}$$

Equation (1.43) loses all terms containing spin operators to become

$$[\boldsymbol{J} - \boldsymbol{L}]^2 = \boldsymbol{J}^2 + \boldsymbol{L}^2 - 2J_z L_z - J_+ L_- - J_- L_+. \tag{2.58}$$

The set of conserved quantum numbers listed earlier looks like that of Hund's case (a).

The presence of the $\hat{L}_\pm$ operators in the rotational term has interesting consequences, namely, it results in the doubling of the rotational energy levels, called Λ-doubling. Let us understand how Λ-doubling arises and how it manifests itself.

When there is a Π state, there is usually a Σ state not far in energy. To see this, imagine the molecular states arising from the interaction of a structureless atom in the ^{1}S state and an atom in a ^{1}P state, a problem analogous to the one considered in Exercise 1.6. The total orbital angular momentum of the electrons in the second atom is $L = 1$. This angular momentum can have the projection 0 or ± 1 on the interatomic axis. The interaction of a ^{1}S atom with a ^{1}P atom thus gives rise to two electronic molecular states: $^1\Sigma$ and $^1\Pi$. Because both of these states approach the same energy in the limit of large interatomic separation, they are generally not very far in energy.

The rotational operator (2.57), due to the terms containing $L_\pm$, couples the Π state and the Σ state. These couplings are weak but important, as they affect the rotational structure of the molecule in the Π state. The rotational states of the molecule are the eigenstates of the 3×3 matrix:

$$\begin{pmatrix} \epsilon_\Pi & 0 & \Omega_{\Sigma\Pi} \\ 0 & \epsilon_\Pi & \Omega_{\Sigma\Pi} \\ \Omega^*_{\Sigma\Pi} & \Omega^*_{\Sigma\Pi} & \epsilon_\Sigma \end{pmatrix},$$

where

$$\epsilon_\Sigma = \Delta_{\Sigma\Pi} + B_\Sigma[J(J+1) + \langle \boldsymbol{L}^2 \rangle_\Sigma] \tag{2.59}$$

$$\epsilon_\Pi = B_\Pi[J(J+1) - 2 + \langle \boldsymbol{L}^2 \rangle_\Pi] \tag{2.60}$$

$$\Omega_{\Sigma\Pi} = -B_e[J(J+1)]^{1/2}\langle \Sigma | L_+ | \Pi \rangle, \tag{2.61}$$

$\langle \boldsymbol{L}^2 \rangle$ denote the diagonal matrix elements of $\boldsymbol{L}^2$ in the corresponding electronic states, $\Delta_{\Sigma\Pi}$ is the electronic energy separation between the states Π and Σ, and B_Π and B_Σ are the rotational constants of the molecule in the corresponding electronic state. We leave B_e in the off-diagonal matrix elements undefined. Since the matrix elements of L_+ are generally undetermined (since L is undefined!), the product of B_e and the matrix elements of L_+ can be treated as an adjustable parameter, which can be tuned, within reason, to reproduce the experimentally observed rotational level structure.

The matrix above is an interesting matrix. The reader may observe that it is of the form

$$\begin{pmatrix} a & 0 & c \\ 0 & a & c \\ c^* & c^* & b \end{pmatrix}$$

and can be transformed to a block-diagonal form

$$\begin{pmatrix} a & 0 & 0 \\ 0 & a & \sqrt{2}c \\ 0 & \sqrt{2}c^* & b \end{pmatrix}$$

by the unitary transformation

$$\begin{pmatrix} \frac{1}{\sqrt{2}} & -\frac{1}{\sqrt{2}} & 0 \\ \frac{1}{\sqrt{2}} & \frac{1}{\sqrt{2}} & 0 \\ 0 & 0 & 1 \end{pmatrix}.$$

This means that we can combine the states $|\Lambda = \pm 1\rangle |JM\Omega = \pm 1\rangle$ to form new eigenstates of the rotational operator (2.57)

$$|JM\tilde{\Omega}\varepsilon\rangle = \frac{1}{\sqrt{2}}[|\Lambda = 1\rangle |JM\Omega = 1\rangle + (-1)^J \varepsilon |\Lambda = -1\rangle |JM\Omega = -1\rangle], \tag{2.62}$$

where $\varepsilon = \pm 1$ and $\tilde{\Omega} = |\Omega|$. Only one of these two states, the state with $\varepsilon(-1)^J = 1$, is coupled to the Σ state. These couplings shift the energy of this state; thus, the two $\varepsilon = \pm$ states (2.62) have slightly different energies. This phenomenon is appropriately termed Λ-doubling.

The rotational states of a Π state molecule should thus be expected to come in Λ-doubled pairs. How big is the energy splitting of the $\varepsilon = \pm$ states? The difference $b - a \approx \Delta_{\Sigma\Pi}$ is the separation energy between the Σ and Π electronic states, which is typically >1000 cm^{-1}. The magnitude of the off-diagonal matrix elements c is $\approx B_e$, typically <50 cm^{-1}. For most molecules, $|c/(b-a)| \ll 1$, and the coupling to the Σ state is a small perturbation, leading to a small energy splitting of the $\varepsilon = \pm$ states, usually much smaller than the energy separation of the rotational energy levels with different J. There are some stray molecules, in which $\Delta_{\Sigma\Pi} \sim 0$, but we will normally not be concerned with these.

By constructing the states (2.62), we get an added bonus. The value of ε, as defined, determines the parity of the state (see Section 1.5). This explains why only one of the two states (2.62) is coupled to the Σ state. The operator (2.57) must conserve parity. The parity of the rotational states of a $^1\Sigma$ molecule is given by $(-1)^J$ (cf., Exercise 2.2). For a given value of J, only one of the two states (2.62) has the same parity.

Λ-doubling is important because it provides states of different parity that lie close, very close, in energy. We know that the dipole moment operator, and consequently the Stark operator, couples states of different parity and immediately conclude that the matrix elements

$$\langle JM\tilde{\Omega}\varepsilon|\boldsymbol{E}\cdot\boldsymbol{d}_0|JM\tilde{\Omega}\varepsilon'\rangle \neq 0, \tag{2.63}$$

when $\varepsilon \neq \varepsilon'$. For fixed values of J, M, and $\tilde{\Omega}$, we are dealing with 2×2 matrices, in which the diagonal matrix elements are very similar, while the off-diagonal matrix elements depend on the magnitude of the electric field. This means that the cross-over from the quadratic to the linear Stark effect in molecules with Λ-doubling occurs at very weak electric fields.

The specific forms of the matrix elements of the Stark operator can be obtained by recognizing that $|JM\tilde{\Omega}\rangle$ represent the Wigner D-functions $D^{J*}_{M\tilde{\Omega}}$ and that the spherical harmonic Y_{10} in Eq. (2.35) is a special case of a Wigner D-function, namely, $Y_{10} = \sqrt{\frac{3}{4\pi}}D^{1*}_{00}$. This can be used to recast the matrix elements $\langle JM\tilde{\Omega}\varepsilon| -\boldsymbol{E}\cdot\boldsymbol{d}_0|J'M'\tilde{\Omega}'\varepsilon'\rangle$ as integrals over products of three D-functions, which yields (see Appendix D and Refs. [9, 10])

$$\begin{aligned}\langle JM\tilde{\Omega}\varepsilon| -\boldsymbol{E}\cdot\boldsymbol{d}_0|J'M'\tilde{\Omega}'\varepsilon'\rangle &= -Ed_0\langle JM\tilde{\Omega}\varepsilon|\cos\theta|J'M'\tilde{\Omega}'\varepsilon'\rangle \\ &= -Ed_0\delta_{\varepsilon,-\varepsilon'}\delta_{MM'}(-1)^{M'-\tilde{\Omega}'}[(2J+1)(2J'+1)]^{1/2} \\ &\times\begin{pmatrix} J & 1 & J' \\ M & 0 & -M'\end{pmatrix}\begin{pmatrix} J & 1 & J' \\ \tilde{\Omega} & 0 & -\tilde{\Omega}'\end{pmatrix}.\end{aligned} \tag{2.64}$$

This is the general expression for the matrix elements of the Stark operator in Hund's case (a) basis. This equation shows, as expected, that there are no couplings between states with different values of M, the projection of $\boldsymbol{J}$ on the electric field direction. Interestingly, Eq. (2.64) implies that the couplings between states with different Ω are also zero, although there is no fundamental symmetry that would require the *body-fixed* projection of $\boldsymbol{J}$ to be conserved (recall that Ω is the projection of $\boldsymbol{J}$ on the interatomic axis). The Stark Hamiltonian conserves $\tilde{\Omega}$ because the vector $\boldsymbol{d}_0$ is tied to the intermolecular axis, pointing from one atom to another.

The main features of the Stark effect in a Λ-doubled system using an example of a $^2\Pi$ molecule is illustrated in Section 2.4.5.

2.4.5 Molecules in a $^2\Pi$ Electronic State

$$^2\Pi : S = 1/2, |\Lambda| = 1, |\Sigma| = 1/2, |\Omega| = 1/2, 3/2.$$

The nonzero electron spin brings in the spin–orbit interactions and adds a bit of complexity by letting $\tilde{\Omega}$ take two values. Just like in the case of $^1\Pi$ molecules, the rotational states of $^2\Pi$ molecules are best described by the parity-adapted basis

$$\begin{aligned} |JM\tilde{\Omega}\epsilon\rangle = {} & \frac{1}{\sqrt{2}}[|\Lambda = 1, \Sigma = \tilde{\Omega} - 1\rangle|JM\tilde{\Omega}\rangle \\ & + \epsilon(-1)^{J-1/2}|\Lambda = -1, \Sigma = -\tilde{\Omega} + 1\rangle|JM - \tilde{\Omega}\rangle]. \end{aligned} \tag{2.65}$$

Notice that the phase factor of the second term has a factor 1/2 tacked on. This makes the power of -1 integer.

The energy level structure of a $^2\Pi$ molecule is determined by the rotational Hamiltonian (1.42) and the spin–orbit interactions. While a separate book can be written on the spin–orbit interactions in molecules, they are very often approximated as

$$\hat{H}_{\text{SO}} = A\boldsymbol{L} \cdot \boldsymbol{S} = A\left[L_z S_z + \frac{1}{2}(S_+ L_- + L_- S_+)\right]. \tag{2.66}$$

where A is an empirical constant. The operator (2.66) contains the diagonal part

$$\langle JM\tilde{\Omega}\epsilon|AL_z S_z|JM\tilde{\Omega}\epsilon\rangle = A(\tilde{\Omega} - 1), \tag{2.67}$$

which separates the $\tilde{\Omega} = 1/2$ and $\tilde{\Omega} = 3/2$ states by the energy gap equal to A. The rotational states of the molecule can thus be divided into a manifold labeled by $\tilde{\Omega} = 1/2$ and a manifold labeled by $\tilde{\Omega} = 3/2$. The remaining terms in Eq. (2.66), just like their twins in Eq. (1.43), couple the $^2\Pi$ state with a $^2\Sigma$ state.

It is inconvenient to have to diagonalize the 3×3 matrix discussed in Section 2.4.5 in order to determine Λ-doubling energies. To avoid dealing with the Σ state, one can introduce an effective Hamiltonian that acts only within the $^2\Pi$ state and separates the states (2.65) of opposite parity by the Λ-doubling energy. Such effective operator needs to do a proper job at describing the dependence of the Λ-doublets on J and $\tilde{\Omega}$. Because the couplings to the Σ state are diagonal in J and M, the effective operator must be diagonal in these quantum numbers as well. Brown and Carrington [4] give a general form (which works for any Π state) of the effective Λ-doubling operator

$$\hat{H}_{\mathrm{LD}} = \frac{1}{2}q(J_+^2 + J_-^2) - \frac{1}{2}(p + 2q)(J_+S_+ + J_-S_-) + \frac{1}{2}(o + p + q)(S_+^2 + S_-^2), \tag{2.68}$$

where the constants $o, p,$ and q are related to the matrix elements that couple the Π state with, not just one, but all the Σ states of the right spin multiplicity. This operator is meant to replace all the terms containing L_+ or L_- in Eqs. (1.43) and (2.66). Note that for the $^2\Pi$ (spin-1/2) state the matrix elements of the operators $S_\pm^2$ vanish.

The effective Hamiltonian for a $^2\Pi$ molecule in an electric field can thus be written as

$$\begin{aligned}\hat{H}_{\mathrm{eff}} = B_e[\boldsymbol{J}^2 - 2J_z^2 + 2L_zS_z - J_+S_- - J_-S_+] \\ + AL_zS_z + \hat{H}_{\mathrm{LD}} - \boldsymbol{E}\cdot\boldsymbol{d}_0,\end{aligned} \tag{2.69}$$

where I have omitted the terms $\boldsymbol{L}^2$ and $\boldsymbol{S}^2$ that would only contribute to zero-point energy and used the equality $J_z = L_z + S_z$. The matrix elements of the $J_\pm$ and $S_\pm$ operators can be readily evaluated in the basis (2.65), as shown in Appendix B.

With this, we have all the pieces needed to compute the Stark shifts of the rotational energy levels of a $^2\Pi$ molecule. Figure 2.5 illustrates the electric field modification of the energy levels of the OH molecule in the ground electronic and vibrational state. The left panels of the figure show the energy levels corresponding to $\tilde{\Omega} = 3/2$ and the right panels show the levels with $\tilde{\Omega} = 1/2$. At zero electric field, the different parity states are often labeled as e states, which have $\epsilon(-1)^{J-1/2} = +1$, and f states, which have $\epsilon(-1)^{J-1/2} = -1$. Notice that the Λ-doubling splitting increases with J and is much greater for the rotational states with $\tilde{\Omega} = 1/2$.

Beyond showing the Stark shifts of the OH molecule, the six panels of Figure 2.5 nicely illustrate the generic features of the dc Stark effect, namely, the splitting of the rotational states into $J + 1/2$ (for integer J it would be $J + 1$) M components, the repulsion of the states of opposite parity (notice how the e and f states go opposite ways) and the transition from the quadratic to the linear Stark shifts (notice how the departure from linearity at low values of the electric field becomes more prominent as the Λ-doubling splitting increases). This is all consistent with the general arguments derived in Section 2.3.

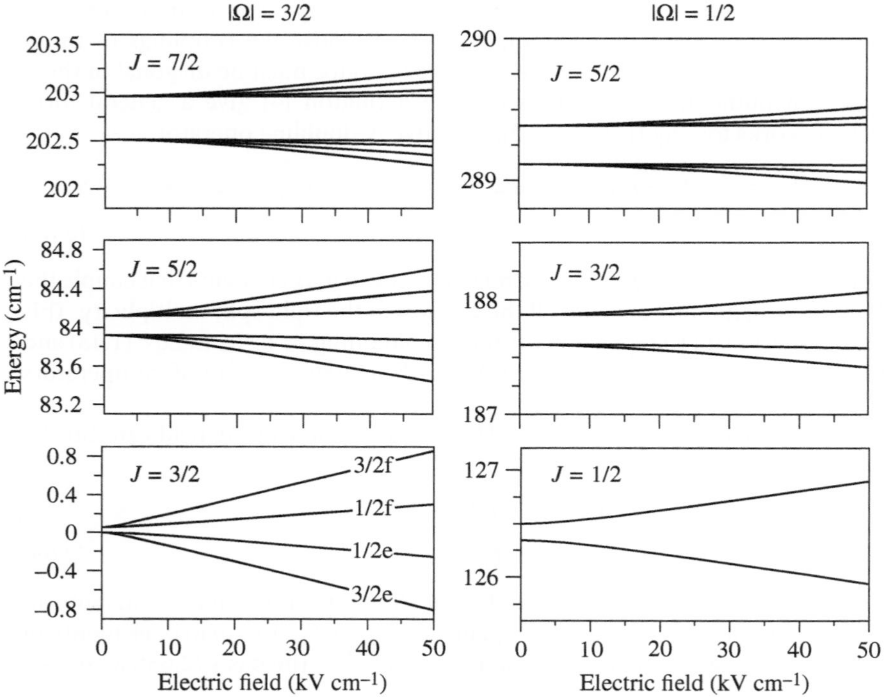

Figure 2.5 The Stark shifts of the rotational energy levels of a diatomic molecule in a $^2\Pi$ electronic state. The energy levels presented are for the molecule OH, which has $B_e = 18.55\ \text{cm}^{-1}, A = -139.273\ \text{cm}^{-1}, q = -0.03877\ \text{cm}^{-1}, p = 0.235608\ \text{cm}^{-1}$ and the dipole moment 1.67 Debye. *Source*: Reproduced with permission from Pavlovic et al. [14]. © 2009, American Chemical Society.

Exercises

2.1 Plot the eigenvalues of a 3×3 matrix

$$\begin{pmatrix} a & d & 0 \\ d & b & d \\ 0 & d & c \end{pmatrix}$$

for $a = 1$, $b = 2$, and $c = 6$ as functions of d in the interval between 0 and 20.

2.2 Show that the spherical harmonics $Y_{lm}(\theta, \phi)$ have parity given by $(-1)^l$.

To do this, we must consider what happens to the spherical polar angles θ and ϕ under the transformation $(x, y, z) \to (-x, -y, -z)$, given the relations between (x, y, z), and (r, θ, ϕ) in Exercise 1.8. After this, all we need to do is analyze how the spherical harmonics change under these transformations of the angles.

2.3 Prove that the matrix elements $\langle Y_{l,m}|Y_{k,q}|Y_{l',m'}\rangle$ are zero if $k + l + l'$ is odd.
This can be done using the properties of the spherical harmonics under the inversion of the coordinate frame (established in the previous exercise) or the properties of $3j$-symbols. Alternatively, this can be proven using the Wigner-Eckart theorem discussed in Section 9.2 and the properties of the $3j$-symbols discussed in Appendix B.

2.4 Assign the quantum number M to each curve in the upper panel of Figure 2.4.
This can be done by inspection, as described at the end of Section 2.4.3.

3

Zeeman Effect

3.1 The Electron Spin

Deriving the effective Hamiltonian for a molecule in a magnetic field is nontrivial matter. Had electrons been spinless, the problem would have been simple. We would then consider each electron as a particle carrying charge $-e$ and moving with velocity $\boldsymbol{v}$ in an electromagnetic field defined by the scalar φ and vector $\boldsymbol{A}$ potentials. The classical Lagrangian for such a particle is [13]

$$L = mv^2/2 + e\varphi - e\boldsymbol{v} \cdot \boldsymbol{A}. \tag{3.1}$$

The generalized momentum can be obtained from the Lagrangian as follows [15]:

$$\boldsymbol{p} = \partial L/\partial \boldsymbol{v} = m\boldsymbol{v} - e\boldsymbol{A}. \tag{3.2}$$

We can see that it is different from the momentum of a free particle by the term $-e\boldsymbol{A}$. The Hamiltonian can be obtained from the Lagrangian using $H = \boldsymbol{v} \cdot \partial L/\partial \boldsymbol{v} - L$, which gives

$$H = \frac{mv^2}{2} - e\varphi. \tag{3.3}$$

Using the expression $m\boldsymbol{v} = \boldsymbol{p} + e\boldsymbol{A}$ from Eq. (3.2), we can write the Hamiltonian in terms of the generalized momentum:

$$H = \frac{1}{2m}(\boldsymbol{p} + e\boldsymbol{A})^2 - e\varphi. \tag{3.4}$$

So, if the electrons had no spin, all we would have to do to write the quantum Hamiltonian of an electron in the presence of an electromagnetic field is to replace $\boldsymbol{p}$ with $-i\hbar\nabla$ in Eq. (3.4). We could then express $\boldsymbol{A}$ in terms of the magnetic field (as we will do below) and obtain the quantum Hamiltonian of the charged particle in a magnetic field.

However, the electrons do have spin and the magnetic field interacts with the electron spin. So the question we need to answer first is, how do the spin-dependent interactions appear in the Hamiltonian for an electron in a magnetic field?

Molecules in Electromagnetic Fields: From Ultracold Physics to Controlled Chemistry, First Edition. Roman V. Krems.

3.1.1 The Dirac Equation

The concept of the electron spin naturally arises in Dirac's relativistic treatment of the electron. The relativistic version of the Schrödinger equation can be obtained by quantizing the energy–momentum relation [16]

$$E^2 = c^2\boldsymbol{p}^2 + m^2c^4 \tag{3.5}$$

by replacing $\boldsymbol{p}$ with $-i\hbar\nabla$ and E with $i\hbar\partial/\partial t$, which yields

$$\frac{1}{c^2}\frac{\partial^2\psi}{\partial^2 t} - \nabla^2\psi + \frac{m^2c^2}{\hbar^2}\psi = 0. \tag{3.6}$$

This equation contains the second time derivative of the wave function and looks like a distant relative of the Schrödinger equation. This second-order differential equation requires the initial conditions for both the wave function and its time derivative. This is undesirable. Quantum mechanics hinges on the belief that knowing the wave function at $t = 0$ is sufficient to determine the time evolution of the wave function (provided the Hamiltonian is known). If the fundamental equation of quantum mechanics contained the second time derivative, a system with a given wave function and a given Hamiltonian could evolve in different ways. The wave function would lose its meaningfulness as a fundamental function determining the state (and fate) of the system.

So we want a fundamental equation that would only contain the first time derivative of ψ. This raises the question, is it possible to obtain a relativistic equation that would have the form $\hat{H}\psi = i\hbar\partial\psi/\partial t$ and have the same solutions as Eq. (3.6)? According to Eq. (3.5), such an equation would have to be linear in momentum $\boldsymbol{p}$. However, since $\boldsymbol{p}$ is a vector quantity, an equation linear in $\boldsymbol{p}$ is generally not invariant under rotations of the coordinate system so it cannot describe a particle in isotropic space. (In other words, we want all terms in the Hamiltonian to be scalar and that would not be the case for an equation linear in $\boldsymbol{p}$.)

To circumvent this problem, Dirac proposed to write the wave function for an electron in the form of a four-component vector (spinor)

$$\psi = \begin{pmatrix} \psi_1 \\ \psi_2 \\ \psi_3 \\ \psi_4 \end{pmatrix},$$

which led him to an equation

$$(c\boldsymbol{\alpha}\cdot\boldsymbol{p} + \beta mc^2)\psi = i\hbar\frac{\partial\psi}{\partial t}, \tag{3.7}$$

that is invariant under space rotations. In order for each component of the wave function to satisfy Eq. (3.6), the quantities $\boldsymbol{\alpha}$ and β must be 4×4 matrices given by the following equations:

$$\boldsymbol{\alpha} = \hat{x}\alpha_x + \hat{y}\alpha_y + \hat{z}\alpha_z$$

with

$$\alpha_x = \begin{pmatrix} 0 & 0 & 0 & 1 \\ 0 & 0 & 1 & 0 \\ 0 & 1 & 0 & 0 \\ 1 & 0 & 0 & 0 \end{pmatrix} \quad \alpha_y = \begin{pmatrix} 0 & 0 & 0 & -i \\ 0 & 0 & i & 0 \\ 0 & -i & 0 & 0 \\ i & 0 & 0 & 0 \end{pmatrix}$$

$$\alpha_z = \begin{pmatrix} 0 & 0 & 1 & 0 \\ 0 & 0 & 0 & -1 \\ 1 & 0 & 0 & 0 \\ 0 & -1 & 0 & 0 \end{pmatrix}$$

and

$$\beta = \begin{pmatrix} 1 & 0 & 0 & 0 \\ 0 & 1 & 0 & 0 \\ 0 & 0 & -1 & 0 \\ 0 & 0 & 0 & -1 \end{pmatrix},$$

where $\hat{x}$, $\hat{y}$, and $\hat{z}$ are the unit vectors pointing along the Cartesian coordinate axes. Equation (3.7) is thus really a set of four coupled equations for the four components of the wave function.

Note that Eq. (3.5) implies the existence of negative and positive energy solutions $E_\pm = \pm\sqrt{c^2\boldsymbol{p}^2 + m^2c^4}$. The negative energy solutions present a problem. In particular, they imply that an electron – as a charged particle – may emit electromagnetic field and lose energy, indefinitely. We know from experience this doesn't happen. As a work-around for this problem, Dirac postulated the existence of the Dirac sea, i.e. a sea of fermions occupying entirely the negative energy solutions of the Dirac equation. Exciting one of these fermions to a positive energy state then creates an electron, while the hole left in the negative-energy "sea" represents a positron.

The Dirac equation can include electromagnetic fields by a simple extension of the momentum $\boldsymbol{p} \to \boldsymbol{p} + e\boldsymbol{A}$ leading to the Hamiltonian[1]

$$\hat{H} = -e\varphi + c\boldsymbol{\alpha}\cdot(\boldsymbol{p} + e\boldsymbol{A}) + \beta mc^2. \tag{3.8}$$

We are interested in the nonreleativistic limit of this equation, i.e. the limit of a slowly moving electron. To find the nonrelativistic Hamiltonian, the procedure [4] is to set $\hat{H} = \hat{H}_{\text{nr}} + mc^2$, assuming that $\hat{H}_{\text{nr}}$ is small compared with mc^2, and square both sides of Eq. (3.8). After dividing both sides of the resulting equation by $2mc^2$ and neglecting all terms with c in the denominator, we obtain (Exercise 3.1)

$$\hat{H}_{\text{nr}} = -\beta e\varphi + \frac{1}{2m_e}[\boldsymbol{\alpha}\cdot(\boldsymbol{p} + e\boldsymbol{A})]^2, \tag{3.9}$$

1 Note that I assume that the charge of the particle is $q = -e$. Thus, the terms that would in the literature appear as $(\boldsymbol{p} - q\boldsymbol{A})$ and $q\varphi$ have the opposite sign.

where we use the rest mass m_e of the electron. If $\boldsymbol{\alpha}$ is expressed as $\boldsymbol{\alpha} = \rho\tilde{\boldsymbol{\sigma}}$, where

$$\rho = \begin{pmatrix} 0 & 0 & 1 & 0 \\ 0 & 0 & 0 & 1 \\ 1 & 0 & 0 & 0 \\ 0 & 1 & 0 & 0 \end{pmatrix}$$

and the Cartesian components of $\tilde{\boldsymbol{\sigma}}$ are

$$\tilde{\sigma}_x = \begin{pmatrix} 0 & 1 & 0 & 0 \\ 1 & 0 & 0 & 0 \\ 0 & 0 & 0 & 1 \\ 0 & 0 & 1 & 0 \end{pmatrix} \quad \tilde{\sigma}_y = \begin{pmatrix} 0 & -i & 0 & 0 \\ i & 0 & 0 & 0 \\ 0 & 0 & 0 & -i \\ 0 & 0 & i & 0 \end{pmatrix}$$

$$\tilde{\sigma}_z = \begin{pmatrix} 1 & 0 & 0 & 0 \\ 0 & -1 & 0 & 0 \\ 0 & 0 & 1 & 0 \\ 0 & 0 & 0 & -1 \end{pmatrix},$$

then Eq. (3.9) can be rewritten (Exercise 3.2) as

$$\hat{H}_{\text{nr}} = \frac{1}{2m_e}(\boldsymbol{p} + e\boldsymbol{A})^2 + \frac{e\hbar}{2m_e}\tilde{\boldsymbol{\sigma}} \cdot \boldsymbol{B} - \beta e\varphi, \tag{3.10}$$

where $\boldsymbol{B} = \nabla \times \boldsymbol{A}$ is the magnetic field vector. See Ref. [4] for more details.

Not coincidentally, the upper and lower blocks of the matrices $\tilde{\sigma}_x$, $\tilde{\sigma}_y$, and $\tilde{\sigma}_z$ are the corresponding 2×2 Pauli matrices for a spin-1/2 particle:

$$\sigma_x = \begin{pmatrix} 0 & 1 \\ 1 & 0 \end{pmatrix} \quad \sigma_y = \begin{pmatrix} 0 & -i \\ i & 0 \end{pmatrix} \quad \sigma_z = \begin{pmatrix} 1 & 0 \\ 0 & -1 \end{pmatrix}. \tag{3.11}$$

The Hamiltonian (3.10) is block-diagonal and contains no couplings between the subspace of (ψ_1, ψ_2) and the subspace of (ψ_3, ψ_4). The part of the Hamiltonian in the (ψ_1, ψ_2)-subspace corresponds to positive energy solutions and that in the (ψ_3, ψ_4)-subspace – to negative energy solutions. Since we are interested in electrons with positive energy, let us ignore the negative energy solutions and rewrite Eq. (3.10) in terms of the Pauli matrices

$$\hat{H}_e = \frac{1}{2m_e}(\boldsymbol{p} + e\boldsymbol{A})^2 + \frac{e\hbar}{2m_e}\boldsymbol{\sigma} \cdot \boldsymbol{B} - e\varphi, \tag{3.12}$$

acting on the 2×2 subspace of the four-component wave function. This is the nonrelativistic limit of the Hamiltonian for an electron in an electromagnetic field.

A comparison of Eqs. (3.4) and (3.12) shows that the Dirac treatment leads to an additional term in the Hamiltonian

$$V_B = \frac{e\hbar}{2m_e}\boldsymbol{\sigma} \cdot \boldsymbol{B} \tag{3.13}$$

that depends on the magnetic field $\boldsymbol{B}$ and is parametrized by the vector quantity $\boldsymbol{\sigma}$. We will write this term as

$$V_B = -\boldsymbol{\mu} \cdot \boldsymbol{B}, \tag{3.14}$$

where

$$\boldsymbol{\mu} = -\frac{e\hbar}{2m_e}\boldsymbol{\sigma} \tag{3.15}$$

is the magnetic dipole moment of the electron. The constant $\mu_B = e\hbar/2m_e$ is the Bohr magneton. The value of the Bohr magneton is 0.5 a.u. or $5.788381804(39) \times 10^{-5}$ eV T^{-1}.

3.2 Zeeman Energy of a Moving Electron

In general, a moving electron possesses an orbital as well as the spin angular momentum. Both angular momenta interact with the external magnetic field. The interaction of the orbital angular momentum with the magnetic field comes from the first term in Eq. (3.12).

The magnetic field is related to the vector potential as $\boldsymbol{B} = \nabla \times \boldsymbol{A}$ [13] (see footnote on page 36). If the magnetic field is uniform (i.e. does not vary with spatial coordinates), we can also write

$$\boldsymbol{A} = \frac{1}{2}\boldsymbol{B} \times \boldsymbol{r}. \tag{3.16}$$

To prove this, consider the vector identity [13]

$$\nabla \times (\boldsymbol{B} \times \boldsymbol{r}) = \boldsymbol{B}(\nabla \cdot \boldsymbol{r}) - (\boldsymbol{B} \cdot \nabla)\boldsymbol{r} + (\boldsymbol{r} \cdot \nabla)\boldsymbol{B} - \boldsymbol{r}(\nabla \cdot \boldsymbol{B}), \tag{3.17}$$

which can be verified directly. Since $\boldsymbol{B}$ is independent of the spatial coordinates, the last two terms vanish. Given that $\nabla \cdot \boldsymbol{r} = 3$, we obtain

$$\nabla \times (\boldsymbol{B} \times \boldsymbol{r}) = 2\boldsymbol{B}, \tag{3.18}$$

which proves Eq. (3.16).

We can now expand the first term in Eq. (3.12) and use Eq. (3.16), to obtain

$$(\boldsymbol{p} + e\boldsymbol{A})^2 \approx \boldsymbol{p}^2 + e(\boldsymbol{p} \cdot \boldsymbol{A} + \boldsymbol{A} \cdot \boldsymbol{p}) = \boldsymbol{p}^2 + e(\boldsymbol{r} \times \boldsymbol{p}) \cdot \boldsymbol{B} = \boldsymbol{p}^2 + e\hbar\boldsymbol{L} \cdot \boldsymbol{B}, \tag{3.19}$$

where $\hbar\boldsymbol{L} = \boldsymbol{r} \times \boldsymbol{p}$ is the orbital angular momentum of the electron.[2]

Reintroducing the electron spin angular momentum as

$$\boldsymbol{S} = \frac{1}{2}\boldsymbol{\sigma}, \tag{3.20}$$

2 Note that, in general, $\boldsymbol{A}$ is a function of coordinates and $\boldsymbol{p} \cdot \boldsymbol{A} - \boldsymbol{A} \cdot \boldsymbol{p} = -i\hbar\, \mathrm{div}\boldsymbol{A} \neq 0$. For the special case of a uniform field, whose vector potential is $\boldsymbol{A} = \frac{1}{2}\boldsymbol{B} \times \boldsymbol{r}$, $\mathrm{div}\boldsymbol{A} = 0$ so $\boldsymbol{p}$ and $\boldsymbol{A}$ commute [8].

we can rewrite Eq. (3.12) in the most suitable form

$$\hat{H}_e = \frac{\boldsymbol{p}^2}{2m_e} + \mu_B(\boldsymbol{L} + 2\boldsymbol{S}) \cdot \boldsymbol{B} - e\varphi. \tag{3.21}$$

The middle term leads to the Zeeman effect in atoms and molecules. It is generally written as

$$\hat{H}_{\text{Zeeman}} = \mu_B(g_L\boldsymbol{L} + g_S\boldsymbol{S}) \cdot \boldsymbol{B}, \tag{3.22}$$

where g_L and g_S are the electron orbital and spin g-factors, respectively. As follows from the above, the Dirac theory predicts the values $g_L = 1$ and $g_S = 2$. Quantum electrodynamics offers a more accurate value of g_S represented by an expansion [17,18]

$$g_S = 2\left(1 + \frac{\alpha}{2\pi} - 2.973\frac{\alpha^2}{\pi^2} + \cdots\right), \tag{3.23}$$

where $\alpha = e^2/(4\pi\epsilon_0)\hbar c \approx 1/137.035999679(94)$ is the fine structure constant. Equation (3.23) gives the value $g_S = 2.00231930436153(53)$. The fundamental constants quoted here and throughout this monograph are taken from the NIST database [19]. The value of g_L may also deviate from one due to nonrelativistic effects and the rotational motion of the molecule (see Section 3.6). In the calculations presented here, we will assume that $g_L = 1$.

3.3 Magnetic Dipole Moment

Equation (3.21) has a friendly form that can be generalized to multi-electron systems. If the space-fixed quantization axis Z is chosen to point in the direction of $\boldsymbol{B}$, the Zeeman Hamiltonian can be written as

$$\hat{H}_{\text{Zeeman}} = \mu_B B(g_L L_Z + g_S S_Z), \tag{3.24}$$

where B is the magnitude of the magnetic field. Recognizing that the Z-components of the *total* orbital and spin angular momenta of N electrons are simply the sums of the Z-components of the individual angular momenta $L_Z = \sum_i^N l_{Z_i}$ and $S_Z = \sum_i^N s_{Z_i}$, we see that all we have to do in order to obtain the Zeeman Hamiltonian for an N-electron system is insert the total angular momentum projections in Eq. (3.24). This suggests that for an ensemble of electrons with total spin $\boldsymbol{S}$ we can define the magnetic moment as $\boldsymbol{\mu}_S = -g_S\mu_B\boldsymbol{S}$ and for an ensemble of electrons with total angular momentum $\boldsymbol{L}$ we can define the magnetic moment $\boldsymbol{\mu}_L = -g_L\mu_B\boldsymbol{L}$. The total magnetic moment can be written in terms of the total angular momentum $\boldsymbol{J} = \boldsymbol{L} + \boldsymbol{S}$

$$\boldsymbol{\mu} = \boldsymbol{\mu}_S + \boldsymbol{\mu}_L = -g_J\mu_B\boldsymbol{J}, \tag{3.25}$$

where g_J is the Landé g-factor given by

$$g_J = \frac{J(J+1) - S(S+1) + L(L+1)}{2J(J+1)} + g_s \frac{J(J+1) + S(S+1) - L(L+1)}{2J(J+1)}. \tag{3.26}$$

To obtain Eq. (3.26), consider the expectation value of the operator $\boldsymbol{\mu} \cdot \boldsymbol{J}$ in the basis of eigenstates of $\boldsymbol{J}^2$ and J_Z, denoted by $|JM\rangle$,

$$\langle JM| - g_J \mu_B \boldsymbol{J}^2 |JM\rangle = -g_J \mu_B J(J+1). \tag{3.27}$$

On the other hand,

$$\langle JM|(\boldsymbol{\mu}_S + \boldsymbol{\mu}_L) \cdot \boldsymbol{J}|JM\rangle = -\mu_B \langle JM|g_S \boldsymbol{S} \cdot \boldsymbol{J} + \boldsymbol{L} \cdot \boldsymbol{J}|JM\rangle \tag{3.28}$$

Recognizing that

$$\boldsymbol{S} \cdot \boldsymbol{J} = \boldsymbol{S}^2 + \boldsymbol{S} \cdot \boldsymbol{L}, \tag{3.29}$$

$$\boldsymbol{L} \cdot \boldsymbol{J} = \boldsymbol{L}^2 + \boldsymbol{S} \cdot \boldsymbol{L}, \tag{3.30}$$

$$\boldsymbol{J}^2 = (\boldsymbol{L} + \boldsymbol{S})^2 = \boldsymbol{L}^2 + \boldsymbol{S}^2 + 2\boldsymbol{S} \cdot \boldsymbol{L}, \tag{3.31}$$

and that

$$\boldsymbol{S}^2|JM\rangle = S(S+1)|JM\rangle, \tag{3.32}$$

$$\boldsymbol{L}^2|JM\rangle = L(L+1)|JM\rangle, \tag{3.33}$$

we obtain the desired result (3.26). Equation (3.25) can be used to calculate the Zeeman shifts of atomic energy levels characterized by the electronic angular momentum $\boldsymbol{J}$, yielding

$$\Delta E_{JM} = -\langle JM|\boldsymbol{\mu} \cdot \boldsymbol{B}|JM\rangle = g_J \mu_B BM. \tag{3.34}$$

For molecules, the total angular momentum $\boldsymbol{J}$ includes not only $\boldsymbol{L}$ and $\boldsymbol{S}$ but also the rotational angular momentum of the nuclei $\boldsymbol{R}$, which makes the calculations more interesting.

Let me conclude this part by a few general observations. Because the magnetic field $\boldsymbol{B} = \nabla \times \boldsymbol{A}$ is defined by the cross product of two vectors, it is a pseudovector, meaning that it does not change under the inversion of the coordinate system. Since scalars remain unchanged under parity transformation, this tells us that the magnetic moment operator in $\boldsymbol{\mu} \cdot \boldsymbol{B}$ must also be invariant under parity inversion. In turn, this means that the spin angular momentum, although not defined as a cross product, must be invariant under parity inversion. We also must conclude that the matrix elements $\langle \psi|\boldsymbol{\mu}|\psi'\rangle$ are nonzero, only if $|\psi\rangle$ and $|\psi'\rangle$ have the same parity. This suggests that the perturbations of the molecular energy levels induced by the magnetic field can be described by first-order perturbation theory and the Zeeman shifts of molecular energy levels at low magnetic fields are linear, in stark contrast with the Stark effect.

3.4 Zeeman Operator in the Molecule-Fixed Frame

The electronic states of molecules are labeled by the projections of electronic angular momenta on the interatomic axis (see Section 1.4). In order to evaluate the matrix elements of the Zeeman Hamiltonian (3.24) sandwiched between molecular states, it is, therefore, desirable to rewrite the Zeeman operator in terms of operators that act on states in the molecule-fixed frame. This can be done by transforming the operators L_Z and S_Z to the molecule-fixed frame whose z-axis is directed along the internuclear axis. As before, we shall use the uppercase X, Y, and Z for the space-fixed coordinates and the lowercase x, y, and z for the molecule-fixed coordinates. In general, we can write

$$L_Z = \alpha_{Z,x} L_x + \alpha_{Z,y} L_y + \alpha_{Z,z} L_z, \tag{3.35}$$

where $\alpha_{R,s}$ are the elements of the direction cosine matrix that provides the relation between the laboratory-fixed and molecule-fixed coordinates (see Appendix C). Using the expressions for the x- and y-components of the angular momentum in terms of the raising and lowering operators defined in Appendix B,

$$L_x = (L_+ + L_-)/2 \tag{3.36}$$

and

$$L_y = i(L_- - L_+)/2, \tag{3.37}$$

Eq. (3.35) can be written in terms of $L_\pm$ and L_z

$$L_Z = \frac{1}{2}\Phi_Z^- L_+ + \frac{1}{2}\Phi_Z^+ L_- + \Phi_Z^z L_z, \tag{3.38}$$

where we introduced $\Phi_Z^\pm = \alpha_{Z,x} \pm i\alpha_{Z,y}$ and $\Phi_Z^z = \alpha_{Z,z}$. The same equation holds for S_Z so the Zeeman operator (3.24) can be written as

$$\hat{H}_{\text{Zeeman}} = \frac{\mu_B B}{2}[\Phi_Z^-(L_+ + g_S S_+) + \Phi_Z^+(L_- + g_S S_-) + 2\Phi_Z^z(L_z + g_S S_z)]. \tag{3.39}$$

When sandwiched between molecular states, the operators L_z, S_z, $L_\pm$, and $S_\pm$ act on the electronic part of the states in the molecule-fixed frame and the operators $\Phi_Z^\pm$ and Φ_Z^z on the rotational states $|JM\Omega\rangle$. We see that the Zeeman operator may couple different electronic states of the molecule through the appearance of $L_\pm$ and $S_\pm$. Our task is now to evaluate the matrix elements $\langle JM\Omega|\Phi_Z^\pm|J'M'\Omega'\rangle$ and $\langle JM\Omega|\Phi_Z^z|J'M'\Omega'\rangle$, which can be obtained using the elements of the direction cosine matrix given in Appendix C, see also Ref. [20]. The nonzero matrix elements are

$$\langle JM\Omega \pm 1|\Phi_Z^{\mp}|JM\Omega\rangle = \frac{[(J \mp \Omega)(J \pm \Omega + 1)]^{1/2}}{J(J+1)} M,$$

$$\langle J + 1M\Omega \pm 1|\Phi_Z^{\mp}|JM\Omega\rangle$$
$$= \mp \frac{[(J \pm \Omega + 1)(J \pm \Omega + 2)(J + M + 1)(J - M + 1)]^{1/2}}{(J+1)[2J+1)(2J+3)]^{1/2}},$$

$$\langle J - 1M\Omega \pm 1|\Phi_Z^{\mp}|JM\Omega\rangle = \pm \frac{[(J \mp \Omega)(J \mp \Omega - 1)(J + M)(J - M)]^{1/2}}{J(2J+1)(2J-1)},$$

$$\langle JM\Omega|\Phi_Z^{z}|JM\Omega\rangle = \frac{\Omega M}{J(J+1)},$$

$$\langle J + 1M\Omega|\Phi_Z^{z}|JM\Omega\rangle$$
$$= \frac{[(J + \Omega + 1)(J - \Omega + 1)(J + M + 1)(J - M + 1)]^{1/2}}{(J+1)[(2J+1)(2J+3)]^{1/2}},$$

$$\langle J - 1M\Omega|\Phi_Z^{z}|JM\Omega\rangle = \frac{[(J + \Omega)(J - \Omega)(J + M)(J - M)]^{1/2}}{J[(2J+1)(2J-1)]^{1/2}}. \quad (3.40)$$

By examining these matrix elements, we can make the following observations:

- The Zeeman operator (3.39) has nonzero diagonal matrix elements. This means that the Zeeman shifts of molecular energy levels can be linear at low magnetic fields.
- The Zeeman operator couples states with $J' = J$ and $J' = J \pm 1$.
- There are no couplings between states with different M. This means that the projection of $\boldsymbol{J}$ on the magnetic field axis is conserved and can be used as a label of the Zeeman levels.
- Unlike in the case of electric-field-induced perturbations, magnetic fields induce couplings between states with different values of Ω so the body-fixed projection of $\boldsymbol{J}$ is not conserved. This indicates that the magnetic moment of the molecule is generally not tied to its internuclear axis.

3.5 Zeeman Shifts of Rotational Levels

We shall now see how Eq. (3.40) can be used to compute the Zeeman shifts for molecules in specific electronic states.

3.5.1 Molecules in a $^2\Sigma$ State

$$^2\Sigma : S = 1/2, \Lambda = 0, \Sigma = \pm 1/2, \Omega = \Sigma, \boldsymbol{J} = \boldsymbol{N} + \boldsymbol{S}.$$

The molecular states of a $^2\Sigma$ molecule are best described by Hund's coupling case (b). The field-free rotational states of the molecule defined in Eq. (2.39) are linear combinations of Hund's case (a) basis states (cf., Eq. (1.39)):

$$|J(NS)M\rangle = \sum_{\Sigma} C^{J\Omega}_{N0S\Sigma} |JM\Omega\rangle|S\Sigma\rangle, \tag{3.41}$$

where $\Omega = \Sigma$ because $\Lambda = 0$. Each molecular state corresponds to well-defined values of N and J. For the rotational ground state, we have $N = 0$ and $J = S = 1/2$. The rotationally excited states correspond to $J = N \pm 1/2$.

It is instructive to examine the parity of the states $|J(NS)M\rangle$. By acting with the parity operator on both sides of Eq. (3.41), we obtain (see Section 1.5)

$$\begin{aligned}\hat{P}|J(NS)M\rangle &= \sum_{\Sigma} C^{J\Omega}_{N0S\Sigma}(-1)^{J-\Omega+S-\Sigma+\lambda}|JM-\Omega\rangle|S-\Sigma\rangle,\\ &= \sum_{\Sigma} C^{J-\Omega}_{N0S-\Sigma}(-1)^{N-\Omega+2S-\Sigma+\lambda}|JM-\Omega\rangle|S-\Sigma\rangle,\\ &= (-1)^{N+\lambda}|J(NS)M\rangle,\end{aligned} \tag{3.42}$$

where $\lambda = 0$ for a Σ^+ state and $\lambda = 1$ for a Σ^- state (see Section 1.5), and we have used the symmetry property of the Clebsch–Gordan coefficients $C^{j_3m_3}_{j_1m_1j_2m_2} = (-1)^{j_1+j_2-j_3}C^{j_3-m_3}_{j_1-m_1j_2-m_2}$. The parity of the states $|J(NS)M\rangle$ is thus determined by the value of N. The parity of the lowest energy rotational state is given by $(-1)^{\lambda}$. These results can be used to draw important conclusions about the structure of the matrix of the Zeeman operator within the subspace of a single $^2\Sigma$ electronic state. Since the Zeeman interaction couples states with $|\Delta J| = 1$ and the same parity, the operator (3.39) couples states of different J only when the two values $J = N \pm 1/2$ correspond to the same value of N. We can also conclude that the ground rotational state is not coupled by the Zeeman operator to other states within the same electronic state. Effectively, this means that in the state with $N = 0$ the electron spin is decoupled from the rotational motion of the molecule and the Zeeman effect should resemble that of an isolated spin-1/2 particle.

It is preferable to use Hund's case (a) basis for the evaluation of the Zeeman shifts of the rotational energy levels. There are two reasons for this. First, the Zeeman interaction (3.39) contains operators $L_\pm$ and $S_\pm$ that couple different electronic states. These couplings are most conveniently treated using a basis containing explicitly the eigenstates of S_z and L_z. Second, the matrix of the Zeeman operator is generally not diagonal in either of the two bases. Therefore, there is no need to prefer Hund's case (b) over case (a) basis.

Because the Zeeman interaction conserves parity, it is useful to work with parity adapted basis states, which for the case of a $^2\Sigma$ molecule are

$$\begin{aligned}|JM,\tilde{\Omega} = 1/2,\varepsilon\rangle = \frac{1}{\sqrt{2}}[&|JM,\Omega = 1/2\rangle|S,\Sigma = 1/2\rangle\\ &+ \varepsilon(-1)^{J-1/2+\lambda}|JM,\Omega = -1/2\rangle|S,\Sigma = -1/2\rangle].\end{aligned} \tag{3.43}$$

The matrix elements of the operator (3.39) in the basis (3.43) can be readily evaluated using Eq. (3.40).

For example, consider a $^2\Sigma$ electronic state well separated in energy from other electronic states.[3] In this case, the matrix elements of $L_\pm$ can be neglected and the Zeeman shifts are determined by the matrix elements of the operator

$$\hat{H}_{\text{Zeeman},^2\Sigma} = \frac{\mu_B g_S B}{2\hbar}[\Phi_Z^- S_+ + \Phi_Z^+ S_- + 2\Phi_Z^z S_z]. \tag{3.44}$$

Note that the matrix elements of L_z are zero. The diagonal matrix elements of the operator (3.44) in the basis (3.43) have the form (Exercise 3.3)

$$\langle JM\tilde{\Omega}|\hat{H}_{\text{Zeeman},^2\Sigma}|JM\tilde{\Omega}\rangle = \frac{\mu_B g_S B}{2}M \times \begin{cases} 1/J & \text{if } \epsilon(-1)^{J-1/2+\lambda} = 1. \\ -1/(J+1) & \text{if } \epsilon(-1)^{J-1/2+\lambda} = -1. \end{cases} \tag{3.45}$$

The rotational ground state corresponds to $J = 1/2$ and $\epsilon = (-1)^\lambda$. The Zeeman shifts of the rotational ground state are thus given by

$$\Delta E_M = \mu_B g_S B M, \tag{3.46}$$

where $M = \pm 1/2$.

In general, the Zeeman energy levels of a $^2\Sigma$ molecule are perturbed by couplings to other electronic states, specifically, $^2\Pi$ states. These couplings are mediated by the $L_\pm$ operators appearing both in the rotational operator (see Section 2.4.4) and in the Zeeman operator (3.39). If the Σ and Π states happen to be close in energy, it is necessary to include the Σ–Π couplings in the calculations and diagonalize the matrix of the rotational and Zeeman interactions in the basis spanning multiple electronic states [20–22]. The matrix elements $\langle\Sigma|L_\pm|\Pi\rangle$ are usually undefined and can be found by fitting the experimental measurements of the Zeeman shifts of the rotational states in the Σ state [20]. The effect of these couplings is determined largely by the relative energies of the Σ and Π electronic states. Note that the same couplings give rise to Λ-doubling of the rotational energy levels of molecules in the Π electronic states discussed in Section 2.4.4.

Friedrich et al. [20] presents an interesting diagram of the Zeeman energy levels of the CaH molecule in the ground $X^2\Sigma^+$ and excited $B^2\Sigma^+$ electronic states, reproduced in Figure 3.1. This diagram was calculated in order to simulate the magnetic energy shifts of the frequencies for the electronic transitions shown by the thin vertical lines. The state $X^2\Sigma^+$ is separated from both $A^2\Pi$ and $B^2\Sigma^+$ by the energy gap of about 15 000 cm^{-1}. However, the A and B states are close in energy, separated only by 1349 cm^{-1} [19]. The Zeeman shifts of the

3 See Section 3.5.3 for an alternative strategy of calculating the Zeeman shifts in an isolated Σ state.

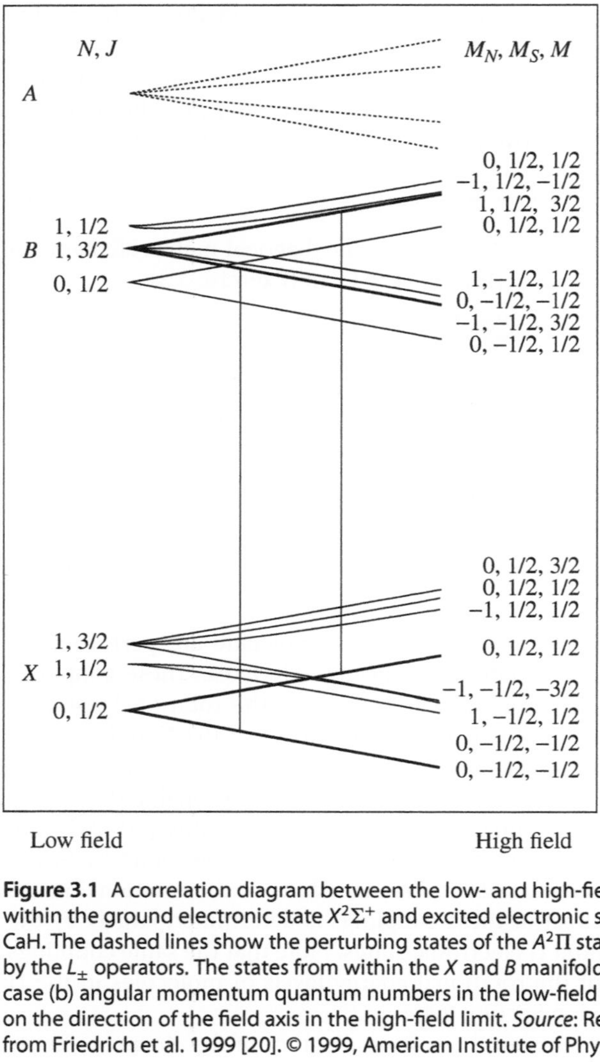

Figure 3.1 A correlation diagram between the low- and high-field limits for states from within the ground electronic state $X^2\Sigma^+$ and excited electronic state $B^2\Sigma^+$ of the molecule CaH. The dashed lines show the perturbing states of the $A^2\Pi$ state coupled to the Σ states by the $L_\pm$ operators. The states from within the X and B manifolds are labeled by the Hund's case (b) angular momentum quantum numbers in the low-field limit and their projections on the direction of the field axis in the high-field limit. *Source*: Reproduced with permission from Friedrich et al. 1999 [20]. © 1999, American Institute of Physics.

rotational states in the $B^2\Sigma^+$ state are, therefore, significantly affected by the $L_\pm$ couplings [20]. On the other hand, the couplings between $X^2\Sigma^+$ and $A^2\Pi$ do not affect the Zeeman shifts of the rotational states in the $X^2\Sigma^+$ state.

Figure 3.1 illustrates the main features of the Zeeman effect in $^2\Sigma$ molecules. In particular, I would like to point out the following:

- The rotational ground state $N = 0$ splits into a doublet of Zeeman levels, one corresponding to the electron spin projection $M_S = +1/2$ and one to

$M_S = -1/2$. The magnetic shifts of these levels are linear at all magnitudes of the magnetic field. Effectively, this means that in the state with $N = 0$ the electron spin is decoupled from the rotational motion of the molecule and the Zeeman effect is the same as that of an isolated spin-1/2 particle.

- In the limit of very weak magnetic fields, the Zeeman shifts of all other rotational states are also linear. As discussed earlier, this is a consequence of the magnetic field being a pseudo-vector, allowing for the magnetic-field-induced couplings to appear in first-order perturbation theory corrections.
- At intermediate fields, some of the rotational states exhibit nonlinear Zeeman effect. This is a consequence of the couplings between states with different J.
- At high magnetic fields, the Zeeman shifts of all rotational states are linear. At high fields, the electron spin becomes decoupled from the rotational motion of the molecule. Therefore, each state can be labeled by well-defined projections of $\boldsymbol{N}$, $\boldsymbol{S}$, and $\boldsymbol{J}$ on the magnetic field axis. At high fields, half of the states are low-field seeking (i.e. exhibiting a positive field gradient of energy) and the other half are high-field seeking (negative field gradient of energy). We will return to the discussion of low-field seekers vs high-field seekers in Chapter 6 devoted to external field traps.

3.5.2 Molecules in a $^2\Pi$ Electronic State

$$^2\Pi : S = 1/2, |\Lambda| = 1, |\Sigma| = 1/2, |\Omega| = 1/2, 3/2.$$

If a $^2\Pi$ molecule is placed in a weak magnetic field, the Zeeman shifts of the rotational states can be approximated by the diagonal elements of the Zeeman operator (3.39) in the basis (2.65), which according to Eqs. (3.39) and (3.40) are

$$\langle JM\tilde{\Omega}, \epsilon | \hat{H}_{\text{Zeeman}} | JM\tilde{\Omega}, \epsilon \rangle = \mu_B BM \frac{\tilde{\Omega}}{J(J+1)} \left[\tilde{\Lambda} + g_S(\tilde{\Omega} - \tilde{\Lambda}) \right], \tag{3.47}$$

where $\tilde{\Omega} = |\Omega|$ and $\tilde{\Lambda} = |\Lambda|$. As in the case of $^2\Sigma$ molecules, the rotational state with a given J splits into $2J + 1$ Zeeman states, which exhibit linear divergence at low magnetic fields. As the field strength increases, the behavior of the Zeeman states become markedly different from the Zeeman shifts in $^2\Sigma$ molecules. In particular, the Zeeman shifts depart from linearity and *all* of the Zeeman states become high-field seeking. This is illustrated in Figure 3.2, which displays the Zeeman energy levels of a CaF molecule in the rotational state with $J = 3/2$ of the $A\ ^2\Pi_{3/2}$ electronic state [23]. The dashed lines show the magnitudes of the diagonal matrix elements (3.47). The figure illustrates that the estimates given by Eq. (3.47) are accurate for Zeeman shifts in the CaF molecule induced by magnetic fields below $\lesssim 0.5$ T. However, at large fields it is necessary to include the off-diagonal matrix elements of the Zeeman operator.

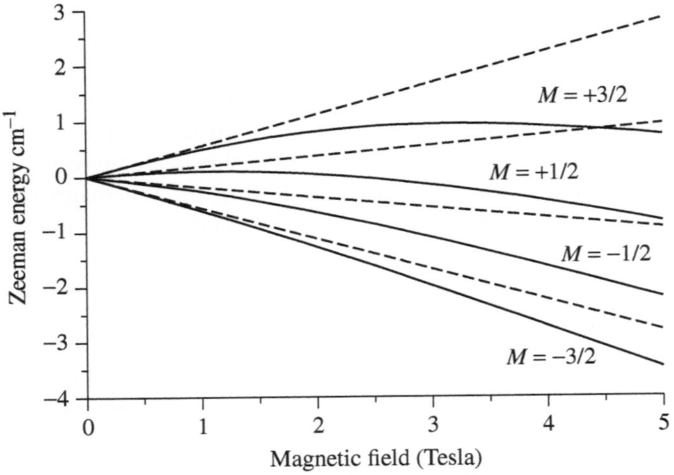

Figure 3.2 Zeeman levels of a CaF molecule in the rotational state characterized by $J = 3/2$ of the $A^2\Pi_{3/2}$ electronic state: Full curves–accurate calculations; dashed lines–the magnitudes of the diagonal matrix elements given by Eq. (3.47). *Source*: Reproduced with permission from Krems et al. 2004 [23]. © 2004, American Institute of Physics.

The Zeeman operator couples states with $|\Delta J| = 0, 1$ and the same parity. These selection rules prohibit couplings between different rotational states in Σ electronic states (see Section 3.5.1), where different rotational states either have different parity or correspond to J quantum numbers differing by more than one. In the case of Π electronic states, each rotational J state offers a doublet of opposite parity states (see Section 2.4.4). As a consequence, the Zeeman operator (3.39) couples different fine structure components of the $^2\Pi$ state (i.e. $^2\Pi_{3/2}$–$^2\Pi_{1/2}$ couplings) and different rotational states of the same parity. This is, of course, in addition to the Π–Σ couplings discussed earlier. The question is, which of these couplings are the most important? The answer is determined by the relative spacing of the rotational, fine structure, and electronic states. For a majority of $^2\Pi$ molecules in low-energy rotational states, the rotational splitting is smaller than the energy gap between the $^2\Pi_{3/2}$ and $^2\Pi_{1/2}$ components. If this is the case, the departure of the Zeeman shifts from linearity is largely determined by the couplings between different rotational states of the same electronic state.

As an illustrative example, consider Figure 3.3, which presents the magnetic shifts of the frequencies for the $(X^2\Sigma^+, N = 0, M = 1/2) \rightarrow (A^2\Pi_{3/2}, J = 3/2, M = 3/2)$ and $(X^2\Sigma^+, N = 0, M = -1/2) \rightarrow (A^2\Pi_{3/2}, J = 3/2, M = -3/2)$ transitions as functions of the magnetic field. The full curves displayed in the

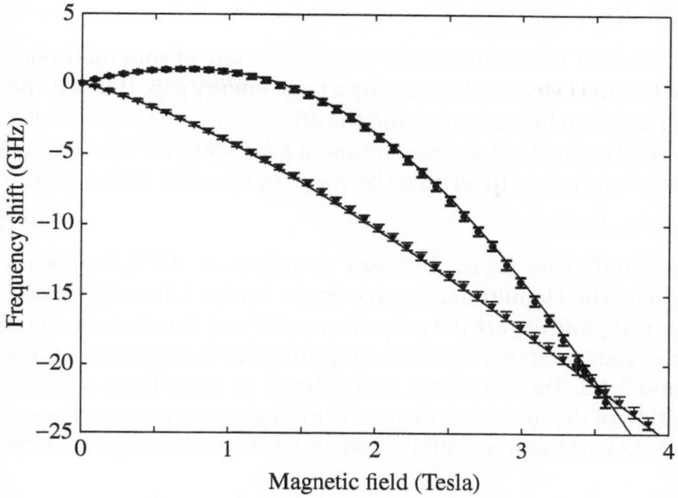

Figure 3.3 Symbols - measured frequency shift for the $(X^2\Sigma^+, N = 0, M = 1/2) \rightarrow (A^2\Pi_{3/2}, J = 3/2, M = 3/2)$ (circles) and $(X^2\Sigma^+, N = 0, M = -1/2) \rightarrow (A^2\Pi_{3/2}, J = 3/2, M = -3/2)$ (triangles) transitions; curves – direction cosine calculations. *Source*: Reproduced with permission from Krems et al. 2004 [23]. © 2004, American Institute of Physics.

figure were obtained by diagonalizing the matrix of the Hamiltonian (3.39) separately in the basis of a single $X^2\Sigma^+$ electronic state in order to obtain the Zeeman shifts of the $|X^2\Sigma^+, N = 0, M\rangle$ states and in the basis of a single $A^2\Pi_{3/2}$ electronic state in order to obtain the Zeeman shifts of the $|A^2\Pi_{3/2}, J = 3/2, M\rangle$ states. The couplings to the $A^2\Pi_{1/2}$ state and other electronic states were neglected. The agreement of this calculation with the spectroscopy measurements shown by the symbols confirms that the curvature of the Zeeman shifts is determined entirely by the couplings between different J states of $^2\Pi_{3/2}$. Interestingly, as a result of the nonlinear Zeeman effect, the two transitions have the same frequency at the magnetic field ~3.5 T. This can be used to probe both Zeeman states $|X^2\Sigma^+, N = 0, M\rangle$ by a single-frequency laser field.

That all Zeeman states of $^2\Pi$ molecules become high-field seeking at high magnetic fields has important implications for magnetic trapping experiments discussed in Chapter 6. It is important to note that the magnetic field magnitude, at which low-field seeking states become high-field seeking, depends on the rotational constant of the molecule and the rotational state J. If the rotational constant is large enough, the low-field seeking Zeeman states may not tip over until the magnetic field reaches very large magnitudes. An example of such molecules is the OH radical in the ground electronic state.

3.5.3 Isolated Σ States

In this section, we consider the Zeeman effect in a $^{2S+1}\Sigma$ state of spin multiplicity $2S+1$ separated from Π electronic states by a large energy gap. If this is the case, the Σ–Π couplings can be neglected and the effective Zeeman operator is reduced to Eq. (3.44). Tracing back the derivation of Eq. (3.39), we can see that this approximation is equivalent to writing the Zeeman operator in the form

$$\hat{H}_{\text{Zeeman},^{2S+1}\Sigma} = \mu_B B g_S S_Z. \tag{3.48}$$

Instead of using Hund's case (a) or (b) bases as in Section 3.5.2, it appears to be easier to evaluate the Hamiltonian matrix in the basis of direct products $|NM_N\rangle|SM_S\rangle$, where M_N and M_S are the projections of $\boldsymbol{N}$ and $\boldsymbol{S}$ on the magnetic field axis. We have already used this basis for computing the Stark perturbations in Sections 2.4.2 and 2.4.3. The advantage of the direct product basis is that it reduces the evaluation of the matrix elements of any angular momentum operators to the Wigner–Eckart theorem, applied separately to each subspace in the direct product.

Any term in the Hamiltonian can be written as a scalar product of tensor operators acting separately on the subspace of $|NM_N\rangle$ states and the subspace of $|SM_S\rangle$ states

$$\hat{V}(N,S) = \sum_q \hat{T}^k_q(N)\hat{T}^k_{-q}(S), \tag{3.49}$$

where k is the rank of the tensors. Depending on the nature of the interactions, k can take different values. For example, the spin–rotation interaction is a scalar product of two rank-1 tensors, while the spin–spin interaction is a scalar product of two rank-2 tensors (cf., Sections 2.4.2 and 2.4.3). The matrix elements of the operator (3.49) are given by

$$\begin{aligned}&\langle NM_N|\langle SM_S|\hat{V}(N,S)|N'M'_N\rangle|SM'_S\rangle\\ &\quad = \sum_q \langle NM_N|\hat{T}^k_q(N)|N'M'_N\rangle\langle SM_S|\hat{T}^k_{-q}(S)|SM'_S\rangle,\end{aligned} \tag{3.50}$$

where [9]

$$\langle NM_N|\hat{T}^k_q(N)|N'M'_N\rangle = (2N+1)^{-1/2} C^{NM_N}_{N'M'_N kq}\langle N||\hat{T}^k||N'\rangle \tag{3.51}$$

and a similar equation can be written for the spin-dependent matrix elements. The last expressions is the Wigner–Eckart theorem discussed in detail in Section 9.1. The double-bar matrix elements are often well known [9] or can be obtained as described in Section 9.3.2. Sections 2.4.2 and 2.4.3 give specific examples of these matrix elements.

The matrix elements of the operator (3.48) in the direct product basis are trivial

$$\langle NM_N|\langle SM_S|\mu_B B g_S S_Z|N'M'_N\rangle|SM'_S\rangle = \delta_{NN'}\delta_{M_N M'_N}\delta_{M_S M'_S}\mu_B B g_S M_S. \tag{3.52}$$

Using this result and the observation that N is a good quantum number for a $^2\Sigma$ state, we immediately obtain the Zeeman shifts given by Eq. (3.46) for the ground rotational state $N = 0$. The evaluation of the Zeeman shifts of the rotationally excited states requires the diagonalization of the matrix of the operators (2.38) and (3.48) in the basis of the direct product states. In the limit of very strong magnetic field, the Zeeman interaction (3.52) becomes dominant and the eigenstates of the Hamiltonian matrix can be labeled by the quantum numbers M_N and M_S, as shown in Figure 3.1. Note that, according to Eq. (2.39), the projection of $\boldsymbol{J}$ on the magnetic field axis is $M = M_N + M_S$.

3.6 Nuclear Zeeman Effect

The preceding discussion completely ignores the interaction of the nuclei with the magnetic field. Being charged particles, the nuclei do interact with an applied magnetic field as molecules execute the rotational motion. Furthermore, many nuclei have nonzero nuclear spin, which gives rise to Zeeman interactions proportional to nuclear magnetic moments. Because of the large mass of the nuclei, these interactions are generally weak. However, if the Zeeman shifts of rotational states need to be known with extremely high precision, it is necessary to consider the effect of an applied magnetic field on the nuclei.

The nonrelativistic interaction of two nuclei A and B and N electrons with a magnetic field can be obtained from the kinetic energy of these $N + 2$ charged particles in an electromagnetic field

$$\hat{T} = \frac{1}{2m_A}(\boldsymbol{P}_A - Z_A e\boldsymbol{A})^2 + \frac{1}{2m_B}(\boldsymbol{P}_B - Z_B e\boldsymbol{A})^2 + \sum_i^N \frac{1}{2m_e}(\boldsymbol{p}_i + e\boldsymbol{A})^2, \tag{3.53}$$

where Z_A and Z_B are the numbers of protons in the nuclei A and B, and e, as before, is the magnitude of the electron charge. Using Eq. (3.19), we can rewrite this equation as

$$\hat{T} = \frac{1}{2m_A}[\boldsymbol{P}_A^2 - Z_A e(\boldsymbol{R}_A \times \boldsymbol{P}_A) \cdot \boldsymbol{B}] + \frac{1}{2m_B}[\boldsymbol{P}_B^2 - Z_B e(\boldsymbol{R}_B \times \boldsymbol{P}_B) \cdot \boldsymbol{B}]$$
$$+ \frac{1}{2m_e}\sum_i^N [\boldsymbol{p}_i^2 + e(\boldsymbol{r}_i \times \boldsymbol{p}_i) \cdot \boldsymbol{B}]. \tag{3.54}$$

Here, the coordinate vectors $\boldsymbol{R}_A$, $\boldsymbol{R}_B$, and $\boldsymbol{r}_i$ are defined with respect to the origin of an arbitrary laboratory-fixed frame. As in Section 1.2.2 and Exercise 1.2, we want to separate out the center-of-mass motion of the molecular system. This requires defining a new frame of reference with the origin at the center of mass of the $N + 2$ particle system. The magnetic field dependent part of the kinetic energy in this frame is [22]

$$\hat{T}_B = -\frac{e\hbar}{2m_e}\chi\frac{Z_A m_B^2 + Z_B m_A^2}{m_A m_B}\boldsymbol{B}\cdot\boldsymbol{R} + \frac{e}{2m_e}(1-\chi)\boldsymbol{B}\cdot\sum_i \boldsymbol{r}_i \times \boldsymbol{p}_i$$
$$-\frac{e}{2m_e}\chi\boldsymbol{B}\cdot\sum_i\sum_{j\neq i}\boldsymbol{r}_i\times\boldsymbol{p}_j + O(\chi^2), \tag{3.55}$$

where $\chi = m_e/(m_A + m_B) \ll 1$ and $\boldsymbol{R}$ is the rotational angular momentum of the nuclei introduced in Section 1.2.2. The leading term on the right-hand side of Eq. (3.55) is the already familiar $\mu_B g_L \boldsymbol{L}\cdot\boldsymbol{B}$ appearing in Eq. (3.19). The last term, when averaged over all electronic degrees of freedom, is proportional to $\mu_B \boldsymbol{L}\cdot\boldsymbol{B}$, and thus produces a small correction to the leading term [22].

Just like an electron, a nucleus α with a nonzero spin $\boldsymbol{I}$ has a magnetic moment $\boldsymbol{\mu}_I = g_\alpha \mu_N \boldsymbol{I}$, where g_α is the nuclear g-factor and μ_N is the nuclear magneton equal to $e\hbar/2m_p$ with m_p being the proton mass. The Zeeman Hamiltonian for a diatomic molecule accounting for the electron and nuclear spin effects thus has the form

$$\hat{H}_{\text{Zeeman}} = \mu_B(g_L\boldsymbol{L} + g_S\boldsymbol{S} + g_R\boldsymbol{R} + g_I^A\boldsymbol{I}_A + g_I^B\boldsymbol{I}_B)\cdot\boldsymbol{B}. \tag{3.56}$$

The g_L and g_R factors derive from Eq. (3.55) and have the magnitude

$$g_L = \left(1 - \frac{m_e}{m_A + m_B}\right) \tag{3.57}$$

$$g_R = -\frac{m_e}{m_A + m_B}\frac{Z_A m_B^2 + Z_B m_A^2}{m_A m_B}. \tag{3.58}$$

The g_I factors are given by $g_I = -\mu_I/(\mu_B I)$ [22].

3.6.1 Zeeman Effect in a $^1\Sigma$ Molecule

As an illustrative example, consider the Zeeman effect in an isolated $^1\Sigma$ state. In this case, the matrix elements of the first two terms in Eq. (3.56) vanish, and the Zeeman shifts are determined by the matrix elements of the $\boldsymbol{R}\cdot\boldsymbol{B}$ and $\boldsymbol{I}\cdot\boldsymbol{B}$ operators. The nonzero nuclear spins lead to hyperfine structure, which needs to be properly accounted for (see Section 1.7). Since the electron spin is absent, the hyperfine structure of $^1\Sigma$ molecules is largely determined by the interaction of the nuclear quadrupole moments with the electric field gradients at the nuclei, the nuclear spin–rotation interaction between the magnetic moments of the nuclei and the magnetic field created by the rotational motion of the molecule, and the nuclear spin–spin interactions. The matrix elements of these interactions as well as the matrix elements of the Zeeman interaction (3.56) can be conveniently evaluated in the basis of direct products $|I_A M_{I_A}\rangle|I_B M_{I_B}\rangle|NM_N\rangle$, where the projections M_{I_A}, M_{I_B}, and M_N are all with respect to the direction

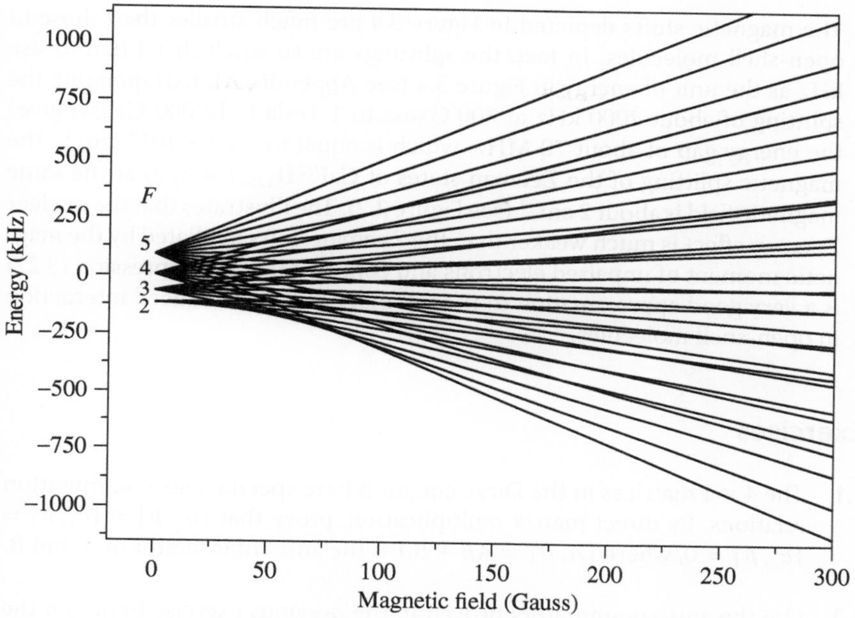

Figure 3.4 Zeeman splitting of the hyperfine energy levels of $^{87}Rb^{133}Cs(^1\Sigma^+)$ in the ground rotational state. *Source*: Adapted with permission from Wallis and Krems 2014 [24]. © 2014, American Physical Society.

of the magnetic field. The diagonalization of the Hamiltonian matrix gives the Zeeman states deriving from the hyperfine states of the molecule at zero magnetic field. Figure 3.4 shows the Zeeman states of the $^{87}Rb^{133}Cs(^1\Sigma^+)$ molecule in the lowest-energy rotational state computed using the interaction constants from Ref. [25].

There are a few general observations to be made from Figure 3.4:

- At zero magnetic field, the rotational ground state of the molecule is split into four hyperfine states, each corresponding to a well-defined quantum number F of the total angular momentum $\boldsymbol{F} = \boldsymbol{I}_A + \boldsymbol{I}_B + \boldsymbol{N}$.
- The Zeeman interaction (3.56) couples states with $|F - F'| = 0$ and 1, leading to nonlinear magnetic shifts of the Zeeman states.
- Each Zeeman state is characterized by a well-defined value of the projection of $\boldsymbol{F}$ on the magnetic field direction. In other words, the projection of the *total* angular momentum on the magnetic field direction is conserved.
- In the limit of high magnetic fields, all Zeeman shifts are linear and each Zeeman state corresponds to well-defined projections of $\boldsymbol{I}_A$ and $\boldsymbol{I}_B$ on the magnetic field axis. This happens when the magnetic-field-induced interactions become much larger than the hyperfine interactions.

- The magnetic shifts depicted in Figure 3.4 are much smaller than those in open-shell molecules. In fact, the splittings are so small that I had to use kHz as the unit of energy in Figure 3.4 (see Appendix A). Extrapolating the splitting of about 2000 kHz at 300 Gauss to 1 Tesla (=10 000 Gauss) gives the energy gap of about 70 MHz, which is equal to $\sim 2.3 \times 10^{-3}$ cm^{-1}. The magnetic splitting of the Zeeman states of CaF($^2\Pi_{3/2}, J = 3/2$) at the same magnetic field is about 2 cm^{-1} (see Figure 3.2). The illustrates that the nuclear Zeeman effect is much weaker than the Zeeman shifts mediated by the magnetic moment of unpaired electrons and confirms that the expression (3.22) is a very good approximation to the full magnetic-field-induced interaction in open-shell molecules.

Exercises

3.1 The 4×4 matrices in the Dirac equation have specific anti-commutation relations. By direct matrix multiplication, prove that $\{\alpha_x, \beta\} = \{\alpha_y, \beta\} = \{\alpha_z, \beta\} = 0$, where $\{A, B\} = AB + BA$ is the anti-commutator of A and B.

3.2 Use the anti-commutators proven in the previous exercise to obtain the nonrelativistic limit (3.9) of the Dirac Hamiltonian (3.8).

3.3 The step from Eq. (3.9) to Eq. (3.10) is important as it leads to the magnetic field-dependent term written in terms of the Pauli matrices. Obtain Eq. (3.10) from Eq. (3.9). To do this, use the vector identities

$$(\tilde{\sigma} \cdot \boldsymbol{A})(\tilde{\sigma} \cdot \boldsymbol{B}) = (\boldsymbol{A} \cdot \boldsymbol{B}) + i\tilde{\sigma}(\boldsymbol{A} \times \boldsymbol{B})$$

and

$$\boldsymbol{A} \times \boldsymbol{p} + \boldsymbol{p} \times \boldsymbol{A} = -i\hbar \nabla \times \boldsymbol{p}.$$

Consult Ref. [4] for more details.

3.4 Obtain Eq. (3.45).

3.5 Calculate the Zeeman shifts of the energy levels for a $^2\Sigma$ molecule in the states with $J = 3/2$. Compare the results with those given by Eq. (3.45).

3.6 Write the spin–rotation and the spin–spin interaction of a molecule in a $^3\Sigma$ electronic states in the form of Eq. (3.49). The explicit operators for the spin–rotation and spin–spin interactions are given in Section 2.4.3.

3.7 Calculate the Zeeman states of the $CaF(A^2\Pi)$ molecule displayed in Figure 3.2. Repeat the calculation with the fictitious rotational constants $B = 2 \times B_{CaF}$ and $B = 55 \times B_{CaF}$, where $B_{CaF} = 0.343\ cm^{-1}$ is the rotational constant of the CaF molecule in the $A^2\Pi$ electronic state.

3.8 Obtain Eq. (3.55) from Eq. (3.54). To do this, introduce the center-of-mass coordinates and assume that the molecule is charge neutral.

4

AC Stark Effect

In this chapter, we will consider the effect of a time-varying (AC) electric field on molecular energy levels. An AC electric field

$$E(t) = \xi \hat{\epsilon} \cos \omega t, \tag{4.1}$$

is characterized by the amplitude ξ, the oscillation direction specified by the unit vector $\hat{\epsilon}$, and the oscillation frequency ω. The response of a molecule to an applied AC electric field (4.1) generally depends on all these parameters.

Equation (4.1) represents the time dependence of the electric field of an electromagnetic wave. Thus, molecules can be subjected to a periodically varying electric field (4.1) if placed, for example, in a field of a laser beam or in a microwave cavity. The electromagnetic fields are often characterized by their intensity. The intensity of a monochromatic electromagnetic wave propagating in free space is given by

$$I = \frac{c\epsilon_0}{2} |\xi|^2, \tag{4.2}$$

where c is the speed of light, ϵ_0 is the vacuum permittivity, and ξ is the amplitude of the oscillating electric field, same as in Eq. (4.1). The intensity is commonly measured in units of W cm^{-2} and represents the amount of power transmitted through a unit surface area perpendicular to the propagation direction. It can also be thought of as the energy density multiplied by the velocity of the propagating wave.

Inserting the values of c and ϵ_0 into Eq. (4.2), we find that a beam of electromagnetic field with a power of 1 W cm^{-2} provides an electric field of up to $\xi \approx 27.5$ V cm^{-1} at the peaks of the oscillations. Is it easy to generate an electromagnetic field with the intensity $I = 1$ W cm^{-2}? While the answer depends on the frequency of the field, note that commercially available laser pointers produce light with power $\lesssim 5 \times 10^{-3}$ W. If focussed onto an area of 1 mm^2, this provides the intensity up to 0.5 W cm^{-2}. In a laboratory setting, it is fairly easy to generate laser fields many times stronger than this. In fact, if pulsed and focused to a very small area, laser light can have a very high intensity of up

Molecules in Electromagnetic Fields: From Ultracold Physics to Controlled Chemistry, First Edition. Roman V. Krems.

to – in the extreme case [26] – 10^{23} W cm^{-2}. This shows that AC electric fields can be made much, much stronger that DC electric fields (which, as mentioned in Chapter 2, are limited to $<5 \times 10^5$ V cm^{-1}).

Assuming that the electric field at any given moment of time is uniform, we can use Eq. (2.7) from Chapter 2 to write the Hamiltonian of a molecular system in an AC electric field as

$$\hat{H} = \hat{H}_0 - \boldsymbol{d} \cdot \hat{\epsilon}\, \xi \cos \omega t, \tag{4.3}$$

where $\hat{H}_0$ is the Hamiltonian of the molecule in the absence of the field and $\boldsymbol{d}$ is the dipole moment operator. Because the field is varying in time, the Hamiltonian is time dependent, so it is necessary to solve the time-dependent Schrödinger equation in order to find how the field modifies the properties of the molecule. The energy of a system with a time-dependent Hamiltonian is not conserved so, for an arbitrary system with the Hamiltonian represented as a sum of a time-independent term $\hat{H}_0$ and a time-dependent term $V(t)$, it does not make sense to discuss shifts of the eigenstates of $\hat{H}_0$ due to $V(t)$. Instead, one typically is concerned with the problem of the population transfer from one eigenstate of $\hat{H}_0$ to another.

However, if the frequency of the oscillating field is far detuned from the frequency of the transitions between the eigenstates of $\hat{H}_0$, i.e.

$$\hbar\omega \gg |\epsilon_\alpha - \epsilon_\beta| \quad \text{or} \quad \hbar\omega \ll |\epsilon_\alpha - \epsilon_\beta|, \tag{4.4}$$

the presence of the field perturbs the molecular energy levels, without inducing much population transfer. These perturbations are the AC Stark shifts. In this chapter, we will discuss how to calculate these shifts.

4.1 Periodic Hamiltonians

The Hamiltonian (4.3) is a periodic function of time, i.e. $\hat{H}(t + T) = \hat{H}(t)$, where $T = 2\pi/\omega$. Can we take advantage of this periodicity?

Consider the time-dependent Schrödinger equation

$$i\hbar \frac{\partial}{\partial t} |\psi(t)\rangle = \hat{H} |\psi(t)\rangle. \tag{4.5}$$

The periodicity of the Hamiltonian should be reflected in the solutions of the Schrödinger equation. But how?

Note that the solutions $|\psi\rangle$ themselves are, in general, not periodic, i.e., in general

$$|\psi(t + T)\rangle \neq |\psi(t)\rangle. \tag{4.6}$$

What we will do now is write these solutions in terms of states that reflect the periodicity of the Hamiltonian.

Let us define an operator $\hat{T}$ that translates time from t to $t+T$

$$\hat{T}|\psi(t)\rangle = |\psi(t+T)\rangle. \tag{4.7}$$

This operator has two properties of importance to us. First, it commutes with the Hamiltonian because the Hamiltonian is invariant under the transformation $t \to t+T$. Second, the operator $\hat{T}$ is unitary because the time evolution determined by the Schrödinger equation preserves the norm of the state vector:

$$\langle\psi|\hat{T}^\dagger\hat{T}|\psi\rangle = \langle\hat{T}\psi|\hat{T}\psi\rangle = 1 = \langle\psi|\psi\rangle \Rightarrow \hat{T}^\dagger\hat{T} = 1. \tag{4.8}$$

Since $\hat{T}$ and $\hat{H}$ commute, we can find a set of solutions $|\psi\rangle$ of Eq. (4.5) that are also the eigenstates of $\hat{T}$:

$$\hat{T}|\psi\rangle = \mathrm{e}^{i\gamma}|\psi\rangle. \tag{4.9}$$

Because $\hat{T}$ is unitary, its eigenvalues must be of the form $\mathrm{e}^{i\gamma}$, where γ is real. This ensures that

$$\langle\psi|\hat{T}^\dagger\hat{T}|\psi\rangle = \langle\psi|\psi\rangle. \tag{4.10}$$

We will write γ as $\gamma = qT/\hbar$ so that

$$\hat{T}|\psi\rangle = \mathrm{e}^{iqT/\hbar}|\psi\rangle. \tag{4.11}$$

Let us now consider states defined as following:

$$|\phi\rangle = \mathrm{e}^{-iqt/\hbar}|\psi\rangle, \tag{4.12}$$

with the same q in the exponent as in the previous equation. If we act with the $\hat{T}$ operator on these states, we will obtain

$$\hat{T}|\phi(t)\rangle = \mathrm{e}^{-iq(t+T)/\hbar}|\psi(t+T)\rangle = \mathrm{e}^{-iqt/\hbar}\mathrm{e}^{-iqT/\hbar}\mathrm{e}^{iqT/\hbar}|\psi(t)\rangle = |\phi(t)\rangle,$$

where we have used Eqs. (4.7) and (4.11).

On the other hand,

$$\hat{T}|\phi(t)\rangle = |\phi(t+T)\rangle, \tag{4.13}$$

so we see that

$$|\phi(t+T)\rangle = |\phi(t)\rangle. \tag{4.14}$$

In other words, the solutions of the time-dependent Schrödinger equation (4.5) can be written as

$$|\psi\rangle = \mathrm{e}^{iqt/\hbar}|\phi\rangle, \tag{4.15}$$

where the states $|\phi\rangle$ are periodic functions of time, with the same period T as the Hamiltonian. Remember also that q is real, which follows from the unitarity of $\hat{T}$, as discussed above.

This is a beautiful result that can be generalized to solve ordinary differential equations with periodic potentials. For example, a similar result can be obtained for a quantum particle in a time-independent periodic potential, such as an electron in a crystal lattice. For such a problem, the same arguments as we have made in this section can be used to obtain the eigenstates of the time-independent Schrödinger equation in the form

$$|\psi(\boldsymbol{r})\rangle = e^{i\boldsymbol{q}\cdot\boldsymbol{r}/\hbar}|\phi(\boldsymbol{r})\rangle, \tag{4.16}$$

where $\boldsymbol{q}$ is the momentum of the particle, $\boldsymbol{r}$ is the spatial coordinate, and the states $|\phi(\boldsymbol{r})\rangle$ have the same spatial periodicity as the lattice potential. We will encounter another example in Section 5.1.2, where we will use a similar approach to obtain the solutions for a quantum pendulum described by the angular Mathieu differential equation with periodic boundary conditions.

Writing the solutions of the differential equations with periodic potentials in the form of Eq. (4.15) is the essence of the Floquet theory. When applied to quantum particles in spatially periodic potentials, the Floquet theory produces the solutions given by Eq. (4.16), which are known as Bloch waves. We will use the Floquet theory to solve the time-dependent Schrödinger equation for molecules in an AC field and analyze the main features of the AC Stark effect.

We should be really excited about Eq. (4.15). It allows us to write the solutions of the time-dependent Schrödinger equation as a product of a simple phase factor and states with known time periodicity. Any periodic function can be represented as a Fourier series, which can be used to reformulate Eq. (4.5) as an eigenvalue problem. This is what we will do in Section 4.2 – we will solve the time-dependent Schrödinger equation by diagonalizing a time-independent matrix called the Floquet matrix.

4.2 The Floquet Theory

In this section, we will apply the Foquet theory more formally to the problem of a molecule in an AC electric field and obtain the corresponding Floquet matrix. We will attempt to understand the physical significance of the Floquet matrix and its eigenvalues and learn how to use the eigenvalues of the Floquet matrix to construct the solutions of the time-dependent Schrödinger equation. We will also understand the meaning of the coefficients q in Eq. (4.15). In a subsequent section, we will quantize the electromagnetic field and see that the Floquet matrix is identical to the Hamiltonian that describes a molecule interacting with an ensemble of photons.

The main theorem of the Floquet theory can be summarized as follows [27]: Consider a first-order differential equation in matrix form

$$\dot{\mathbf{f}} = \mathbf{L}(t)\mathbf{f} \tag{4.17}$$

with a linear operator $\mathbf{L}$ satisfying $\mathbf{L}(t) = \mathbf{L}(t + T)$. The solutions to this equation can be written as

$$\mathbf{f}(t) = e^{\lambda t}\mathbf{g}(t), \tag{4.18}$$

where $\mathbf{g}(t) = \mathbf{g}(t + T)$ is a column matrix of periodic functions of t with the same period T. If there are N linearly independent solutions to Eq. (4.17), there are N Floquet exponents λ that satisfy

$$e^{\lambda_1 T}e^{\lambda_2 T}\cdots e^{\lambda_N T} = \exp\left\{\int_0^T \mathrm{Tr}[\mathbf{L}(\tau)]\mathrm{d}\tau\right\}. \tag{4.19}$$

Note that this formulation is not restricted to first-order differential equations. The operator L can contain higher-order derivative operators. Note also that Eq. (4.15) is a special case of Eq. (4.18). In general, λ can be real or complex, depending on the form of the operator L. For the time-dependent Schrödinger equation, as we have seen in Section 4.1, λ must be purely imaginary.

We will now apply the Floquet's theorem to the Schrödinger equation with the Hamiltonian (4.3). Much of the subsequent discussion is based on Ref. [28].

Consider N orthonormal states $|\psi_\beta(t)\rangle$ satisfying the Schrödinger equation

$$i\hbar\frac{\partial}{\partial t}|\psi_\beta(t)\rangle = \hat{H}|\psi_\beta(t)\rangle. \tag{4.20}$$

In the absence of the electric field, β would be a label of a field-free molecular state. In the presence of a field, different field-free molecular states are coupled. However, we can still expand each of the states $|\psi_\beta\rangle$ in terms of the field-free molecular states. The field-free molecular states are the eigenstates of the Hamiltonian $\hat{H}_0$:

$$\hat{H}_0|\chi_\alpha\rangle = \epsilon_\alpha|\chi_\alpha\rangle. \tag{4.21}$$

Each of these eigenstates evolves with time as any stationary state

$$|\chi_\alpha\rangle = e^{-i\epsilon_\alpha t/\hbar}|\phi_\alpha\rangle, \tag{4.22}$$

where $|\phi_\alpha\rangle$ are time independent.

Using the expansion

$$|\psi_\beta(t)\rangle = \sum_{\alpha=1}^{N} c_\alpha^\beta(t)|\chi_\alpha\rangle = \sum_{\alpha=1}^{N} a_\alpha^\beta(t)|\phi_\alpha\rangle, \tag{4.23}$$

we can rewrite Eq. (4.20) in matrix form as

$$i\hbar\frac{\mathrm{d}}{\mathrm{d}t}\mathrm{F}(t) = \mathbf{H}(t)\mathbf{F}(t), \tag{4.24}$$

where **F** and **H** are the square matrices with the elements

$$\mathbf{F} = \begin{pmatrix} a_1^1 & \cdot & a_1^N \\ \cdot & \cdot & \cdot \\ \cdot & \cdot & \cdot \\ a_N^1 & \cdot & a_N^N \end{pmatrix} \tag{4.25}$$

and

$$H_{\alpha,\beta} = \epsilon_\alpha \delta_{\alpha,\beta} - \langle \phi_\alpha | \boldsymbol{d} \cdot \hat{\epsilon} | \phi_\beta \rangle \, \xi \cos \omega t. \tag{4.26}$$

Note that the phase factors $e^{-i\epsilon_\alpha t/\hbar}$ are absorbed into the coefficients a_α^β.

Because the states $|\psi_\beta(t)\rangle$ are orthonormal, the matrix **F** is unitary. One can readily verify that $\mathbf{F}^\dagger \mathbf{F} = 1$. We can define the matrix

$$\mathbf{U}(t, t_0) = \mathbf{F}(t)\mathbf{F}^\dagger(t_0). \tag{4.27}$$

This square matrix is the *time-evolution operator*. We call it that because it transforms $\mathbf{F}(t_0)$ into $\mathbf{F}(t)$:

$$\mathbf{F}(t) = \mathbf{U}(t, t_0)\mathbf{F}(t_0). \tag{4.28}$$

So far, we have just rewritten the time-dependent Schrödinger equation in the form of Eq. (4.17). The different columns of the matrix **F** correspond to the different solutions $\mathbf{f}(t)$ in Eq. (4.17). We will now apply the Floquet's theorem to the solutions of this equation.

The Floquet's theorem says that the solutions of Eq. (4.24) can be written in the form

$$\mathbf{F}(t) = \boldsymbol{\Phi}(t)e^{\lambda t}, \tag{4.29}$$

where $\boldsymbol{\Phi}(t)$ is a matrix of periodic functions of time and λ is a diagonal time-independent matrix. The different elements of the matrix λ are the different Floquet exponents $\lambda_1, \lambda_2, \dots, \lambda_N$ corresponding to the different solutions of the Schrödinger equation (different columns of the matrix **F**). We will now exploit the periodicity of the matrix elements of $\boldsymbol{\Phi}(t)$ and find the elements of the matrix λ.

First, we note that

$$\mathbf{F}(t + T) = \boldsymbol{\Phi}(t)e^{\lambda(t+T)} = \mathbf{F}(t)e^{\lambda T}. \tag{4.30}$$

Because $\mathbf{F}(t)$ is unitary at all times, this equation shows that the matrix $\lambda \propto i\mathbf{Q}$, where **Q** is a Hermitian matrix. We will write $\lambda = -i\mathbf{Q}/\hbar$ so that

$$\mathbf{F}(t) = \boldsymbol{\Phi}(t)e^{-\frac{i}{\hbar}\mathbf{Q}t}. \tag{4.31}$$

Since **Q** is Hermitian and diagonal, all of its matrix elements are real. This is a special property of the solutions of the Schrödinger equation. We should have expected this result based on the discussion in Section 4.1. There, we showed that the solutions of the Schrödinger equation can be written in the form of

Eq. (4.15), where q is real. The reality of q follows from the unitarity of $\hat{T}$, which follows from the fact that the time evolution governed by the Schrödinger equation does not change the norm of the state vector. This is another way to say that $\mathbf{F}(t)$ – as defined by Eq. (4.25) – must remain unitary at all times.

Since the matrix elements of $\mathbf{\Phi}(t)$ are all periodic functions, which have period $T = 2\pi/\omega$, each of them can be expanded in a Fourier series

$$\Phi_{\alpha,\beta} = \sum_n e^{in\omega t} F^n_{\alpha,\beta}, \tag{4.32}$$

where $F^n_{\alpha,\beta}$ are the expansion coefficients. The elements of the matrix $\mathbf{F}$ can thus be written as

$$F_{\alpha,\beta}(t) = e^{-iq_\beta t/\hbar} \sum_n F^n_{\alpha,\beta} e^{in\omega t}, \tag{4.33}$$

where q_β are the elements of the diagonal matrix $\mathbf{Q}$. As can be seen in Eq. (4.26), the off-diagonal matrix elements of $\mathbf{H}(t)$ are also periodic functions so they can likewise be expanded in a Fourier series to yield

$$H_{\alpha,\beta\neq\alpha}(t) = \sum_n H^n_{\alpha,\beta} e^{in\omega t}. \tag{4.34}$$

The diagonal matrix elements of the Hamiltonian are time independent. However, we will also write

$$H_{\alpha,\alpha}(t) = \sum_n H^n_{\alpha,\alpha}\ e^{in\omega t}, \tag{4.35}$$

for the diagonal elements, assuming that $H^{n=0}_{\alpha,\alpha} = \epsilon_\alpha$ and $H^{n\neq0}_{\alpha,\alpha} = 0$.

By inserting all these Fourier expansions into the Schrödinger equation (4.24), we obtain the following recursive equations:

$$\sum_n F^n_{\alpha,\beta}(-\hbar\omega n + q_\beta) e^{in\omega t} e^{-iq_\beta t/\hbar} = e^{-iq_\beta t/\hbar} \sum_\gamma \sum_k \sum_l H^k_{\alpha,\gamma} F^l_{\gamma,\beta} e^{i(k+l)\omega t}. \tag{4.36}$$

We can now drop the phase factor $e^{-iq_\beta t/\hbar}$. We also note that the last term has a sum over k and l. We can set $n = k + l$ and replace the sum over k and l with the sum over k and n, which will simplify our equations to

$$\sum_n F^n_{\alpha,\beta}(-\hbar\omega n + q_\beta) e^{in\omega t} = \sum_\gamma \sum_k \sum_n H^k_{\alpha,\gamma} F^{n-k}_{\gamma,\beta} e^{in\omega t}. \tag{4.37}$$

Dropping the sum over n and the phase factors $e^{in\omega t}$ leads to a one-line equation:

$$F^n_{\alpha,\beta}(-\hbar\omega n + q_\beta) = \sum_\gamma \sum_k H^k_{\alpha,\gamma} F^{n-k}_{\gamma,\beta}. \tag{4.38}$$

We can make another variable substitution $p = n - k$. To minimize the number of indexes, I will rename p to k and write the resulting equation as

$$\sum_{\gamma}\sum_{k}[H_{\alpha,\gamma}^{(n-k)} + \hbar\omega n\delta_{nk}\delta_{\alpha\gamma}]F_{\gamma,\beta}^{k} = q_{\beta}F_{\alpha,\beta}^{n}. \tag{4.39}$$

This is an eigenvalue equation for the Floquet exponents q_β and the coefficients $F_{\gamma,\beta}^{k}$. To see this more clearly, let us introduce the basis of states $|\alpha n\rangle$, where the Greek letter labels the molecular states and the Roman letter the Fourier components. Then, Eq. (4.39) can be written as sums representing a matrix-vector product

$$\sum_{\gamma}\sum_{k}\langle\alpha n|\hat{H}_{\mathrm{F}}|\gamma k\rangle F_{\gamma,\beta}^{k} = q_{\beta}F_{\alpha,\beta}^{n}. \tag{4.40}$$

Note that the states $|\alpha n\rangle$ are orthonormal and the matrix of $\hat{H}_F$ in the basis of these states is Hermitian.

4.2.1 Floquet Matrix

The operator $\hat{H}_{\mathrm{F}}$ is the Floquet Hamiltonian. The physical meaning of the states $|\alpha n\rangle$ will be more transparent in Section 4.8, where we will quantize the electromagnetic field and show that the Hamiltonian of a molecule immersed in a bath of photons is exactly the same as the Floquet matrix. In that context, the states $|\alpha n\rangle$ are the products of the molecular states α and the *photon number* states. For now, we will refer to these states as the "bare product" states, implying that they represent the products of "bare," i.e. field-free, molecular states and the states $|n\rangle$ representing the different Fourier components.

The diagonal matrix elements of $\hat{H}_{\mathrm{F}}$ in the basis $|\alpha n\rangle$ are

$$\langle\alpha n|\hat{H}_{\mathrm{F}}|\alpha n\rangle = \epsilon_{\alpha} + \hbar\omega n. \tag{4.41}$$

The off-diagonal matrix elements are

$$\langle\alpha n|\hat{H}_{\mathrm{F}}|\beta k\rangle = H_{\alpha\beta}^{(n-k)}\delta_{n,k\pm1}, \tag{4.42}$$

where

$$H_{\alpha\beta}^{(n-k=\pm1)} = -\langle\phi_{\alpha}|\boldsymbol{d}\cdot\hat{\boldsymbol{e}}|\phi_{\beta}\rangle\frac{\xi}{2} = \hbar\Omega/2, \tag{4.43}$$

and

$$\Omega \equiv -\frac{\langle\phi_{\alpha}|\boldsymbol{d}\cdot\hat{\boldsymbol{e}}|\phi_{\beta}\rangle}{\hbar}\xi. \tag{4.44}$$

Note that the selection rule in Eq. (4.42) is determined by the specific form of the time-dependent term in the Hamiltonian (4.3). The Fourier expansion of the cosine is

$$\cos\omega t = \frac{1}{2}\{\mathrm{e}^{i\omega t} + \mathrm{e}^{-i\omega t}\}, \tag{4.45}$$

which means that only the terms $H^n_{\alpha,\beta}$ with $n = \pm 1$ survive in the expansion (4.34).

In addition, the matrix elements $\langle \phi_\alpha | \boldsymbol{d} \cdot \hat{\epsilon} | \phi_\beta \rangle$ are nonzero only if the states $|\phi_\alpha\rangle$ and $|\phi_\beta\rangle$ have opposite parity (see Section 2.2). This is the selection rule that is determined by the nature of the molecular states $|\phi_\alpha\rangle$. This selection rule along with Eq. (4.42) makes the Floquet Hamiltonian matrix very sparse.

The sums over n in the Fourier expansions (4.33) and (4.34) extend to infinity. Therefore, the matrix of the Floquet Hamiltonian is infinitely dimensional. If the Hilbert space of molecular states is restricted only to two states α and β, the matrix of the Floquet Hamiltonian has the following form:

$$\begin{pmatrix} \cdot\ \cdot & \cdot & \cdot & \cdot & \cdot & \cdot & \cdot & \cdot & \cdot \\ \cdot\ \epsilon_\beta - 2\hbar\omega & \hbar\Omega/2 & 0 & 0 & 0 & 0 & 0 & 0 & \cdot \\ \cdot\ \hbar\Omega^*/2 & \epsilon_\alpha - \hbar\omega & 0 & 0 & \hbar\Omega/2 & 0 & 0 & 0 & \cdot \\ \cdot\ 0 & 0 & \epsilon_\beta - \hbar\omega & \hbar\Omega^*/2 & 0 & 0 & 0 & 0 & \cdot \\ \cdot\ 0 & 0 & \hbar\Omega^*/2 & \epsilon_\alpha & 0 & 0 & \hbar\Omega/2 & 0 & \cdot \\ \cdot\ 0 & \hbar\Omega^*/2 & 0 & 0 & \epsilon_\beta & \hbar\Omega/2 & 0 & 0 & \cdot \\ \cdot\ 0 & 0 & 0 & 0 & \hbar\Omega^*/2 & \epsilon_\alpha + \hbar\omega & 0 & 0 & \cdot \\ \cdot\ 0 & 0 & 0 & \hbar\Omega^*/2 & 0 & 0 & \epsilon_\beta + \hbar\omega & \hbar\Omega/2 & \cdot \\ \cdot\ 0 & 0 & 0 & 0 & 0 & 0 & \hbar\Omega^*/2 & \epsilon_\alpha + 2\hbar\omega & \cdot \\ \cdot\ \cdot & \cdot & \cdot & \cdot & \cdot & \cdot & \cdot & \cdot & \cdot \end{pmatrix}. \tag{4.46}$$

4.2.2 Time Evolution Operator

Let $|\lambda_\beta\rangle$ be the eigenvector of $\hat{H}_{\rm F}$ corresponding to eigenvalue q_β. This eigenvector can be expanded in states $|\alpha n\rangle$. As follows from Eq. (4.40), the expansion coefficients are the coefficients $F^n_{\alpha,\beta}$:

$$|\lambda_\beta\rangle = \sum_\alpha \sum_n |n\alpha\rangle F^n_{\alpha,\beta}. \tag{4.47}$$

The coefficients $F^n_{\alpha,\beta}$ are thus the projections of the "bare product" states $|\alpha n\rangle$ on the eigenstates of the Floquet matrix

$$F^n_{\alpha,\beta} = \langle \alpha n | \lambda_\beta \rangle. \tag{4.48}$$

The elements (4.33) of the matrix $\mathbf{F}(t)$ can be written in terms of these projections as

$$F_{\alpha,\beta}(t) = \sum_n \langle \alpha n | \lambda_\beta \rangle e^{in\omega t} e^{-iq_\beta t/\hbar}, \tag{4.49}$$

and the elements of the time-evolution matrix $\mathbf{U}(t, t_0) = \mathbf{F}(t)\mathbf{F}^\dagger(t_0)$ are

$$U_{\alpha,\beta}(t, t_0) = \sum_\gamma F_{\alpha,\gamma} F^*_{\gamma,\beta} \tag{4.50}$$

$$= \sum_n \sum_k \sum_\gamma \langle \alpha n | \lambda_\gamma \rangle e^{-iq_\gamma (t-t_0)/\hbar} e^{i(n-k)\omega t} \langle \lambda_\gamma | \beta k \rangle. \tag{4.51}$$

Since the eigenvectors $|\lambda_\gamma\rangle$ form a complete set, we can use the identity

$$e^{-i\hat{H}_F/\hbar} = \sum_\gamma |\lambda_\gamma\rangle e^{-iq_\gamma/\hbar}\langle\lambda_\gamma|, \tag{4.52}$$

to write the time-evolution operator as

$$U_{\alpha,\beta}(t,t_0) = \sum_n \sum_k \langle\alpha n|e^{-i\hat{H}_F(t-t_0)/\hbar}|\beta k\rangle e^{i(n-k)\omega t}. \tag{4.53}$$

The time-evolution operator brings the system from state $\mathbf{F}(t_0)$ to state $\mathbf{F}(t)$. Equation (4.53) shows that the time-evolution matrix can be expressed in terms of the amplitudes for the transitions between different states $|\alpha n\rangle$. We can use the Floquet matrix to calculate the time evolution of the system initially in a particular molecular state. Equation (4.53) shows that we do not need to integrate the differential equation or even compute the eigenvalues of the Floquet matrix. All we need to know are the matrix elements of the Floquet Hamiltonian. Note that we will need to diagonalize the Floquet Hamiltonian if we want to compute the Stark shifts. This will be discussed in the subsequent sections.

4.2.3 Brief Summary of Floquet Theory Results

Let us summarize a few important observations based on Section 4.2.2:

- The Floquet Hamiltonian is Hermitian, and the states $|\alpha n\rangle$ are orthonormal.
- The diagonal elements of the Floquet Hamiltonian are the field-free molecular energies ϵ_α shifted by multiples of $\hbar\omega$.
- The Floquet Hamiltonian matrix is an infinite series of blocks differing only by multiples of $\hbar\omega$.
- Reflecting the periodic structure of the Hamiltonian matrix, the eigenvalues of the Floquet Hamiltonian have periodic structure. Specifically, if q is a particular eigenvalue of $\hat{H}_F$, so is $q + n\hbar\omega$, where n is any integer.
- The time-evolution matrix (4.53) is expressed in terms of the amplitudes for the transitions between different "bare product" states $|\alpha n\rangle$.
- The eigenvalues of the Floquet Hamiltonian represent the energy of the combined molecule-field system. When $\hbar\omega \ll \epsilon_\beta - \epsilon_\alpha$, the eigenvalues of the Floquet Hamiltonian matrix can be correlated with the energies of the molecule in the absence of the field. Thus, they represent the AC Stark shifts of molecular states.
- The eigenvalues of the Floquet Hamiltonian are related to the Hamiltonian (4.26) by

$$\sum_\alpha q_\alpha = \frac{1}{T}\int_0^T \mathrm{Tr}(\mathbf{H})\mathrm{d}t. \tag{4.54}$$

Equation (4.54) follows directly from Eq. (4.19), but it is also instructive to derive it. This derivation is borrowed from the lecture notes of Ward, a mathematics professor at the University of British Columbia.

Consider a small time interval $\varepsilon = t - t_0$. Then, we can expand $\mathbf{F}(t)$ in a Taylor series arount t_0

$$\mathbf{F}(t) = \mathbf{F}(t_0) + \varepsilon\frac{\mathrm{d}}{\mathrm{d}t}\mathbf{F}(t_0) + \cdots = \mathbf{F}(t_0) - \varepsilon\frac{i}{\hbar}\mathbf{H}\mathbf{F}(t_0) + \cdots$$
$$\simeq \left[\mathbf{I} - \varepsilon\frac{i}{\hbar}\mathbf{H}\right]\mathbf{F}(t_0), \tag{4.55}$$

where $\mathbf{I}$ is the identity matrix. From this equation,

$$\det[\mathbf{F}(t)] = \det\left[\mathbf{I} - \varepsilon\frac{i}{\hbar}\mathbf{H}\right]\det[\mathbf{F}(t_0)], \tag{4.56}$$

where we used the fact that the determinant of a matrix product is equal to the product of the determinants of the individual matrices.

The determinant of the sum in the last equation can be expanded as

$$\det\left[\mathbf{I} - \varepsilon\frac{i}{\hbar}\mathbf{H}\right] \approx 1 - \varepsilon\frac{i}{\hbar}\mathrm{Tr}(\mathrm{H}). \tag{4.57}$$

It can be easily verified that all other terms in the determinant on the left-hand side will contain higher powers of ε.

Expanding the determinant $W(t) = \det[\mathbf{F}(t)]$ in a Taylor series, we obtain

$$W(t) = W(t_0) + \varepsilon\frac{\mathrm{d}}{\mathrm{d}t}W(t_0) + \cdots . \tag{4.58}$$

This shows that

$$\frac{\mathrm{d}}{\mathrm{d}t}W(t_0) = -\frac{i}{\hbar}\mathrm{Tr}(\mathbf{H})W(t_0). \tag{4.59}$$

Since t_0 is arbitrary, this equality must be satisfied at all times

$$\frac{\mathrm{d}}{\mathrm{d}t}W(t) = -\frac{i}{\hbar}\mathrm{Tr}(\mathbf{H})W(t). \tag{4.60}$$

The integration of this equation gives

$$W(t_1) = W(t_0)\exp\left\{-\frac{i}{\hbar}\int_{t_0}^{t_1}\mathrm{Tr}(\mathbf{H})\mathrm{d}t\right\}. \tag{4.61}$$

Finally, using Eq. (4.28), we have

$$\det[\mathbf{F}(t_1)] = \det[\mathbf{U}(t_1, t_0)]\det[\mathbf{F}(t_0)], \tag{4.62}$$

which together with Eq. (4.61) yields

$$\det[\mathbf{U}(t_1, t_0)] = \exp\left\{-\frac{i}{\hbar}\int_{t_0}^{t_1}\mathrm{Tr}(\mathbf{H})\mathrm{d}t\right\}. \tag{4.63}$$

We now note that by Eqs. (4.31) and (4.30), the matrices $\mathbf{U}(t_0 + T, t_0)$ and $\exp(-i\mathbf{Q}T/\hbar)$ are related by a unitary transformation

$$\mathbf{U}(t_0 + T, t) = \mathbf{F}(t_0)\mathrm{e}^{-\frac{i}{\hbar}\mathbf{Q}T}\mathbf{F}^{\dagger}(t_0), \tag{4.64}$$

so they have the same eigenvalues. This and Eq. (4.63) lead to the relation (4.54).

Note that, for the problem considered here, the diagonal elements of the Hamiltonian are time independent. Therefore, Eq. (4.54) means that

$$\sum_{\alpha} q_{\alpha} = \mathrm{Tr}(\mathbf{H}). \tag{4.65}$$

The same result can be obtained by noting that the unitary transformation does not change the trace of a square matrix and that the trace of the Floquet matrix is the same as the trace of the original Hamiltonian, because the frequency-dependent terms enter the diagonal elements of the Floquet Hamiltonian matrix with alternating sign so they cancel each other. Equation (4.65) can be used to verify the accuracy of the numerical calculations of the eigenvalues q_{α} of the Floquet matrix.

4.3 Two-Mode Floquet Theory

What if a molecule is placed in *two* superimposed AC fields with different frequencies? This is a very common situation. One example of this problem is an optical centrifuge for molecules discussed in Section 5.6.

The Hamiltonian of a molecule in two superimposed AC fields can be written as

$$\hat{H} = \hat{H}_0 - \boldsymbol{d} \cdot \hat{\boldsymbol{e}}_1\, \xi \cos \omega_1 t - \boldsymbol{d} \cdot \hat{\boldsymbol{e}}_2\, \xi \cos \omega_2 t. \tag{4.66}$$

The presence of the second time-dependent term spoils the nice periodicity of the Hamiltonian (4.3) and makes the Floquet analysis more complicated, though not impossible [29, 30].

The key in this analysis is to introduce a parameter ω, such that [29]

$$\omega_1 = N_1 \omega, \tag{4.67}$$

and

$$\omega_2 = N_2 \omega, \tag{4.68}$$

where N_1 and N_2 are integers. Note that ω can be chosen arbitrarily small so that Eqs. (4.67) and (4.68) hold to within arbitrary precision. Since N_1 and N_2 are integers, the Hamiltonian (4.66) is a periodic function of time with period $T = 2\pi/\omega$. This is wonderful because this means that we can apply the Floquet theory described in Section 4.2 to the Hamiltonian (4.66).

Using the arguments in Section 4.2, we can write the diagonal matrix elements of the Floquet Hamiltonian matrix for a molecule in two superimposed AC fields as follows:

$$\langle \alpha n|\hat{H}_{\mathrm{F}}|\alpha n\rangle = \epsilon_\alpha + \hbar\omega n. \tag{4.69}$$

The off-diagonal matrix elements are

$$\langle \alpha n|\hat{H}_{\mathrm{F}}|\beta k\rangle = \sum_{i=1}^{2} V_{\alpha\beta}^{(i)}(\delta_{n-k,N_i} + \delta_{n-k,-N_i}), \tag{4.70}$$

where

$$V_{\alpha\beta}^{(i)} = -\langle \phi_\alpha|\boldsymbol{d}\cdot\hat{\epsilon}_i|\phi_\beta\rangle\frac{\xi}{2} = \hbar\Omega_i/2. \tag{4.71}$$

This Hamiltonian matrix depends on the artifical parameter ω, which makes it inconvenient. What we want instead is to express the Floquet matrix in terms of the "real" field frequencies ω_1 and ω_2.

From the previous section, we know that the Floquet matrix is an infinite series of blocks differing by multiples of $\hbar\omega$ in the diagonal matrix elements. This matrix will necessarily contain blocks corresponding to $n\hbar\omega$ with such values of n that

$$n\omega = n_1\omega_1 + n_2\omega_2, \tag{4.72}$$

where $n_1 = 0, 1, \ldots, \infty$ and $n_2 = 0, 1, \ldots, \infty$. Consider the state $|\alpha n\rangle$ corresponding to the value $n = N_1 n_1 + N_2 n_2$ that is the value of n from Eq. (4.72). By Eq. (4.70), this state will be coupled to state $|\beta k\rangle$, for which

$$k = (n_1 \pm 1)N_1 + n_2 N_2 \tag{4.73}$$

or

$$k = n_1 N_1 + (n_2 \pm 1)N_2. \tag{4.74}$$

Therefore, instead of using the states $|\alpha n\rangle$, we can use the basis of states $|\alpha n_1 n_2\rangle$. The elements of the Floquet matrix in this basis are

$$\langle \alpha n_1 n_2|\hat{H}_{\mathrm{F}}|\alpha n_1 n_2\rangle = \epsilon_\alpha + n_1\hbar\omega_1 + n_2\hbar\omega_2. \tag{4.75}$$

The off-diagonal matrix elements are

$$\langle \alpha n_1 n_2|\hat{H}_{\mathrm{F}}|\beta k_1 k_2\rangle = V_{\alpha\beta}^{(1)}\delta_{n_2 k_2}\delta_{n_1-k_1,\pm 1} + V_{\alpha\beta}^{(2)}\delta_{n_2-k_2,\pm 1}\delta_{n_1,k_1}. \tag{4.76}$$

The eigenvalues of the Floquet matrix with the elements (4.75) and (4.76) are the same as some of the eigenvalues of the matrix with the elements (4.69) and (4.70). If shifted by multiples of $\hbar\omega$, these eignevalues cover the entire eigenspectrum of the Floquet Hamiltonian for a molecule in a superposition of two AC fields.

4.4 Rotating Wave Approximation

Consider again a two-level system with the Hamiltonian (4.3) and the energy difference between the two molecular states $\epsilon_\beta - \epsilon_\alpha = \hbar\omega_0$. We can define the stationary states

$$|g\rangle = |\phi_\alpha\rangle,$$
$$|e\rangle = e^{-i\omega_0 t}|\phi_\beta\rangle, \tag{4.77}$$

which are the solutions of the time-dependent Schrödinger equation with the time-independent Hamiltonian $\hat{H}_0$. In Section 4.2, the time-dependent exponents of the stationary states were absorbed into the expansion coefficients $a(t)$ in Eq. (4.23).

The matrix of the Hamiltonian (4.3) in the basis of $|g\rangle$ and $|e\rangle$ defined by Eq. (4.77) is

$$\hat{H} = \begin{pmatrix} 0 & \hbar\Omega e^{-i\omega_0 t}\cos\omega t \\ \hbar\Omega^* e^{i\omega_0 t}\cos\omega t & \hbar\omega_0 \end{pmatrix}. \tag{4.78}$$

Using the Euler's formula, the off-diagonal matrix elements can be written as

$$\langle g|\hat{H}|e\rangle = \frac{\hbar\Omega}{2}\{e^{i(\omega-\omega_0)t} + e^{-i(\omega+\omega_0)t}\},$$
$$\langle e|\hat{H}|g\rangle = \frac{\hbar\Omega^*}{2}\{e^{i(\omega+\omega_0)t} + e^{-i(\omega-\omega_0)t}\}. \tag{4.79}$$

If $\omega \approx \omega_0$, then the terms with $\omega_0 - \omega$ vary with time much slower than the exponentials with $\omega + \omega_0$. By the time the terms with $\omega_0 - \omega$ complete one period of oscillation, the exponentials with $\omega + \omega_0$ oscillate many times, averaging to zero.

The rotating wave approximation neglects the rapidly oscillating terms, leading to the Hamiltonian

$$\hat{H} = \begin{pmatrix} 0 & \frac{\hbar\Omega}{2}e^{i\Delta t} \\ \frac{\hbar\Omega^*}{2}e^{-i\Delta t} & \hbar\omega_0 \end{pmatrix}, \tag{4.80}$$

where $\Delta = \omega - \omega_0$ is the detuning of the frequency of the electric field from the molecular transition frequency.

If written in the basis of states $|\phi_\alpha\rangle$ and $|\phi_\beta\rangle$, the Hamiltonian (4.80) becomes

$$\hat{H} = \begin{pmatrix} 0 & \frac{\hbar\Omega}{2}e^{i\omega t} \\ \frac{\hbar\Omega^*}{2}e^{-i\omega t} & \hbar\omega_0 \end{pmatrix}. \tag{4.81}$$

The off-diagonal elements of this Hamiltonian matrix permit only one term in the Fourier expansion (4.34), instead of two for the cosine function, leading to the Floquet matrix

$$\begin{pmatrix} \ddots & \vdots & \vdots & \vdots & \vdots & \vdots & \vdots & \vdots & \vdots & \\ \cdot & \epsilon_\beta - 2\hbar\omega & \hbar\Omega/2 & 0 & 0 & 0 & 0 & 0 & 0 & \cdot \\ \cdot & \hbar\Omega^*/2 & \epsilon_\alpha - \hbar\omega & 0 & 0 & 0 & 0 & 0 & 0 & \cdot \\ \cdot & 0 & 0 & \epsilon_\beta - \hbar\omega & \hbar\Omega/2 & 0 & 0 & 0 & 0 & \cdot \\ \cdot & 0 & 0 & \hbar\Omega^*/2 & \epsilon_\alpha & 0 & 0 & 0 & 0 & \cdot \\ \cdot & 0 & 0 & 0 & 0 & \epsilon_\beta & \hbar\Omega/2 & 0 & 0 & \cdot \\ \cdot & 0 & 0 & 0 & 0 & \hbar\Omega^*/2 & \epsilon_\alpha + \hbar\omega & 0 & 0 & \cdot \\ \cdot & 0 & 0 & 0 & 0 & 0 & 0 & \epsilon_\beta + \hbar\omega & \hbar\Omega/2 & \cdot \\ \cdot & 0 & 0 & 0 & 0 & 0 & 0 & \hbar\Omega^*/2 & \epsilon_\alpha + 2\hbar\omega & \cdot \\ & \vdots & \vdots & \vdots & \vdots & \vdots & \vdots & \vdots & \vdots & \ddots \end{pmatrix}. \tag{4.82}$$

Compare this matrix to the full Floquet matrix given in Eq. (4.46). As we can see, the rotating wave approximation uncouples the 2×2 blocks of the Floquet Hamiltonian matrix corresponding to different multiples of $\hbar\omega$ and reduces the need to diagonalize the infinitely dimensional Floquet matrix to the problem involving the diagonalization of a 2×2 matrix.

The justification for the rotating wave approximation is provided by the form of the matrix (4.46). Consider, for example, the "bare product" state $|\beta, n = 0\rangle$, which is coupled by $\cos \omega t$ to the states $|\alpha, n = 1\rangle$ and $|\alpha, n = -1\rangle$. If

$$\hbar\omega \approx \epsilon_\beta - \epsilon_\alpha,$$

the diagonal matrix elements

$$\langle \beta, n = 0 | \hat{H}_{\mathrm{F}} | \beta, n = 0 \rangle \text{ and } \langle \alpha, n = +1 | \hat{H}_{\mathrm{F}} | \alpha, n = +1 \rangle$$

are close in magnitude, while the diagonal matrix elements

$$\langle \beta, n = 0 | \hat{H}_{\mathrm{F}} | \beta, n = 0 \rangle \text{ and } \langle \alpha, n = -1 | \hat{H}_{\mathrm{F}} | \alpha, n = -1 \rangle$$

are very different. The coupling $\langle \beta, n = 0 | \hat{H}_{\mathrm{F}} | \alpha, n = -1 \rangle$ is therefore much less effective than the coupling between the states $|\beta, n = 0\rangle$ and $|\alpha, n = +1\rangle$. In general, the states $|\beta, n = 0\rangle$ and $|\alpha, n = 1\rangle$ will always be closer in energy than the states $|\beta, n = 0\rangle$ and $|\alpha, n = -1\rangle$, as illustrated in Figure 4.1. Therefore, the rotating wave approximation may be valid even if the detuning Δ is large, providing Ω – the coupling strength – is sufficiently small.

The diagonalization of the 2×2 blocks of the Floquet matrix

$$\hat{H} = \begin{pmatrix} \hbar\omega_0 & \hbar\Omega/2 \\ \hbar\Omega^*/2 & \hbar(\omega_0 + \Delta) \end{pmatrix} \tag{4.83}$$

yields two energies of the field-dressed molecule (see Section 2.3)

$$E_{1,2} = \hbar\omega_0 + \frac{\hbar\Delta}{2} \pm \hbar\frac{\sqrt{|\Omega|^2 + \Delta^2}}{2}. \tag{4.84}$$

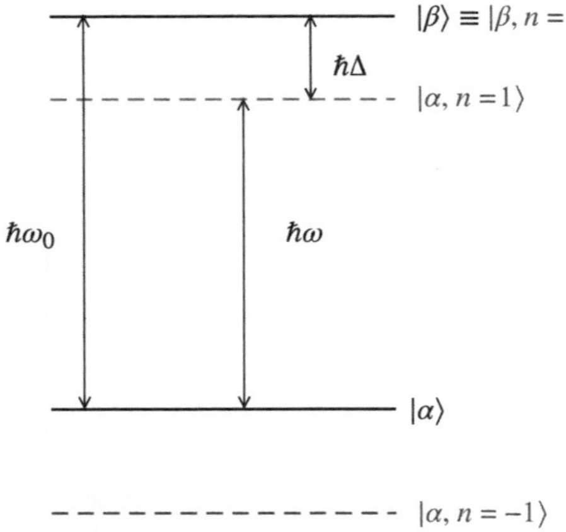

Figure 4.1 A schematic diagram of the field-free molecular energy levels $|\alpha\rangle$ and $|\beta\rangle$ shown by solid lines and the states $|\alpha, n = 1\rangle$ and $|\alpha, n = -1\rangle$ shown by dashed lines. The time-dependent operator $V(t) \propto \cos \omega t$ couples $|\beta, n = 0\rangle$ with both $|\alpha, n = 1\rangle$ and $|\alpha, n = -1\rangle$. The rotating wave approximation eliminates the coupling to $|\alpha, n = -1\rangle$.

It is instructive to consider the limit of small Ω, in particular, when $|\Omega| \ll |\Delta|$. If Δ is negative (the case of *red-detuned* electric field), the state $|\alpha, n = 1\rangle$ is lower in energy than $|\beta\rangle$. Then, by Eqs. (2.24) and (2.25),

$$E_\alpha = \hbar\omega_0 - \hbar|\Delta| - \frac{\hbar|\Omega|^2}{4\Delta} \tag{4.85}$$

and

$$E_\beta = \hbar\omega_0 + \frac{\hbar|\Omega|^2}{4\Delta}. \tag{4.86}$$

So the state α is shifted down and the state β up by the amount of energy $\hbar|\Omega|^2/4\Delta$. If the detuning is positive (the case of *blue-detuned* electric field), the eigenvalues are

$$E_\alpha = \hbar\omega_0 + \hbar\Delta + \frac{\hbar|\Omega|^2}{4\Delta} \tag{4.87}$$

and

$$E_\beta = \hbar\omega_0 - \frac{\hbar|\Omega|^2}{4\Delta}, \tag{4.88}$$

i.e. the state α is shifted up and the state β down by the same amount of energy. Note that the energy shifts are $\propto |\Omega|^2$ and $\propto 1/|\Delta|$. This captures the essence of the AC Stark effect induced by a weak, far-detuned field.

4.5 Dynamic Dipole Polarizability

While Eqs. (4.85)–(4.88) give the AC Stark shifts of a two-level system under conditions when the rotating wave approximation is valid, molecules are inherently multilevel systems, possessing manifolds of closely spaced vibrational and rotational energy levels. This means two things: (i) more often than not, we must abandon the two-level picture and (ii) the rotating wave approximation, if used, must be treated with a lot of caution. In this section, we will go beyond the two-level picture and consider how the multilevel energy structure of rotational, vibrational and electronic states determines the interaction of molecules with AC electric fields.

In principle, the energy levels of a molecule in an off-resonant time-varying electric field can be accurately computed by diagonalizing the Floquet Hamiltonian matrix evaluated in the basis spanning all relevant molecular states. Often, however, diagonalizing the Floquet matrix is like shooting at a fly with a cannon. The accurate results – and physical insights – can often be obtained by considering the relative effects of different off-diagonal elements and by uncoupling blocks of the Floquet matrix.

For example, the structure of the Floquet matrix is telling us that the state $|\beta, n=0\rangle$ is directly coupled by the field to the states $|\alpha, n\pm 1\rangle$:

$$|\beta, n=0\rangle \overset{\hbar\omega}{\longleftrightarrow} |\alpha, n\pm 1\rangle. \tag{4.89}$$

Since the state $|\alpha, n=1\rangle$ is coupled to state $|\beta, n=2\rangle$, the latter may have a second-order effect on the state $|\beta, n=0\rangle$:

$$|\beta, n=0\rangle \overset{\hbar\omega}{\longleftrightarrow} |\alpha, n+1\rangle \overset{\hbar\omega}{\longleftrightarrow} |\beta, n+2\rangle. \tag{4.90}$$

Similarly, the states $|\beta, n=-2\rangle$ and $|\beta, n=0\rangle$ are indirectly coupled via $|\alpha, n=-1\rangle$. As will be clear from Section 4.8.3, the couplings between $|\beta, n=0\rangle$ and $|\beta, n=\pm 2\rangle$ correspond to *two-photon* couplings. At low field strengths, the effect of these couplings on the energy of the molecular state $|\beta\rangle$ is generally smaller than the effect of direct couplings $|\beta, n=0\rangle \leftrightarrow |\alpha, n\pm 1\rangle$. The two-photon couplings will generally become important at strong fields and/or when the states coupled by the second-order couplings are in near resonance. Section 4.6 will discuss the case when such two-photon couplings have to be taken into account. In the present section, we will assume that the indirect couplings involving two or more photons can be neglected, i.e. we will assume that the energy shifts of molecular states are determined only by the direct couplings $|\beta, n=0\rangle \leftrightarrow |\alpha, n\pm 1\rangle$. In this case, the effect of the multilevel structure of molecules on their response to an electromagnetic field may be conveniently described with the help of the electric dipole polarizability.

4.5.1 Polarizability Tensor

Consider a specific state $|\beta, n=0\rangle$ corresponding to the molecular state $|\phi_\beta\rangle$. We will use perturbation theory to compute the Stark shift of this molecular state. Because the diagonal matrix elements of the field-induced interaction vanish $\langle\phi_\beta|\boldsymbol{d}\cdot\hat{\epsilon}|\phi_\beta\rangle = 0$, there is no first-order correction to the energy and the effect of the field will appear in second order.

The second-order correction to the energy of this state due to the field-induced perturbation is

$$\Delta E_\beta(\omega) = \frac{\xi^2}{4}\sum_\gamma \sum_{k=\pm 1} \frac{|\langle \phi_\beta | \boldsymbol{d}\cdot\hat{\epsilon}|\phi_\gamma\rangle|^2}{\epsilon_\beta - \epsilon_\gamma - k\hbar\omega}, \tag{4.91}$$

where the first sum is over all molecular states that are directly coupled to state $|\phi_\beta\rangle$. The quantum number n is not important here. As discussed in Section 4.2.3, the energy shift of the state $|\beta, n=0\rangle$ will be exactly the same as the energy shift of any state $|\beta, n\rangle$.

Note that, in addition to the sum over γ, there are 2×9 terms in Eq. (4.91), since each of $\boldsymbol{d}\cdot\hat{\epsilon} = d_X\hat{\epsilon}_X + d_Y\hat{\epsilon}_Y + d_Z\hat{\epsilon}_Z$. Note also that, for any two vectors $\boldsymbol{a}$ and $\boldsymbol{b}$, the product $(\boldsymbol{a}\cdot\boldsymbol{b})^*(\boldsymbol{a}\cdot\boldsymbol{b})$ can be written in vector-matrix-vector form:

$$(\boldsymbol{a}\cdot\boldsymbol{b})^*(\boldsymbol{a}\cdot\boldsymbol{b}) = (a_x^* \;\; a_y^* \;\; a_z^*)\begin{pmatrix} b_x^*b_x & b_x^*b_y & b_x^*b_z \\ b_y^*b_x & b_y^*b_y & b_y^*b_z \\ b_z^*b_z & b_z^*b_y & b_z^*b_z \end{pmatrix}\begin{pmatrix} a_x \\ a_y \\ a_z \end{pmatrix}. \tag{4.92}$$

The square matrix in the middle represents a rank-2 tensor defined by its nine components.

With this in mind, Eq. (4.91) can be written as the expectation value

$$\Delta E_\beta(\omega) = -\frac{\xi^2}{4} \times \langle \phi_\beta|\hat{\epsilon}^* \cdot \left[\sum_\gamma \left\{\frac{1}{\epsilon_\gamma - \epsilon_\beta - \hbar\omega} + \frac{1}{\epsilon_\gamma - \epsilon_\beta + \hbar\omega}\right\} \boldsymbol{d}^*|\phi_\gamma\rangle\langle\phi_\gamma|\boldsymbol{d}\right] \cdot\hat{\epsilon}|\phi_\beta\rangle \tag{4.93}$$

or

$$\Delta E_\beta(\omega) = -\frac{\xi^2}{4}\langle\phi_\beta|\hat{\epsilon}^*\cdot\boldsymbol{\alpha}\cdot\hat{\epsilon}|\phi_\beta\rangle. \tag{4.94}$$

The operator $\boldsymbol{\alpha}$ as defined by Eq. (4.94) is a rank-2 tensor

$$\boldsymbol{\alpha} = \begin{pmatrix} \alpha_X^*\alpha_X & \alpha_X^*\alpha_Y & \alpha_X^*\alpha_Z \\ \alpha_Y^*\alpha_X & \alpha_Y^*\alpha_Y & \alpha_Y^*\alpha_Z \\ \alpha_Z^*\alpha_Z & \alpha_Z^*\alpha_Y & \alpha_Z^*\alpha_Z \end{pmatrix}, \tag{4.95}$$

whose nine components are

$$\alpha_{\sigma,\sigma'}(\omega) = \sum_\gamma \left\{\frac{1}{\epsilon_\gamma - \epsilon_\beta - \hbar\omega} + \frac{1}{\epsilon_\gamma - \epsilon_\beta + \hbar\omega}\right\} d_\sigma^*|\phi_\gamma\rangle\langle\phi_\gamma|d_{\sigma'}, \tag{4.96}$$

where σ and σ' denote one of $\{X, Y, Z\}$ and d_σ are the Cartesian components of the dipole moment operator in the space-fixed coordinate frame. This tensor is the dynamic polarizability operator that acts on the space of $|\phi_\beta\rangle$. Note that $\boldsymbol{\alpha}$ depends on the field frequency ω through the terms in the denominators.

The expectation value $\langle\phi_\beta|\boldsymbol{\alpha}(\omega)|\phi_\beta\rangle$ is the ω-dependent polarizability of the molecule in the state $|\phi_\beta\rangle$.

The polarizability is defined as the proportionality coefficient between the electric field vector and the dipole moment induced by this field [31]. To show that $\boldsymbol{\alpha}$ as defined by Eq. (4.94) is indeed the polarizability tensor, we need to prove that the dipole moment of the molecule $\langle\boldsymbol{d}\rangle$ induced by the electric field

$$\boldsymbol{E} = \xi\hat{\epsilon}\cos\omega t \tag{4.97}$$

is [32]

$$\langle\boldsymbol{d}\rangle = \langle\boldsymbol{\alpha}\rangle\cdot\boldsymbol{E}. \tag{4.98}$$

The state of a molecule in the presence of the electric field (4.1) is a coherent superposition of the molecular states

$$|\psi_\beta\rangle = \sum_\gamma F_{\gamma,\beta}(t)|\phi_\gamma\rangle, \tag{4.99}$$

where $F_{\gamma,\beta}(t)$ are given by the Fourier expansion (4.33). In first order,

$$F^k_{\gamma,\beta} = \frac{\langle\phi_\gamma|\boldsymbol{d}\cdot\hat{\epsilon}|\phi_\beta\rangle}{\epsilon_\beta - \epsilon_\gamma - k\hbar\omega}\frac{\xi}{2} \tag{4.100}$$

with $\gamma \neq \beta$ and $k = \pm 1$. The expectation value of the dipole moment operator is (to first order)

$$\langle\psi_\beta|\boldsymbol{d}|\psi_\beta\rangle = \frac{\xi}{2}\sum_\gamma\sum_{k=\pm1}\frac{\langle\phi_\gamma|\boldsymbol{d}\cdot\hat{\epsilon}|\phi_\beta\rangle}{\epsilon_\beta - \epsilon_\gamma - k\hbar\omega}\langle\phi_\beta|\boldsymbol{d}|\phi_\gamma\rangle e^{ik\omega t}, \tag{4.101}$$

which in the limit of far-off-resonant field $|\epsilon_\beta - \epsilon_\gamma| \gg \hbar\omega$ can be written as

$$\langle\psi_\beta|\boldsymbol{d}|\psi_\beta\rangle = \xi\cos\omega t\sum_\gamma\frac{\langle\phi_\gamma|\boldsymbol{d}\cdot\hat{\epsilon}|\phi_\beta\rangle}{\epsilon_\beta - \epsilon_\gamma}\langle\phi_\beta|\boldsymbol{d}|\phi_\gamma\rangle, \tag{4.102}$$

which is the same as

$$\langle\psi_\beta|\boldsymbol{d}|\psi_\beta\rangle = \langle\phi_\beta|\boldsymbol{\alpha}|\phi_\beta\rangle\cdot\boldsymbol{E}. \tag{4.103}$$

Equation (4.102) shows that the induced dipole moment of a molecule in an oscillating electric field oscillates with the frequency of the field.

4.5.2 Dipole Polarizability of a Diatomic Molecule

Let us now examine the electric dipole polarizability of a diatomic molecule in a particular state $|\phi_\beta\rangle \Rightarrow\rangle|v\rangle|n\rangle|JMK\epsilon\rangle$, where, as before, $|n\rangle$ denotes the electronic state of the molecule, $|v\rangle$ is the vibrational state and $|JMK\rangle$ is the rotational state.

The diagonal matrix elements of the components of the polarizability operator (4.96) are

$$\langle \upsilon nJMK\varepsilon|\alpha_{\sigma,\sigma'}(\omega)|\upsilon nJMK\varepsilon\rangle = \sum_{\upsilon' n' J' M' K' \varepsilon'} \left\{ \frac{1}{\epsilon_\gamma - \epsilon_\beta - \hbar\omega} + \frac{1}{\epsilon_\gamma - \epsilon_\beta + \hbar\omega} \right\} \times \langle \upsilon nJMK\varepsilon|d_\sigma^*|\upsilon' n' J' M' K' \varepsilon'\rangle\langle \upsilon' n' J' M' K' \varepsilon'|d_{\sigma'}|\upsilon nJMK\varepsilon\rangle, \tag{4.104}$$

where ϵ_β is the energy of the state $|\upsilon nJMK\varepsilon\rangle$ and ϵ_γ denotes the energy of all other molecular states $|\upsilon' n' J' M' K' \varepsilon'\rangle$ directly coupled by the electric dipole operator to the state $|\upsilon nJMK\varepsilon\rangle$.

The Cartesian components of the dipole moment vector can be written in terms of spherical harmonics as (see Ref. [10] and Appendix E)

$$d_X = \sqrt{\frac{2\pi}{3}} d\{Y_{1,-1}(\theta,\phi) - Y_{1,+1}(\theta,\phi)\},$$
$$d_Y = i\sqrt{\frac{2\pi}{3}} d\{Y_{1,-1}(\theta,\phi) + Y_{1,+1}(\theta,\phi)\},$$
$$d_Z = \sqrt{\frac{4\pi}{3}} dY_{1,0}(\theta,\phi), \tag{4.105}$$

where the angles θ and ϕ specify the orientation of the dipole moment vector in the space-fixed coordinate frame and $d = |\boldsymbol{d}|$. We should remember that the dipole moment operator

$$\boldsymbol{d} = \sum_i q_i \boldsymbol{r}_i \tag{4.106}$$

depends on the electronic as well as nuclear coordinates.

Using Eqs. (4.105) and the Wigner–Eckart theorem (see Section 9.1), the matrix elements in Eq. (4.104) can be expressed in terms of the matrix elements of the spherical harmonics [33]

$$\langle \upsilon nJMK\varepsilon|dY_{1,q}|\upsilon' n' J' M' K' \varepsilon'\rangle = \delta'_{\varepsilon,-\varepsilon}\langle \upsilon n|d|\upsilon' n'\rangle\langle JMK|Y_{1,q}|J'M'K'\rangle = \delta_{\varepsilon,-\varepsilon'}(-1)^{J-M} D_{n\upsilon JK;n'\upsilon' J'K'} \begin{pmatrix} J & 1 & J' \\ -M & q & M' \end{pmatrix}, \tag{4.107}$$

so we can write

$$\langle \upsilon nJMK\varepsilon|\alpha_{\sigma,\sigma'}(\omega)|\upsilon nJMK\varepsilon\rangle = \sum_{\upsilon' n' J' K' \varepsilon'} (M,M')\text{-independent terms} \times \sum_q \sum_{q'} \delta^{q,q'}_{\sigma,\sigma'} \sum_{M'} \begin{pmatrix} J & 1 & J' \\ -M & q & M' \end{pmatrix} (-1)^{J+J'-M-M'} \begin{pmatrix} J' & 1 & J \\ -M' & q' & M \end{pmatrix}. \tag{4.108}$$

As determined by Eq. (4.105), the sum over q is restricted to terms with $q = \pm 1$ if $\sigma = X$ or Y and to $q = 0$ if $\sigma = Z$. The same holds for q' and σ'. The phase factor $\delta_{\sigma,\sigma'}^{q,q'} = \pm 1$ is determined by Eq. (4.105). Equation (4.108) can be used to analyze the J, M dependence of the polarizability.

Example 4.1 ***Polarizability of the $J = 0$ State***
For example, consider the electric dipole polarizability of a molecule in the state with $J = 0$. For such a state, the sum in Eq. (4.108) reduces to

$$\sum_q \sum_{q'} \alpha_{\sigma,\sigma'}^{q,q'} \sum_{M'} \begin{pmatrix} 0 & 1 & 1 \\ 0 & q & M' \end{pmatrix} (-1)^{1-M'} \begin{pmatrix} 1 & 1 & 0 \\ -M' & q' & 0 \end{pmatrix}$$

$$= \begin{cases} \frac{2}{\sqrt{3}} & \text{if } \sigma = \sigma' = X \text{ or } Y, \\ 0 & \text{if } \sigma \neq \sigma', \\ \frac{1}{\sqrt{3}} & \text{if } \sigma = \sigma' = Z. \end{cases} \quad (4.109)$$

We thus find that for the molecule in a $J = 0$ state,

- the expectation value of $\alpha_{\sigma,\sigma'}$ vanishes unless $\sigma = \sigma'$; and
- $\langle \alpha_{XX} \rangle = \langle \alpha_{YY} \rangle = \langle \alpha_{ZZ} \rangle$.

In other words, molecules in a $J = 0$ state behave as a spherically symmetric object, equally polarizable along any direction.[1]

4.5.3 Rotational vs Vibrational vs Electronic Polarizability

To gain more insight into the dipole polarizability of diatomic molecules, consider the matrix elements of the polarizability operator in Hund's case (a) basis $|v\rangle|\eta\Sigma\Lambda\rangle|JM\Omega\rangle$, which have the following explicit form:

$$\begin{aligned} &\langle v\eta\Sigma\Lambda JM\Omega|\alpha_{\sigma,\sigma'}|v\eta\Sigma\Lambda JM\Omega\rangle \\ &= \sum_{v'\eta'J'M'\Omega'\Lambda'} \left\{ \frac{1}{\epsilon_\gamma - \epsilon_\beta - \hbar\omega} + \frac{1}{\epsilon_\gamma - \epsilon_\beta + \hbar\omega} \right\} \\ &\times \langle v\eta\Sigma\Lambda JM\Omega|d_\sigma|v'\eta'\Sigma\Lambda'J'M'\Omega'\rangle \\ &\times \langle v'\eta'\Sigma\Lambda'J'M'\Omega'|d_{\sigma'}|v\eta\Sigma\Lambda JM\Omega\rangle, \end{aligned} \quad (4.110)$$

1 Note that here X, Y, and Z are the axes of a space-fixed coordinate system. If a molecule could be held fixed in space and oriented along an axis perpendicular to the electric field vector, the polarizability tensor would have no off-diagonal elements, but the diagonal components would be $\langle \alpha_{XX} \rangle = \langle \alpha_{YY} \rangle \neq \langle \alpha_{ZZ} \rangle$, reflecting the symmetry of the diatomic molecule. In general, the number of independent components of the polarizability tensor depends on the symmetry of the molecule. For a completely nonsymmetric molecule, the dynamic polarizability tensor has nine independent components. The polarizability tensor for a linear molecule, whether homonuclear or heteronuclear, has two independent components. For molecules with cubic symmetry, there is only one independent component. See Ref. [31] for a table summarizing the number of independent components of the polarizability tensor for molecules of various symmetries.

with ϵ_β and ϵ_γ now denoting the energies of the corresponding Hund's case (a) states. Note that the matrix of the electric dipole moment operator is diagonal in spin quantum numbers S and Σ. Because $\Omega' = \Sigma + \Lambda'$, the sum over Ω' is redundant.

As we discussed in Section 1.5, Hund's case (a) states are not the eigenstates of the parity operator so they do not correspond to physical molecular states. However, the proper molecular states can be written as superpositions of Hund's case (a) states, as, for example, in Eq. (2.62), so the polarizability of a molecule in a particular ro-vibrational state with proper parity can be computed by combining the matrix elements given here.

Our first task is to evaluate the matrix elements of the Cartesian components d_σ of the dipole moment operator. The molecular states $|\eta S\Sigma\Lambda\rangle$ are defined in the molecule-fixed coordinate frame. Therefore, in order to evaluate the matrix elements in Eq. (4.110), it is necessary to rewrite the dipole moment components d_X, d_Y, and d_Z in terms of the operators defined in the molecule-fixed frame. This can be achieved by first using Eq. (4.105) to write the Cartesian components of the dipole moment in terms of the spherical harmonics in the space-fixed frame and then rotating the spherical harmonics to the molecule-fixed frame, as described in Appendix D. In other words, we must replace the spherical harmonics in Eq. (4.105) with

$$Y_{1,q}(\theta,\phi) = \sum_p D^1_{p,q}(\alpha,\beta,\gamma) Y_{1,p}(\theta',\phi'), \tag{4.111}$$

where the angles θ' and ϕ' are the spherical polar angles specifying the orientation of the dipole moment in the *molecule-fixed* coordinate frame and the D-functions $D^1_{p,q}$ depend on the Euler angles describing the rotation of the molecule-fixed coordinate frame in the space-fixed frame (see Appendix D). Here, q is the projection on the space-fixed Z-axis and p is the projection on the molecule-fixed z-axis, chosen, as before, along the internuclear axis. The values of both q and p are, of course, restricted to 0 and ± 1.

Equation (4.111) is nice, because in it $Y_{1,p}(\theta',\phi')$ are the spherical harmonics that act as operators in the molecule-fixed frame. The D-functions in Eq. (4.111) act on the rotational part of the molecular states $|JM\Omega\rangle$.

Recognizing that the matrix elements $\langle JM\Omega|D^1_{p,q}|J'M'\Omega'\rangle$ are given by the integrals over a product of three D-functions (see Eq. (1.37) and Appendix D), the matrix elements of the dipole moment operators can be written as

$$\begin{aligned}\langle v\eta S\Sigma\Lambda JM\Omega|dY_{1,q}(\theta,\phi)|v'\eta' S\Sigma\Lambda' J'M'\Omega\rangle &= \langle v\eta|d|v'\eta'\rangle \\ &\times \sum_p \langle S\Sigma\Lambda|Y_{1p}(\theta',\phi')|S\Sigma\Lambda'\rangle \\ &\times [(2J+1)(2J'+1)]^{1/2}\begin{pmatrix} J & 1 & J' \\ -M & q & M'\end{pmatrix}\begin{pmatrix} J & 1 & J' \\ -\Omega & p & \Omega'\end{pmatrix},\end{aligned} \tag{4.112}$$

where $\langle v\eta|d|v'\eta'\rangle$ are the transition dipole moment matrix elements.

Now comes an essential part. We will split the sum over the quantum numbers v', η', Λ', J' in Eq. (4.110) into four parts [31, 33–35]:

- The sum over the terms diagonal in η, Λ, and v, yielding the *rotational polarizability* α^r.
- The sum over the terms diagonal in η and Λ but off-diagonal in v, yielding the *vibrational polarizability* α^v.
- The sum over the terms diagonal in Λ but off-diagonal in η, giving the *electronic polarizability* parallel to the internuclear axis $\alpha^e_{\parallel}$.
- The sum over the terms off-diagonal in Λ, giving the *electronic polarizability* perpendicular to the internuclear axis $\alpha^e_{\perp}$.

To make this more transparent, consider the Stark effect in a diatomic molecule induced by the time-dependent electric field (4.1) oscillating along the Z-axis, i.e. with $\hat{\epsilon} = \hat{Z}$. The energy shift (4.94) of a particular state $|\phi_\beta\rangle \Rightarrow |v\eta S\Sigma JM\Omega\rangle$ is then given by

$$\Delta E_\beta(\omega) = -\frac{\xi^2}{4}\langle\phi_\beta|\hat{Z}\cdot\boldsymbol{\alpha}\cdot\hat{Z}|\phi_\beta\rangle = -\frac{\xi^2}{4}\langle\phi_\beta|\alpha_{ZZ}|\phi_\beta\rangle$$
$$= -\frac{\xi^2}{4}(\alpha^r + \alpha^v + \alpha^e_{\parallel} + \alpha^e_{\perp}), \tag{4.113}$$

where

$$\alpha^r(\omega) = \sum_{J'}[(2J+1)(2J'+1)]\begin{pmatrix} J & 1 & J' \\ -M & 0 & M \end{pmatrix}^2 \begin{pmatrix} J & 1 & J' \\ -\Omega & 0 & \Omega \end{pmatrix}^2$$
$$\times |\langle S\Sigma\Lambda|Y_{10}|S\Sigma\Lambda\rangle|^2 |\langle v\eta|d|v\eta\rangle|^2$$
$$\times \left\{\frac{1}{\epsilon_{\eta\Lambda vJ'} - \epsilon_{\eta\Lambda vJ} - \hbar\omega} + \frac{1}{\epsilon_{\eta\Lambda vJ'} - \epsilon_{\eta\Lambda vJ} + \hbar\omega}\right\}, \tag{4.114}$$

$$\alpha^v(\omega) = \sum_{J'}\sum_{v'\neq v}[(2J+1)(2J'+1)]\begin{pmatrix} J & 1 & J' \\ -M & 0 & M \end{pmatrix}^2 \begin{pmatrix} J & 1 & J' \\ -\Omega & 0 & \Omega \end{pmatrix}^2$$
$$\times |\langle S\Sigma\Lambda|Y_{10}|S\Sigma\Lambda\rangle|^2 |\langle v\eta|d|v'\eta\rangle|^2$$
$$\times \left\{\frac{1}{\epsilon_{\eta\Lambda v'J'} - \epsilon_{\eta\Lambda vJ} - \hbar\omega} + \frac{1}{\epsilon_{\eta\Lambda v'J'} - \epsilon_{\eta\Lambda vJ} + \hbar\omega}\right\}, \tag{4.115}$$

$$\alpha^e_{\parallel}(\omega) = \sum_{J'}\sum_{v'\neq v}\sum_{\eta'\neq\eta}[(2J+1)(2J'+1)]\begin{pmatrix} J & 1 & J' \\ -M & 0 & M \end{pmatrix}^2 \begin{pmatrix} J & 1 & J' \\ -\Omega & 0 & \Omega \end{pmatrix}^2$$
$$\times |\langle S\Sigma\Lambda|Y_{10}|S\Sigma\Lambda\rangle|^2 |\langle v\eta|d|v'\eta'\rangle|^2$$
$$\times \left\{\frac{1}{\epsilon_{\eta'\Lambda v'J'} - \epsilon_{\eta\Lambda vJ} - \hbar\omega} + \frac{1}{\epsilon_{\eta'\Lambda v'J'} - \epsilon_{\eta\Lambda vJ} + \hbar\omega}\right\}, \tag{4.116}$$

$$\alpha_{\perp}^{e}(\omega) = \sum_{J'} \sum_{v' \neq v} \sum_{\eta' \neq \eta} \sum_{\Lambda' \neq \Lambda} \sum_{p=\pm 1} [(2J+1)(2J'+1)] \begin{pmatrix} J & 1 & J' \\ -M & 0 & M \end{pmatrix}^2 \begin{pmatrix} J & 1 & J' \\ -\Omega & p & \Omega' \end{pmatrix}^2$$
$$\times |\langle S\Sigma\Lambda | Y_{1p} | S\Sigma\Lambda' \rangle|^2 |\langle v\eta | d | v'\eta' \rangle|^2$$
$$\times \left\{ \frac{1}{\epsilon_{\eta'\Lambda' v' J'} - \epsilon_{\eta\Lambda vJ} - \hbar\omega} + \frac{1}{\epsilon_{\eta'\Lambda' v' J'} - \epsilon_{\eta\Lambda vJ} + \hbar\omega} \right\}. \quad (4.117)$$

Note that the sum in Eq. (4.114) is over the rotational states of the same electronic and vibrational state, the sum in Eq. (4.115) is over the vibrational and rotational states of the same electronic state and the sums in the other two equations are over the electronic, vibrational, and rotational states. The last equation includes only the terms with $p = \pm 1$. Using the relations analogous to Eq. (4.105) in the molecule-fixed coordinate frame, it is easy to see that the $p = \pm 1$ components of the spherical harmonics correspond to the x and y components of the dipole moment, both of which are perpendicular to the interatomic axis, hence the label. The rotational and vibrational polarizabilities are both parallel to the interatomic axis.

The relative contributions of the rotational, vibrational, and electronic polarizabilities depend on the frequency of the electric field, which determines the magnitude of the denominators in Eqs. (4.114)–(4.117). For example, the polarizability of molecules in a microwave field, for which $\hbar\omega$ is similar to the rotational excitation energy, is dominated by the rotational polarizability, whereas the polarizability of molecules subject to an optical laser field is dominated by the electronic polarizability.

4.6 Molecules in an Off-Resonant Laser Field

Consider a molecule placed in an optical field far detuned from any electronic transition. It is useful to specialize Eq. (4.113) to this important and often-encountered case. The frequency of an optical field is generally much larger than the frequency of rotational or vibrational transitions

$$\hbar\omega \gg |\epsilon_{\eta\Lambda v' J'} - \epsilon_{\eta\Lambda vJ}| > |\epsilon_{\eta\Lambda vJ'} - \epsilon_{\eta\Lambda vJ}|. \quad (4.118)$$

This tells us that the contributions of the rotational (4.114) and vibrational (4.115) polarizabilities are very small. For far-detuned fields, we can also approximate the denominators in Eqs. (4.116) and (4.117) as

$$\frac{1}{\epsilon_{\eta'\Lambda' v' J'} - \epsilon_{\eta\Lambda vJ} \pm \hbar\omega} \approx \frac{1}{\epsilon_{\eta'\Lambda' vJ} - \epsilon_{\eta\Lambda vJ} \pm \hbar\omega}, \quad (4.119)$$

thus removing the dependence on v' and J' from these expressions. Thus, we have the following expression for the Stark shifts induced by a weak off-resonant laser field

$$\Delta E_\beta(\omega) = -\frac{\xi^2}{4}(\tilde{\alpha}_\parallel^e + \tilde{\alpha}_\perp^e) \tag{4.120}$$

with $\tilde{\alpha}_\parallel^e$ and $\tilde{\alpha}_\perp^e$ given by Eqs. (4.116) and (4.117), in which the denominators are replaced by

$$\frac{1}{\epsilon_{\eta'\Lambda'v'J'} - \epsilon_{\eta\Lambda vJ} \pm \hbar\omega} \Rightarrow \frac{1}{\epsilon_{\eta'\Lambda'} - \epsilon_{\eta\Lambda} \pm \hbar\omega}. \tag{4.121}$$

As the intensity of the laser field is increased, we may have to revisit the single-photon approximation we made to write Eq. (4.113). At some field strength, Eq. (4.113) and, consequently, Eq. (4.120) become invalid. In particular, the interaction of the molecule with the optical field induces second-order couplings between different rotational states of a given electronic state. Because the energy separation between the rotational states is relatively small, these couplings may strongly mix the rotational states, while not significantly mixing different electronic states. To treat such a case, it is desirable to obtain an effective operator for the molecule–field interaction that would include perturbatively the couplings between different electronic states and act on the space of the rotational states to allow for a nonpertrubative treatment of the rotational part of the problem.

To derive such an operator, consider a two-atom system fixed in three-dimensional space (aka "clamped nuclei") and subject to an electric field oscillating along the Z-axis. The potential energy of such system depends explicitly on the intermolecular distance R and the angle θ between the electric field direction Z and the intermolecular axis z. Since the vibrational and rotational motion of the molecule is absent, the potential energy is given by a simplified version of Eq. (4.110)

$$V(\theta, R) = -\frac{\xi^2}{4}\sum_{\eta'\Lambda'}\left\{\frac{1}{\epsilon_{\eta'\Lambda'} - \epsilon_{\eta\Lambda} - \hbar\omega} + \frac{1}{\epsilon_{\eta'\Lambda'} - \epsilon_{\eta\Lambda} + \hbar\omega}\right\}$$
$$\times \langle\eta\Lambda|d_Z|\eta'\Lambda'\rangle\langle\eta'\Lambda'|d_Z|\eta\Lambda\rangle. \tag{4.122}$$

By Eqs. (4.105) and (4.111), the Z-component of the dipole moment operator is

$$d_Z = d\frac{4\pi}{3}\sum_{p=-1}^{1} Y_{1,p}^*(\theta, \phi)Y_{1,p}(\theta', \phi'), \tag{4.123}$$

where I made the substitution $D^1_{p,0}(\phi,\theta,\chi) = \sqrt{\frac{4\pi}{3}}Y^*_{1,p}(\theta,\phi)$ [9]. Using the explicit expressions for the spherical harmonics

$$Y_{10}(\theta) = \sqrt{\frac{3}{4\pi}}\cos\theta,$$
$$Y_{1\pm1}(\theta,\phi) = \mp\frac{1}{2}\sqrt{\frac{3}{2\pi}}\sin\theta e^{\pm i\phi}, \tag{4.124}$$

we can write Eq. (4.122) as

$$V(\theta,R) = -\frac{\xi^2}{4}(\alpha_{\parallel}(R)\cos^2\theta + \alpha_{\perp}(R)\sin^2\theta), \tag{4.125}$$

where

$$\alpha_{\parallel}(R) = \sum_{\eta'}\left\{\frac{1}{\epsilon_{\eta'\Lambda'}-\epsilon_{\eta\Lambda}-\hbar\omega} + \frac{1}{\epsilon_{\eta'\Lambda'}-\epsilon_{\eta\Lambda}+\hbar\omega}\right\} \times |\langle\eta|d|\eta'\rangle|^2|\langle\eta\Lambda|Y_{10}(\theta')|\eta'\Lambda\rangle|^2 \tag{4.126}$$

and

$$\alpha_{\perp}(R) = -\frac{1}{2}\sum_{\eta'\Lambda'}\sum_{p=\pm1}\left\{\frac{1}{\epsilon_{\eta'\Lambda'}-\epsilon_{\eta\Lambda}-\hbar\omega} + \frac{1}{\epsilon_{\eta'\Lambda'}-\epsilon_{\eta\Lambda}+\hbar\omega}\right\} \times |\langle\eta|d|\eta'\rangle|^2|\langle\eta\Lambda|Y_{1p}(\theta',\phi')|\eta'\Lambda'\rangle|^2. \tag{4.127}$$

At weak fields, the Stark shifts of the molecule in a particular ro-vibrational state $|vJ\rangle$ are given by the diagonal matrix elements of the potential (4.125) in the basis of ro-vibrational states. Noting that (see Appendix D)

$$\begin{aligned}\langle JM\Omega|\cos^2\theta|JM\Omega\rangle &= \sum_{J'M'\Omega'}\langle JM\Omega|\cos\theta|J'M'\Omega'\rangle\langle J'M'\Omega'|\cos\theta|JM\Omega\rangle \\ &= \sum_{J'}[(2J+1)(2J'+1)]\begin{pmatrix} J & 1 & J' \\ -M & 0 & M\end{pmatrix}^2\begin{pmatrix} J & 1 & J' \\ -\Omega & 0 & \Omega\end{pmatrix}^2\end{aligned} \tag{4.128}$$

and that

$$\begin{aligned}\langle JM\Omega|\sin^2\theta|JM\Omega\rangle &= \frac{4\pi}{3}\langle JM\Omega|Y_{11}Y_{1-1}|JM\Omega\rangle \\ &= \sum_{J'M'\Omega'}\langle JM\Omega|Y_{11}|J'M'\Omega'\rangle\langle J'M'\Omega'|Y_{1-1}|JM\Omega\rangle \\ &= \sum_{J'}\sum_{\Omega'}[(2J+1)(2J'+1)] \\ &\quad\times\begin{pmatrix} J & 1 & J' \\ -M & 0 & M\end{pmatrix}^2\begin{pmatrix} J & 1 & J' \\ -\Omega & 1 & \Omega'\end{pmatrix}^2,\end{aligned} \tag{4.129}$$

we see that the diagonal matrix elements are given by Eq. (4.120). For higher fields, we must diagonalize the matrix of the Hamiltonian including the field-induced potential (4.125) in the basis of the ro-vibrational states.

The polarizbailities $\alpha_{\parallel}$ and $\alpha_{\perp}$ can be expanded in Taylor series in R, yielding

$$\alpha_{\parallel}(R) = \alpha_{\parallel,\mathrm{eq}} + \alpha'_{\parallel,\mathrm{eq}}r + \frac{1}{2}\alpha''_{\parallel,\mathrm{eq}}r^2 + \cdots , \tag{4.130}$$

$$\alpha_{\perp}(R) = \alpha_{\perp,\mathrm{eq}} + \alpha'_{\perp,\mathrm{eq}}r + \frac{1}{2}\alpha''_{\perp,\mathrm{eq}}r^2 + \cdots , \tag{4.131}$$

where $r = (R - R_e)/R_e$ is the relative displacement from the equilibrium distance of the diatomic molecule. If the rigid rotor approximation is used, the potential energy of the molecule in an oscillating electric field is thus [36]

$$V_{\alpha}(\theta) = -\frac{\xi^2}{4}(\alpha_{\parallel,\mathrm{eq}}\cos^2\theta + \alpha_{\perp,\mathrm{eq}}\sin^2\theta). \tag{4.132}$$

This equation is often written in a more convenient form

$$V_{\alpha}(\theta) = -\frac{\xi^2}{4}(\alpha_{\parallel,\mathrm{eq}} - \Delta\alpha\sin^2\theta) \tag{4.133}$$

or

$$V_{\alpha}(\theta) = -\frac{\xi^2}{4}(\alpha_{\perp,\mathrm{eq}} + \Delta\alpha\cos^2\theta), \tag{4.134}$$

where $\Delta\alpha = \alpha_{\parallel,\mathrm{eq}} - \alpha_{\perp,\mathrm{eq}}$ is the polarizability anisotropy. For linear molecules, $\alpha_{\parallel,\mathrm{eq}}$ is almost always larger than $\alpha_{\perp,\mathrm{eq}}$ so the polarizability anisotropy, as defined, is positive. The first term in Eqs. (4.133) and (4.134) can be incorporated into the zero point energy. Equations (4.133) and (4.134) illustrate that the rotational states are coupled by the polarizability anisotropy.

4.7 Molecules in a Microwave Field

The frequency of a microwave field is comparable with the energy of the rotational splittings

$$\hbar\omega \approx |\epsilon_{\eta\Lambda vJ'} - \epsilon_{\eta\Lambda vJ}|. \tag{4.135}$$

In the limit of weak field, we can still use Eq. (4.113). Since the denominators in Eqs. (4.115)–(4.117) are much bigger than the denominators in Eq. (4.114), we can usually neglect all terms but α^r in Eq. (4.113). The rotational polarizability is easy to calculate since

$$|\langle S\Sigma\Lambda|Y_{10}|S\Sigma\Lambda\rangle|^2|\langle v\eta|d|v\eta\rangle|^2 \equiv d_0^2, \tag{4.136}$$

where d_0 is the dipole moment of the molecule in the vibrational state $|v\rangle$ of the electronic state $|\eta S\Sigma\Lambda\rangle$ that appeared already in Eq. (2.33). This is simply the dipole moment operator integrated over all electronic degrees of freedom and over the vibrational motion of the molecule. However, because the rotational splittings are small, even moderate magnitudes of microwave fields require nonperturbative treatment.

Nonperturbative treatment means we have to diagonalize the matrix of the Floquet Hamiltonian. In most cases, it is sufficient to restrict the basis of molecular states $|\phi_\beta\rangle$ to the rotational states of the rigid rotor. In this case, ϵ_α in Eq. (4.41) are the rotational energies of the molecule. If the field is polarized along a space-fixed Z-axis, the off-diagonal elements of the Floquet matrix are

$$\frac{\hbar\Omega}{2} = -\frac{\xi d_0}{2}\langle JM\tilde{\Omega}\varepsilon|\cos\theta|J'M\tilde{\Omega}'\varepsilon'\rangle \tag{4.137}$$

with the matrix elements of $\cos\theta$ given by Eq. (2.64). Although the Floquet matrix is an infinitely dimensional matrix, in practice, we need to focus on a particular state $|\beta n\rangle$ and restrict the Floquet basis to N coupled states with energy above and N coupled states with energy below that of $|\beta n\rangle$. The convergence of the eigenvalue correlating with $|\beta n\rangle$ should then be sought by increasing N.

Figure 4.2 shows an example of field-dressed states of a $^1\Sigma$ molecule in an off-resonant microwave field. One can see that the field-dressed states form an infinite series of replicas separated by multiples of $\hbar\omega$ [37]. The field-dressed state correlating with zero of energy in the limit of zero field describes the Stark shift of the rotational ground state $|N=0\rangle$. In the presence of the field, this state is no longer ground state. In fact, there is no ground state for a molecule in an electromagnetic field. When colliding with atoms or with each other, molecules

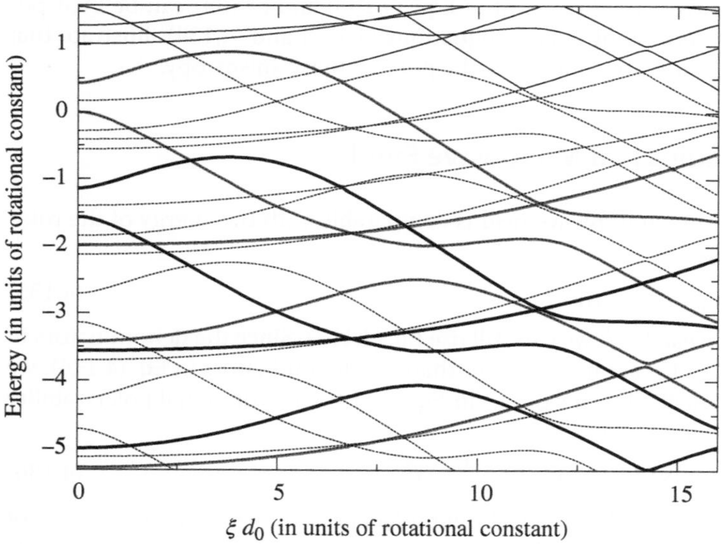

Figure 4.2 Stark shifts of the rotational energy levels of a $^1\Sigma$ molecule in an off-resonant microwave field.

prepared in this field-dressed state may undergo collisional relaxation to lower energy field-dressed states [38]. As we shall see in Section 4.8.3, this process corresponds to collision-induced absorption of microwave photons.

4.8 Molecules in a Quantized Field

An alternative approach to calculating the Stark shifts of molecules in an electromagnetic field is based on quantizing the field [39, 40]. Under some conditions, this approach is completely equivalent to the Floquet theory discussed in Section 4.2; under other conditions, there are some differences. We shall compare the two theories after introducing the quantum treatment of fields.

The goal of this section is to write the Hamiltonian of a molecule in an electromagnetic field in the form

$$\hat{H} = \hat{H}_0 + \hat{H}_\mathrm{f} + \hat{H}_{\mathrm{m-f}}, \tag{4.138}$$

where $\hat{H}_0$ is the field-free Hamiltonian of the molecule in Eqs. (3.4), and (4.23), $\hat{H}_\mathrm{f}$ is the Hamiltonian of the field and $\hat{H}_{\mathrm{m-f}}$ is the interaction of the molecule with the quantized field. The eigenstates of this Hamiltonian will represent the energy of the combined molecule–field system.

4.8.1 Field Quantization

Much of this section is based on the excellent discussion in Ref. [40]. As before (see Section 2.1), we will describe the field by a combination of the vector $\boldsymbol{A}$ and scalar φ potentials. The electric $\boldsymbol{E}$ and magnetic $\boldsymbol{B}$ fields of an electromagnetic field are given in terms of these potentials as follows [13, 41]:

$$\boldsymbol{E} = -\nabla\varphi - \frac{\partial \boldsymbol{A}}{\partial t}, \tag{4.139}$$

$$\boldsymbol{B} = \nabla \times \boldsymbol{A}. \tag{4.140}$$

The relationship between the electric and magnetic fields in vacuum is provided by the Maxwell's equations

$$\begin{aligned}
\nabla \times \boldsymbol{E} &= -\frac{\partial \boldsymbol{B}}{\partial t},\\
\nabla \times \boldsymbol{B} &= \frac{1}{c^2}\frac{\partial \boldsymbol{E}}{\partial t},\\
\nabla \cdot \boldsymbol{E} &= 0,\\
\nabla \cdot \boldsymbol{B} &= 0,
\end{aligned} \tag{4.141}$$

which follow from Eqs. (4.139) and (4.140).

As is well known [13], both the scalar and the vector potentials are not uniquely defined, meaning that many different potentials φ and $\boldsymbol{A}$ yield the same magnetic $\boldsymbol{B}$ and electric $\boldsymbol{E}$ fields. For example, we can see that the potentials

$$\boldsymbol{A}' = \boldsymbol{A} + \nabla f(\boldsymbol{r}, t), \tag{4.142}$$

$$\varphi' = \varphi - \frac{\partial f(\boldsymbol{r}, t)}{\partial t}, \tag{4.143}$$

where $f(\boldsymbol{r}, t)$ is any scalar function of coordinates and time, give the same fields $\boldsymbol{B}$ and $\boldsymbol{E}$. We are going to fix the potentials by setting $\varphi = 0$ (assuming the absence of charges) and

$$\nabla \cdot \boldsymbol{A} = 0. \tag{4.144}$$

The last equation is known as the Coulomb gauge.

Taking the curl of the magnetic field and using the Maxwell's equations, we can write

$$\nabla \times \nabla \times \boldsymbol{A} = -\frac{1}{c^2}\frac{\partial^2 \boldsymbol{A}}{\partial t^2}. \tag{4.145}$$

On the other hand, we can write for any vector $\boldsymbol{A}$ the following identity [13]:

$$\nabla \times \nabla \times \boldsymbol{A} = -\Delta \boldsymbol{A} + \nabla(\nabla \cdot \boldsymbol{A}), \tag{4.146}$$

which can be readily verified by writing out the components of the vector on the left-hand side. Using the Coulomb gauge (4.144), we thus obtain the wave equation for the vector potential in free space [39–41]:

$$\nabla^2 \boldsymbol{A}(\boldsymbol{r}, t) = \frac{1}{c^2}\frac{\partial^2 \boldsymbol{A}(\boldsymbol{r}, t)}{\partial t^2}. \tag{4.147}$$

To solve this equation, we multiply both sides by $\mathrm{e}^{i\boldsymbol{k}\cdot\boldsymbol{r}}$ and integrate over the spatial coordinates

$$\int \nabla^2 \boldsymbol{A}(\boldsymbol{r}, t)\mathrm{e}^{i\boldsymbol{k}\cdot\boldsymbol{r}}\mathrm{d}^3 r = \frac{1}{c^2}\frac{\partial^2}{\partial t^2}\int \boldsymbol{A}(\boldsymbol{r}, t)\mathrm{e}^{i\boldsymbol{k}\cdot\boldsymbol{r}}\mathrm{d}^3 r. \tag{4.148}$$

Equation (4.148) can be written as [42]

$$(-ik)^2 \tilde{\boldsymbol{A}}(\boldsymbol{k}, t) = \frac{1}{c^2}\frac{\partial^2}{\partial t^2}\tilde{\boldsymbol{A}}(\boldsymbol{k}, t), \tag{4.149}$$

where

$$\tilde{\boldsymbol{A}}(\boldsymbol{k}, t) = \frac{1}{(2\pi)^{3/2}}\int \boldsymbol{A}(\boldsymbol{r}, t)\mathrm{e}^{i\boldsymbol{k}\cdot\boldsymbol{r}}\mathrm{d}^3 r \tag{4.150}$$

is the Fourier transform of $\boldsymbol{A}(\boldsymbol{r}, t)$.

Equation (4.149) can be derived by integrating the left-hand side of Eq. (4.148) by parts. However, it should also be familiar from basic quantum mechanics reminding us that the operator $-i\hbar\partial/\partial x$ in ordinary space becomes a multiplication by momentum k_x in momentum space. Incidentally, this also means that Eq. (4.144) is equivalent to

$$\boldsymbol{k} \cdot \tilde{\boldsymbol{A}}(\boldsymbol{k}, t) = 0, \tag{4.151}$$

demonstrating that $\tilde{\boldsymbol{A}}(\boldsymbol{k}, t)$ is always orthogonal to $\boldsymbol{k}$.

The inverse Fourier transform gives

$$\boldsymbol{A}(\boldsymbol{r}, t) = \frac{1}{(2\pi)^{3/2}} \int \tilde{\boldsymbol{A}}(\boldsymbol{k}, t) \mathrm{e}^{-i\boldsymbol{k}\cdot\boldsymbol{r}} \mathrm{d}^3 k. \tag{4.152}$$

Assume now that the field is confined to a cubic box (cavity) of volume V. This discretizes the allowed values of $\boldsymbol{k}$ and replaces the integral over k-space by a sum

$$\boldsymbol{A}(\boldsymbol{r}, t) = \sum_{\boldsymbol{k}} \tilde{\boldsymbol{A}}_{\boldsymbol{k}}(t) \mathrm{e}^{-i\boldsymbol{k}\cdot\boldsymbol{r}}, \tag{4.153}$$

where each of the coefficients $\tilde{\boldsymbol{A}}_{\boldsymbol{k}}(t)$ satisfies Eq. (4.149) and I absorbed the factors containing π into the expansion coefficients. Because the electric and magnetic fields – and hence the vector potential – are real, we have

$$\boldsymbol{A}^*(\boldsymbol{r}, t) = \boldsymbol{A}(\boldsymbol{r}, t), \tag{4.154}$$

which means that for every $|\boldsymbol{k}|$, we have

$$\tilde{\boldsymbol{A}}_{\boldsymbol{k}} \mathrm{e}^{-i\boldsymbol{k}\cdot\boldsymbol{r}} + \tilde{\boldsymbol{A}}_{-\boldsymbol{k}} \mathrm{e}^{i\boldsymbol{k}\cdot\boldsymbol{r}} = \tilde{\boldsymbol{A}}^*_{\boldsymbol{k}} \mathrm{e}^{i\boldsymbol{k}\cdot\boldsymbol{r}} + \tilde{\boldsymbol{A}}^*_{-\boldsymbol{k}} \mathrm{e}^{-i\boldsymbol{k}\cdot\boldsymbol{r}}, \tag{4.155}$$

which means that

$$\tilde{\boldsymbol{A}}_{\boldsymbol{k}} = \tilde{\boldsymbol{A}}^*_{-\boldsymbol{k}}. \tag{4.156}$$

This suggests that each coefficient $\tilde{\boldsymbol{A}}_{\boldsymbol{k}}$ can be written as a sum

$$\tilde{\boldsymbol{A}}_{\boldsymbol{k}}(t) = \boldsymbol{a}_{\boldsymbol{k}}(t) + \boldsymbol{a}^*_{-\boldsymbol{k}}(t). \tag{4.157}$$

The decomposition (4.157) clearly satisfies Eq. (4.156).

Using Eq. (4.157), we can rewrite the expansion (4.153) of the vector field as

$$\boldsymbol{A}(\boldsymbol{r}, t) = \sum_{\boldsymbol{k}} [\boldsymbol{a}_{\boldsymbol{k}}(t) \mathrm{e}^{i\boldsymbol{k}\cdot\boldsymbol{r}} + \boldsymbol{a}^*_{\boldsymbol{k}}(t) \mathrm{e}^{-i\boldsymbol{k}\cdot\boldsymbol{r}}]. \tag{4.158}$$

Note that $\boldsymbol{a}_{\boldsymbol{k}}(t)$ and $\boldsymbol{a}^*_{\boldsymbol{k}}(t)$ – just like $\tilde{\boldsymbol{A}}_{\boldsymbol{k}}$ – are vector quantities.

Equation (4.151) tells us that each vector $\tilde{\boldsymbol{A}}_{\boldsymbol{k}}$ – and hence $\boldsymbol{a}_{\boldsymbol{k}}$ – lives in a plane perpendicular to $\boldsymbol{k}$. We can define two basis vectors $\boldsymbol{\epsilon}_{\boldsymbol{k},1}$ and $\boldsymbol{\epsilon}_{\boldsymbol{k},2}$ in this plane and expand $\boldsymbol{a}_{\boldsymbol{k}}$ as

$$\boldsymbol{a}_{\boldsymbol{k}}(t) = \sum_{s=1,2} a_{\boldsymbol{k},s}(t) \boldsymbol{\epsilon}_{\boldsymbol{k},s}. \tag{4.159}$$

For $\boldsymbol{\epsilon}_{\boldsymbol{k},1}$ and $\boldsymbol{\epsilon}_{\boldsymbol{k},2}$ to be basis vectors, we must require that

$$\boldsymbol{\epsilon}^*_{\boldsymbol{k},1} \cdot \boldsymbol{\epsilon}_{\boldsymbol{k},2} = \boldsymbol{\epsilon}_{\boldsymbol{k},1} \cdot \boldsymbol{k} = \boldsymbol{\epsilon}_{\boldsymbol{k},2} \cdot \boldsymbol{k} = 0 \tag{4.160}$$

and

$$\epsilon_{k,1}^* \cdot \epsilon_{k,1} = \epsilon_{k,2}^* \cdot \epsilon_{k,2} = 1. \tag{4.161}$$

Obviously, there is an infinite set of $\epsilon_{k,1}$ and $\epsilon_{k,2}$ basis vectors. They can be simply the unit vectors $\hat{X}$ and $\hat{Y}$ or they can be defined, for example, as $\epsilon_{k,1} = -(\hat{X} + i\hat{Y})/\sqrt{2}$ and $\epsilon_{k,1} = (\hat{X} - i\hat{Y})/\sqrt{2}$. See Appendix E for the connection with the spherical basis. For the sake of generality, we will assume that these vectors are complex.

We now have the following expansion for the vector potential:

$$\boldsymbol{A}(\boldsymbol{r}, t) = \sum_{\boldsymbol{k}} \sum_{s=1,2} [a_{\boldsymbol{k},s}(t)\epsilon_{\boldsymbol{k},s}\mathrm{e}^{i\boldsymbol{k}\cdot\boldsymbol{r}} + a_{\boldsymbol{k},s}^*(t)\epsilon_{\boldsymbol{k},s}^*\mathrm{e}^{-i\boldsymbol{k}\cdot\boldsymbol{r}}]. \tag{4.162}$$

If we insert this expansion into the original wave equation (4.147) and equalize the coefficients in front of the same $\epsilon_{\boldsymbol{k},s}\mathrm{e}^{i\boldsymbol{k}\cdot\boldsymbol{r}}$, we will obtain the differential equations

$$-k^2 a_{\boldsymbol{k},s}(t) = \frac{1}{c^2}\frac{\partial^2}{\partial t^2} a_{\boldsymbol{k},s}(t), \tag{4.163}$$

which can be integrated to yield

$$a_{\boldsymbol{k},s}(t) = a_{\boldsymbol{k},s}\mathrm{e}^{-i\omega_k t} \tag{4.164}$$

and

$$a_{\boldsymbol{k},s}^*(t) = a_{\boldsymbol{k},s}^*\mathrm{e}^{i\omega_k t}. \tag{4.165}$$

Here, $k^2 = \boldsymbol{k} \cdot \boldsymbol{k}$ and $\omega_k = ck$. This leads to the following expansion of the vector potential

$$\boldsymbol{A}(\boldsymbol{r}, t) = \sum_{\boldsymbol{k}} \sum_{s=1,2} [a_{\boldsymbol{k},s}\epsilon_{\boldsymbol{k},s}\mathrm{e}^{i(\boldsymbol{k}\cdot\boldsymbol{r}-\omega_k t)} + a_{\boldsymbol{k},s}^*\epsilon_{\boldsymbol{k},s}^*\mathrm{e}^{-i(\boldsymbol{k}\cdot\boldsymbol{r}-\omega_k t)}]. \tag{4.166}$$

Using Eqs. (4.140) and (4.139) and keeping in mind the assumption that there are no electric charges in the cavity (i.e. $\varphi = 0$), we can write the magnetic and electric fields for each $\boldsymbol{k}$-mode and polarization s as

$$\boldsymbol{B}_{\boldsymbol{k},s} = i\boldsymbol{k} \times [a_{\boldsymbol{k},s}\epsilon_{\boldsymbol{k},s}\mathrm{e}^{i(\boldsymbol{k}\cdot\boldsymbol{r}-\omega_k t)} - a_{\boldsymbol{k},s}^*\epsilon_{\boldsymbol{k},s}^*\mathrm{e}^{-i(\boldsymbol{k}\cdot\boldsymbol{r}-\omega_k t)}], \tag{4.167}$$

$$\boldsymbol{E}_{\boldsymbol{k},s} = i\omega_k[a_{\boldsymbol{k},s}\epsilon_{\boldsymbol{k},s}\mathrm{e}^{i(\boldsymbol{k}\cdot\boldsymbol{r}-\omega_k t)} - a_{\boldsymbol{k},s}^*\epsilon_{\boldsymbol{k},s}^*\mathrm{e}^{-i(\boldsymbol{k}\cdot\boldsymbol{r}-\omega_k t)}]. \tag{4.168}$$

If the electric field $\boldsymbol{E}_{\boldsymbol{k},s}$ is oscillating along a particular axis, $\epsilon_{\boldsymbol{k},s}$ can be chosen to be real, so the last equation can also be written as

$$\boldsymbol{E}_{\boldsymbol{k},s} = -\epsilon_{\boldsymbol{k},s}2\omega_k|a_{\boldsymbol{k},s}|\sin(\boldsymbol{k}\cdot\boldsymbol{r} - \omega_k t + \varphi). \tag{4.169}$$

The total energy of the field is [13, 40]

$$H_{\mathrm{f}} = \frac{\epsilon_0}{2}\int_V (|\boldsymbol{E}|^2 + c^2|\boldsymbol{B}|^2)\mathrm{d}^3r. \tag{4.170}$$

Substituting Eqs. (4.167) and (4.168) into Eq. (4.170) and using the orthogonality properties of $e^{ik\cdot r}$ and the polarization vectors, we obtain after the integration over the entire volume V the Hamiltonian of the field:

$$H_{\mathrm{f}} = \sum_{k}\sum_{s=1,2} \epsilon_0 \omega_k^2 V (a_{k,s}^* a_{k,s} + a_{k,s} a_{k,s}^*). \tag{4.171}$$

The goal now is to replace the coefficients $a_{k,s}$ and $a_{k,s}^*$ with quantum mechanical operators. One way to do this is to express the Hamiltonian in terms of the canonically conjugate variables. One can then replace the conjugate variables with quantum operators, which will satisfy the canonical commutation relation.

In classical mechanics, the variables q and p are canonically conjugate variables if they satisfy the Hamilton's equations:

$$\frac{dq}{dt} = \frac{\partial H}{\partial p}, \tag{4.172}$$

$$\frac{dp}{dt} = -\frac{\partial H}{\partial q}, \tag{4.173}$$

where H is the Hamiltonian.

Consider, for example, a one-dimensional harmonic oscillator. The Hamiltonian of the harmonic oscillator can be written in terms of the momentum p, coordinate q, mass m, and oscillation frequency ω as follows:

$$H_{\mathrm{HO}} = \frac{p^2}{2m} + \frac{m\omega^2}{2} q^2. \tag{4.174}$$

It is easy to see that p and q satisfy Eqs. (4.172) and (4.173) and are, therefore, the canonically conjugate variables. As a consequence, the quantum operators for coordinate and momentum have the well-known commutation relation.

There is an infinite number of canonically conjugate variables, related to each other by a canonical transformation.[2] For example, the coordinate and momentum can be transformed as $\tilde{p} = p/\sqrt{m}$ and $\tilde{q} = \sqrt{m}q$, leading to

$$H_{\mathrm{HO}} = \frac{1}{2}(\tilde{p}^2 + \omega^2 \tilde{q}^2). \tag{4.175}$$

It can be easily seen that $\tilde{p}$ and $\tilde{q}$ are also the canonically conjugate variables.

Equation (4.175) can be considered to represent the absolute square of a complex number $a = (\tilde{p} + i\omega\tilde{q})/\sqrt{2}$. We now note that the field Hamiltonian (4.171)

---

2 Canonical transformations are an important class of transformations, which necessarily preserve the Hamilton's equations (4.173). As discussed in *Mechanics* by Landau and Lifschitz [15], one example of a canonical transformation is $P = -q$ and $Q = p$, which is simply renaming the generalized coordinates as momenta and vice versa. It is easy to see that this transformation preserves the Hamilton's equations. This clearly illustrates that the quantities q and p in the Hamilton's equations should not be always be associated with coordinates and momenta. Rather, they are canonically conjugate variables, whose Poisson brackets are $[q_i, q_j] = 0$, $[p_i, p_j] = 0$ and $[p_i, q_j] = \delta_{ij}$. The Poisson brackets "deform" to commutators upon transition to quantum mechanics.

is given by a sum of absolute squares $a^*_{k,s}a_{k,s}$. This suggest that we can write the field Hamiltonian as a sum of terms in the form of Eq. (4.175) by defining

$$a_{k,s}(t) = \frac{1}{\sqrt{4\epsilon_0 V\omega_k^2}}(\omega_k Q_{k,s} + iP_{k,s}), \tag{4.176}$$

where $Q_{k,s}$ and $P_{k,s}$ are some real scalar quantities. With this substitution, the total energy of the field can be written as

$$H_f = \sum_k \sum_s \frac{1}{2}(P^2_{k,s} + \omega_k^2 Q^2_{k,s}). \tag{4.177}$$

Note that by construction (given Eq. (4.164)), we have

$$Q_{k,s} = \frac{\sqrt{4\epsilon_0 V\omega_k^2}}{\omega_k}|a_{k,s}|\cos(\omega_k t - \phi) \tag{4.178}$$

and

$$P_{k,s} = -\sqrt{4\epsilon_0 V\omega_k^2}|a_{k,s}|\sin(\omega_k t - \phi), \tag{4.179}$$

so that

$$\frac{dP_{k,s}}{dt} = -\omega_k^2 Q_{k,s} \tag{4.180}$$

and

$$\frac{dQ_{k,s}}{dt} = P_{k,s}. \tag{4.181}$$

We thus see that the variables $P_{k,s}$ and $Q_{k,s}$ satisfy the Hamilton's equations

$$\frac{\partial H_f}{\partial Q_{k,s}} = -\frac{dP_{k,s}}{dt}, \qquad \frac{\partial H_f}{\partial P_{k,s}} = \frac{dQ_{k,s}}{dt}, \tag{4.182}$$

so they are the canonically conjugate variables. This means that, just like when we quantize the harmonic oscillator, we can replace $P_{k,s}$ and $Q_{k,s}$ with time-independent Hermitian operators

$$Q_{k,s} \rightarrow \hat{Q}_{k,s}, \qquad P_{k,s} \rightarrow \hat{P}_{k,s}, \tag{4.183}$$

such that

$$[\hat{Q}_{k,s}, \hat{P}_{k,s}] = i\hbar. \tag{4.184}$$

Since the different $(\boldsymbol{k}, s)$-modes are noninteracting, we also have

$$[\hat{Q}_{k,s}, \hat{P}_{k',s'}] = 0 \tag{4.185}$$

for $\boldsymbol{k} \neq \boldsymbol{k}'$ and $s \neq s'$.

With the replacements (4.183), the coefficients $a_{k,s}(t)$ in Eq. (4.176) become operators

$$a_{k,s} = \frac{1}{\sqrt{4\epsilon_0 V\omega_k^2}}(\omega_k \hat{Q}_{k,s} + i\hat{P}_{k,s}). \tag{4.186}$$

The next step is to define the operators [40]

$$\hat{a}_{k,s} = \frac{1}{\sqrt{2\hbar\omega_k}}(\omega_k \hat{Q}_{k,s} + i\hat{P}_{k,s}), \tag{4.187}$$

$$\hat{a}^\dagger_{k,s} = \frac{1}{\sqrt{2\hbar\omega_k}}(\omega_k \hat{Q}_{k,s} - i\hat{P}_{k,s}). \tag{4.188}$$

These definitions allow us to write

$$a_{k,s} = \sqrt{\frac{\hbar}{2\epsilon_0 V\omega_k}}\hat{a}_{k,s}, \tag{4.189}$$

$$a^*_{k,s} = \sqrt{\frac{\hbar}{2\epsilon_0 V\omega_k}}\hat{a}^\dagger_{k,s}, \tag{4.190}$$

yielding the quantum operator for the vector field

$$\hat{A}(r) = \sum_k \sum_{s=1,2} \sqrt{\frac{\hbar}{2\epsilon_0 V\omega_k}}[\hat{a}_{k,s}\epsilon_k e^{ik\cdot r} + \hat{a}^\dagger_{k,s}\epsilon^*_k e^{-ik\cdot r}], \tag{4.191}$$

and the electric and magnetic fields in the operator form

$$\hat{B}(r) = \sum_k \sum_{s=1,2} i\sqrt{\frac{\hbar}{2\epsilon_0 V\omega_k}} k \times [\hat{a}_{k,s}\epsilon_k e^{ik\cdot r} - \hat{a}^\dagger_{k,s}\epsilon^*_k e^{-ik\cdot r}], \tag{4.192}$$

$$\hat{E}(r) = \sum_k \sum_{s=1,2} i\sqrt{\frac{\hbar\omega_k}{2\epsilon_0 V}}[\hat{a}_{k,s}\epsilon_k e^{ik\cdot r} - \hat{a}^\dagger_{k,s}\epsilon^*_k e^{-ik\cdot r}]. \tag{4.193}$$

Note that the operators (4.191)–(4.193) are all Hermitian.

The Hamiltonian operator of the electromagnetic field can be obtained by substituting Eqs. (4.192) and (4.193) into Eq. (4.170), yielding

$$\hat{H}_\mathrm{f} = \frac{1}{2}\sum_k \sum_{s=1,2} \hbar\omega_k(\hat{a}_{k,s}\hat{a}^\dagger_{k,s} + \hat{a}^\dagger_{k,s}\hat{a}_{k,s}). \tag{4.194}$$

The commutation properties of the operators $\hat{a}_{k,s}$ and $\hat{a}^\dagger_{k,s}$ are determined by the commutation properties (4.184) and (4.185) of $\hat{Q}$ and $\hat{P}$. We can easily verify that

$$[\hat{a}_{k,s}, \hat{a}^\dagger_{k',s'}] = \delta_{k,k'}\delta_{s,s'}. \tag{4.195}$$

This commutation relation defines bosonic operators in quantum field theory so we discover than photons (quanta of electromagnetic field) are bosons!

Equation (4.195) can be used to rewrite the field Hamiltonian (4.194) as

$$\hat{H}_{\mathrm{f}} = \sum_{k} \sum_{s} \hbar\omega_k \left(\hat{a}^{\dagger}_{k,s}\hat{a}_{k,s} + \frac{1}{2} \right). \tag{4.196}$$

This is the familiar Hamiltonian of an ensemble of quantum harmonic oscillators. By analogy with the harmonic oscillator problem, $\hat{a}^{\dagger}_{k,s}\hat{a}_{k,s}$ is the (photon) number operator with the eigenstates

$$\hat{a}^{\dagger}_{k,s}\hat{a}_{k,s}|n\rangle = n|n\rangle, \tag{4.197}$$

representing a number (Fock) state of n photons with momentum $\boldsymbol{k}$ and polarization s. The operators $\hat{a}^{\dagger}_{k,s}$ and $\hat{a}_{k,s}$ are the photon creation and annihilation operators, respectively. Acting on the photon number states, these operators produce

$$\hat{a}^{\dagger}_{k,s}|n\rangle = \sqrt{n+1}|n+1\rangle, \tag{4.198}$$

$$\hat{a}_{k,s}|n\rangle = \sqrt{n}|n-1\rangle. \tag{4.199}$$

4.8.2 Interaction of Molecules with Quantized Field

To simplify the notation, consider a monochromatic, linearly polarized field. The operator of the electric field (4.193) is then simply

$$\hat{\boldsymbol{E}}(\boldsymbol{r}) = i\sqrt{\frac{\hbar\omega}{2\epsilon_0 V}}\boldsymbol{\epsilon}[\hat{a}\mathrm{e}^{i\boldsymbol{k}\cdot\boldsymbol{r}} - \hat{a}^{\dagger}\mathrm{e}^{-i\boldsymbol{k}\cdot\boldsymbol{r}}]. \tag{4.200}$$

As mentioned earlier, we shall assume that the wavelength λ of the field is much longer that the size of the molecule. If the molecule is placed at the origin of the coordinates,

$$\mathrm{e}^{i\boldsymbol{k}\cdot\boldsymbol{r}} = \mathrm{e}^{i(\omega/c)\hat{\boldsymbol{k}}\cdot\boldsymbol{r}} = \mathrm{e}^{i(2\pi/\lambda)\hat{\boldsymbol{k}}\cdot\boldsymbol{r}} \approx 1. \tag{4.201}$$

So we can omit the exponents in Eq. (4.200). The operator for the molecule–field interaction can then be written as[3]

$$\hat{H}_{\mathrm{m-f}} = i\sqrt{\frac{\hbar\omega}{2\epsilon_0 V}}\boldsymbol{d}\cdot\boldsymbol{\epsilon}[\hat{a} - \hat{a}^{\dagger}].$$

This operator acts on the space of the products of molecular states $|\phi_\alpha\rangle$ and the photon number states $|n\rangle$.

3 The use of the long-wavelength approximation is essential here. It is only when this approximation is used that we can write the molecule–field interaction as $\sim \boldsymbol{d}\cdot\boldsymbol{E}$. This interaction arises in the theory of electromagnetism when the so-called Göppert–Mayer gauge is used and the spatial dependence of the vector field is neglected. That's why the approximation (4.201) is often referred to as the electric dipole approximation. If the spatial dependence of the fields cannot be neglected, one should use the Coulomb gauge and write the molecule–field interactions in terms of the matrix elements of the $\boldsymbol{p}\cdot\boldsymbol{A}$ operator; see Ref. [39] for details.

The nonzero matrix elements of the photon creation and annihilation operators are

$$\langle n-1|\hat{a}|n\rangle = \sqrt{n}, \tag{4.202}$$

$$\langle n+1|\hat{a}^{\dagger}|n\rangle = \sqrt{n+1}. \tag{4.203}$$

4.8.3 Quantized Field vs Floquet Theory

The Hamiltonian of the combined molecule–field system can now be written as

$$\hat{H} = \hat{H}_0 + \hbar\omega\hat{a}^{\dagger}\hat{a} + \hat{H}_{\text{m-f}}, \tag{4.204}$$

where I shifted the zero point energy in the field Hamiltonian to omit the factor 1/2. Be reminded that

$$\hat{H}_0|\phi_\alpha\rangle = \epsilon_\alpha|\phi_\alpha\rangle \tag{4.205}$$

and

$$\hbar\omega\hat{a}^{\dagger}\hat{a}|n\rangle = n\hbar\omega|n\rangle. \tag{4.206}$$

The matrix of the Hamiltonian (4.204) in the basis of direct products $|\phi_\alpha\rangle|n\rangle$ has the same form as the matrix of the Floquet Hamiltonian given by Eqs. (4.41) and (4.42). Just like there is an infinite number of terms in the Fourier expansion (4.33), there is an infinite number of the photon number states $|n\rangle$. One difference concerns the range of n. Although the Fourier expansions (4.33) and (4.34) include terms from $-\infty$ to $+\infty$, the photon number states correspond to positive values of n only. This difference is essential in the limit of zero number of photons. Few-photon fields must be treated quantum mechanically.

If the number of photons in the field is large, we can introduce the mean number of photons $\bar{n}$ and write the photon number states as $|n\rangle \Rightarrow |\bar{n}+p\rangle$, where $p \ll \bar{n}$ is a variable index. The energy of the field can be defined with respect to the energy of $\bar{n}$ photons, allowing p to be both positive and negative.

The diagonal elements of the molecule–field Hamiltonian (4.204) are exactly the same as the diagonal elements of the Floquet Hamiltonian matrix (4.41). The off-diagonal matrix elements of the Hamiltonian (4.204) are given by

$$\langle \bar{n}+p-1|\langle\phi_\alpha|\hat{H}|\phi_\beta\rangle|\bar{n}+p\rangle = i\omega_0\langle\phi_\alpha|\boldsymbol{d}\cdot\boldsymbol{\epsilon}|\phi_\beta\rangle\sqrt{\bar{n}+p}\sqrt{\frac{\hbar\omega}{2\epsilon_0 V}} \tag{4.207}$$

and have the same selection rules as the off-diagonal matrix elements of the Floquet Hamiltonian (4.42). The analogy is even more transparent if we replace $\sqrt{\bar{n}+p} \approx \sqrt{\bar{n}}$ and make the definition

$$\frac{\hbar\Omega_{\alpha,\beta}}{2} = i\langle\phi_\alpha|\boldsymbol{d}\cdot\boldsymbol{\epsilon}|\phi_\beta\rangle\sqrt{\frac{\hbar\omega}{2\epsilon_0 V}}\sqrt{\bar{n}}. \tag{4.208}$$

Given that the mean number of photons $\bar{n}$ is related to the intensity of the electric field ξ,

$$\bar{n} = \frac{\xi^2}{2}\frac{\epsilon_0 V}{\hbar\omega}, \tag{4.209}$$

we can rewrite Eq. (4.208) as follows:

$$\frac{\hbar\Omega_{\alpha,\beta}}{2} = \frac{i}{2}\langle\phi_\alpha|\boldsymbol{d}\cdot\epsilon|\phi_\beta\rangle\xi. \tag{4.210}$$

Thus, $\Omega_{\alpha,\beta}$ is the same (to within a phase factor) as Ω defined in Eq. (4.44). This means that, in the limit of a large number of photons, the eigenvalues of the Hamiltonian (4.204) are the same as the eigenvalues of the Floquet Hamiltonian.

The quantum theory of fields allows for a better interpretation of the field-dressed states discussed in Section 4.7. The energies shown in Figure 4.2 correspond to coherent superpositions of different molecular states $|\phi_\alpha\rangle$ and different photon number states $|n\rangle$. The manifolds of states separated by multiples of $\hbar\omega$ can be labeled by $\bar{n}$. Transitions between field-dressed states having the same dependence on Ω but separated by multiples of $\hbar\omega$ correspond to absorption or emission of photons. These transitions can be induced by collisions of field-dressed molecules with atoms or with other molecules [38]. These processes are termed collision-induced absorption or emission of photons. To see this more clearly, consider the manifolds of states shown in Figure 4.2 in gray (corresponding to $\bar{n}$) and in black (corresponding to $\bar{n}-1$). The transition from the highest energy "gray" state to the highest energy "black" state does not change the state of the molecule, but removes one photon from the field.

Exercises

4.1 Consider an AC (alternating current) electric circuit with the current

$$I(t) = A\cos\omega t.$$

The electric power of a circuit is given by $P(t) = I^2(t)$. Determine the relation between the average power over one oscillation cycle and the amplitude A of the current.

4.2 Construct the matrix of the Floquet Hamiltonian for a two-level system subject to a time-dependent perturbation $V(t) = A(\cos wt)^2$.
A good starting point for this problem could be to consider the Fourier expansion of $V(t)$ and determine which of the Fourier components are nonzero. This will determine which of the off-diagonal matrix elements of the Floquet matrix are nonzero.

4.3 Prove Eq. (4.149) by using the Fourier transform of the gradient of a function.

4.4 Show that $\nabla \cdot \boldsymbol{A} = 0$ is equivalent to $\boldsymbol{k} \cdot \tilde{\boldsymbol{A}} = 0$, where $\tilde{\boldsymbol{A}}$ is the Fourier transform of $\boldsymbol{A}$.

4.5 Compute the energy levels of a rigid rotor in a linearly polarized microwave field shown in Figure 4.2 by diagonalizing the corresponding Floquet matrix. What happens to the energy levels as the rotational constant goes to zero?

4.6 Compute the energy levels of a rigid rotor in a circularly polarized microwave field. Compare the results with the energy levels shown in Figure 4.2.

5

Molecular Rotations Under Control

The microscopic interactions of molecules depend strongly on the relative orientation of the interaction partners. For example, two polar molecules approaching with their dipole moments head-to-tail are attracted, while the head-to-head orientation results in repulsion. Making molecular dipoles to approach head-to-tail or head-to-head could be used for controlling molecular reaction dynamics. However, molecules in a thermal gas undergo random rotational motion averaging out the relative orientations of molecules before they collide. Wouldn't it be wonderful if the rotational motion could be constrained or molecules could be made to rotate synchronously?! In this chapter, I will show how molecule–field interactions can be employed to do exactly that.

Restricting molecular rotations in order to achieve control over chemical dynamics has long been a research goal in physical chemistry. It proved to be a challenge. While the theory of how a rigid molecule interacts with an electric field was understood a long time ago [43], the control over random molecular rotations had been considered out of reach of experiments. The effect of an external electric field on molecular rotations was believed to be too small to be of practical use [44]. This belief is well exemplified in the 1976 Science paper written by one of the protagonists of this field [45]. The paper contains a Section entitled "Brute force – how not to orient molecules" [45]. The author estimated that the field strength required to restrict the rotational motion of HCl molecules must be 12 000 000 V cm^{-1} (!). This field is certainly out of reach of laboratory experiments.

Controlling rotations with external fields thus sounded indeed like an impossible task. However, if molecules are produced with low rotational temperature or, somehow, cooled rotationally, their rotational motion becomes much less recalcitrant. At low temperatures, the energy gap between the rotational states is small, which makes molecules more sensitive to perturbations coupling rotational states. So it was realized by several authors some 25 years ago [46–48]

Molecules in Electromagnetic Fields: From Ultracold Physics to Controlled Chemistry, First Edition. Roman V. Krems.

that the rotational motion actually can be restricted by an external DC electric field if molecules are first rotationally cooled. At about the same time started a slew of theoretical and experimental work showing that molecular rotations can be effectively controlled by aligning molecules with intense laser fields. For a relatively comprehensive list of references up to year 2013, see our recent review article [1].

5.1 Orientation and Alignment

Consider a classical dipole $\boldsymbol{d}$ undergoing the rotational motion in free space. If placed in a static electric field $\boldsymbol{E}$, the rotational motion of the dipole is hindered and the dipole tumbles. The cause of the tumbling is the interaction of the dipole with the electric field

$$V(\cos\theta) = -\boldsymbol{d}\cdot\boldsymbol{E} = -dE\cos\theta.$$

Since the minimum of this interaction energy occurs at $\cos\theta = 1$, the dipole is forced to orient itself along the electric field vector.

In this section, we will discuss the quantum mechanical description of the orientation effect. We will also consider the effect of time-varying electric fields leading to the *alignment* of molecules. We need to distinguish orientation from alignment. An ensemble of vectors is said to be oriented, if all vectors lie along a particular axis and point in the same direction (↑↑). The vectors are aligned if they lie along a particular axis, but point in both directions with equal probability (↕↕).

We also need to distinguish the orientation and alignment of the molecular axis from the orientation and alignment of angular momentum. Consider an ensemble of molecules prepared in a state with the rotational angular momentum J. The angular momentum is oriented with respect to the quantization axis, if there exists a population disbalance between states $|J,+M\rangle$ and $|J,-M\rangle$. For example, if molecules are prepared in a single state $|JM\rangle$, the angular momentum is oriented. The angular momentum is aligned, if there exists a population disbalance between states of different $|M|$ but the probability of populating the $|J,+M\rangle$ and $|J,-M\rangle$ states is the same. On the other hand, the orientation of the molecular dipole moment is specified by the angle θ. For linear molecules, the dipole moment lies along the molecular axis. As shown in Figure 5.1, a linear molecule oriented at an angle θ with respect to the z-axis can be in either or both of the states $|J,+M\rangle$ and $|J,-M\rangle$. Thus, an *oriented* linear molecule can be in a state of *aligned* angular momentum. The reverse is also possible.

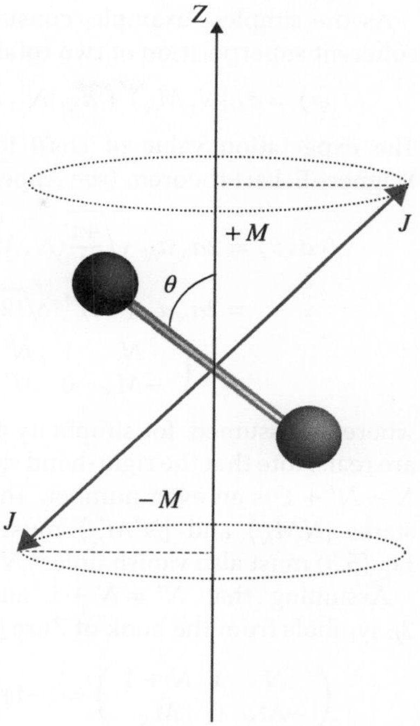

Figure 5.1 The rotational angular momentum of a linear molecule is perpendicular to the molecular axis. Here, we assume that $\boldsymbol{J} = \boldsymbol{R}$, as is the case for a molecule in a $^1\Sigma$ electronic state. A molecule with the molecular axis oriented at an angle θ with respect to the z-axis can be in a superposition of angular momentum states with $+M$ and $-M$ projections, representing an aligned angular momentum state.

5.1.1 Orienting Molecular Axis in Laboratory Frame

For a molecule prepared in state $|\psi\rangle$, the degree of orientation of the molecular axis with respect to a laboratory-fixed quantization axis can be quantified by the expectation value $\langle\cos\theta\rangle = \langle\psi|\cos\theta|\psi\rangle$. The closer the expectation value to 1, the more oriented the molecule. If the molecular axis is allowed to sample any orientations in three-dimensional space, $\langle\cos\theta\rangle = 0$. Recalling that

$$\cos\theta = \sqrt{\frac{4\pi}{3}}Y_{1,0}(\theta) \tag{5.1}$$

and that the spherical harmonic of rank 1 is odd under the inversion of the coordinate system, we see that, *if* $|\psi\rangle$ *is a state of well-defined parity*, $\langle\cos\theta\rangle = 0$. We conclude that in order to orient the molecular axis along a space-fixed axis, it is necessary to prepare molecules in a superposition of different parity states. As we saw in Chapter 2, this is exactly what happens when molecules are placed in a DC electric fields.

As the simplest example, consider a $^1\Sigma$ diatomic molecule prepared in a coherent superposition of two rotational states of different parity

$$|\psi\rangle = a_N|N, M_N\rangle + a_{N'}|N', M'_N\rangle. \tag{5.2}$$

The expectation value of $\cos\theta$ for this state can be calculated using the Wigner–Eckart theorem (see Appendix E and Section 9.1)

$$\begin{aligned}\langle\cos\theta\rangle &= 2a_N a_{N'}\sqrt{\frac{4\pi}{3}}\langle N, M_N|Y_{1,0}|N', M'_N\rangle \\ &= 2a_N a_{N'}(-1)^{M_N}\sqrt{(2N+1)(2N'+1)} \\ &\quad\times\begin{pmatrix} N & 1 & N' \\ -M_N & 0 & M'_N \end{pmatrix}\begin{pmatrix} N & 1 & N' \\ 0 & 0 & 0 \end{pmatrix},\end{aligned} \tag{5.3}$$

where we assumed, for simplicity of notation, that the coefficients a_N and $a_{N'}$ are real. Note that the right-hand side must vanish unless $M_N = M'_N$ and unless $N + N' + 1$ is an even number. The latter reflects the fact that the rotational states $|NM_N\rangle$ and $|N'M'_N\rangle$ must have different parity. The 3j-symbols in Eq. (5.3) must also vanish unless $N' = N \pm 1$.

Assuming that $N' = N + 1$ and using the explicit expression for the 3j-symbols from the book of Zare [9]

$$\begin{pmatrix} N & 1 & N+1 \\ -M_N & 0 & M_N \end{pmatrix} = (-1)^{N-M_N-1}\left[\frac{2(N+M_N+1)(N-M_N+1)}{(2N+3)(2N+2)(2N+1)}\right]^{1/2}, \tag{5.4}$$

we can rewrite Eq. (5.3) as

$$\langle\cos\theta\rangle = 2a_N a_{N'}\left[\frac{(N+1)^2 - M_N^2}{(2N+3)(2N+1)}\right]^{1/2}. \tag{5.5}$$

It is evident from this equation that the maximum value of $\langle\cos\theta\rangle$ occurs when $M_N = 0$. Since the maximum value of $a_N a_{N'} = a_N\sqrt{1-a_N^2}$ occurs at $a_N = 1/\sqrt{2}$, the maximum value of $\langle\cos\theta\rangle$ for $M_N = 0$ is

$$\langle\cos\theta\rangle = \left[\frac{(N+1)(N+1)}{(2N+3)(2N+1)}\right]^{1/2} \approx \frac{1}{2}. \tag{5.6}$$

This is the maximum degree of orientation one can achieve by superimposing two angular momentum states of a $^1\Sigma$ molecule. Note that the value of $\langle\cos\theta\rangle$ can be positive or negative, depending on the relative sign of the coefficients a_N and $a_{N'}$.

In general, when a rigid rotor is placed in a strong DC electric field, the eigenstates of the molecule are superpositions of multiple rotational states,

$$|\psi\rangle_{M_N} = \sum_N a_N|N, M_N\rangle. \tag{5.7}$$

The coefficients a_N, of course, depend on M_N and should generally be labeled by M_N but I omit this label for ease of notation. The expectation value of the orientation cosine can then be written as

$$\langle \cos\theta \rangle = \sum_N 2a_N a_{N+1} \left[\frac{(N+M_N+1)(N-M_N+1)}{(2N+3)(2N+1)} \right]^{1/2}, \tag{5.8}$$

where we again assumed that all expansion coefficients a_N are real. If M_N is zero, the factor in the square brackets is approximately 1/2 so we can write

$$\langle \cos\theta \rangle \approx \sum_N a_N a_{N+1}. \tag{5.9}$$

This shows that mixing more than two states leads to better orientation. For example, for an equal superposition of three states with $N = 0, 1,$ and 2, the value of the orientation cosine is $\langle \cos\theta \rangle \approx 2/3$, and for an equal superposition of four states it is $\langle \cos\theta \rangle \approx 3/4$, provided all the expansion coefficients are positive. If the coefficients a_N are complex, the factor $a_N a_{N+1}$ in Eq. (5.9) must be replaced with $\mathrm{Re}[a_N a_{N+1}]$. Figure 5.2 illustrates the accuracy of Eq. (5.9) and the typical behavior of the expectation value of the direction cosine with the electric field magnitude.

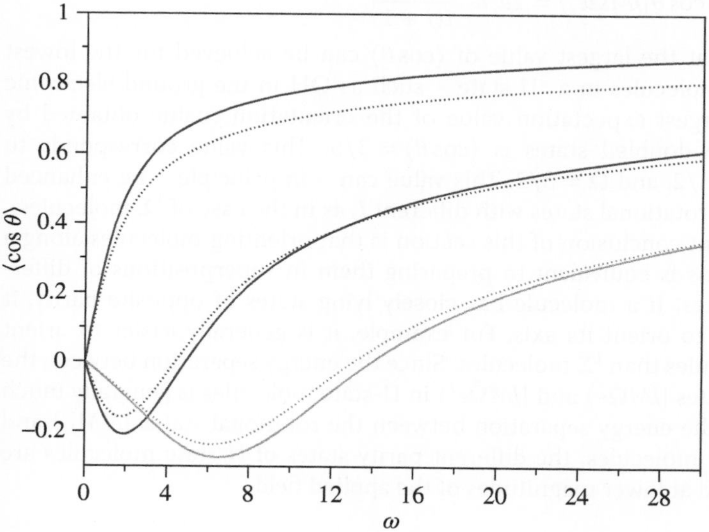

Figure 5.2 The expectation value of the orientation angle cosine of a rigid rotor in a DC electric field vs the dimensionless parameter $\omega = dE/B_e$ for three lowest energy states with $M_N = 0$. The full lines show the results computed with Eq. (5.8) and the dotted lines with Eq. (5.9). Typical values of ω for simple diatomic molecules are in the range of 0–12. The values of $\langle \cos\theta \rangle$ are shown for higher values of ω to illustrate that the limit of $\langle \cos\theta \rangle = 1$ is approached slowly, even for the lowest energy state.

Similar results can be derived for molecules in other electronic states. Consider, for example, a molecule in an electronic state of Π symmetry. As discussed in Chapter 2, the rotational states of molecules in a Π electronic state are characterized by Λ-doubling. The interaction of a Π-state molecule with a DC electric field is determined by the matrix elements given in Eq. (2.64). For a Π-state molecule in a superposition of two Λ-doubled states with the same J, M, and $\tilde{\Omega}$

$$|\psi\rangle = a_\varepsilon |JM\tilde{\Omega}\varepsilon\rangle + a_{\varepsilon'} |JM\tilde{\Omega}\varepsilon'\rangle, \tag{5.10}$$

the expectation value of $\cos\theta$ is (cf., Eq. (2.64))

$$\langle JM\tilde{\Omega}\varepsilon|\cos\theta|JM\tilde{\Omega}\varepsilon'\rangle = 2a_\varepsilon a_{\varepsilon'}\delta_{\varepsilon,-\varepsilon'}(-1)^{M-\tilde{\Omega}}(2J+1) \times \begin{pmatrix} J & 1 & J \\ M & 0 & -M \end{pmatrix} \begin{pmatrix} J & 1 & J \\ \tilde{\Omega} & 0 & -\tilde{\Omega} \end{pmatrix}. \tag{5.11}$$

The $3j$-symbols participating in this equation can be written explicitly as [9]

$$\begin{pmatrix} J & 1 & J \\ M & 0 & -M \end{pmatrix} = (-1)^{J-M} \frac{2M}{[(2J+2)(2J+1)(2J)]^{1/2}}. \tag{5.12}$$

This allows us to rewrite Eq. (5.11) as

$$\langle JM\tilde{\Omega}\varepsilon|\cos\theta|JM\tilde{\Omega}\varepsilon'\rangle = 2a_\varepsilon a_{\varepsilon'} \frac{M\tilde{\Omega}}{J(J+1)}.$$

This shows that the largest value of $\langle\cos\theta\rangle$ can be achieved for the lowest J-states. For molecules in a $^2\Pi$ state – such as OH in the ground electronic state – the largest expectation value of the orientation cosine obtained by mixing two Λ-doubled states is $\langle\cos\theta\rangle \approx 3/5$. This value corresponds to $J = 3/2, M = 3/2$, and $\tilde{\Omega} = 3/2$. This value can – in principle – be enhanced by mixing the rotational states with different J, as in the case of $^1\Sigma$ molecules.

An important conclusion of this section is that orienting molecules along a space-fixed axis is equivalent to preparing them in superpositions of different parity states. If a molecule has closely lying states of opposite parity, it will be easier to orient its axis. For example, it is generally easier to orient Π-state molecules than $^1\Sigma$ molecules. Since the energy separation between the Λ-doubled states $|JM\tilde{\Omega}\varepsilon\rangle$ and $|JM\tilde{\Omega}\varepsilon'\rangle$ in Π-state molecules is generally much smaller than the energy separation between the rotational states $|NM_N\rangle$ and $|N'M_N\rangle$ in $^1\Sigma$ molecules, the different parity states of Π-state molecules are strongly mixed at lower magnitudes of the applied field.

5.1.2 Quantum Pendulum

A rigid rotor for molecules in a $^1\Sigma$ electronic state in a strong DC electric field behaves a lot like a quantum pendulum.[1] The Hamiltonian of a plane pendulum

1 For representative references discussing pendular states of molecules in an electric field, see [47–57].

of mass m and length L

$$\hat{H} = -\frac{\hbar^2}{2mL^2}\frac{\mathrm{d}^2}{\mathrm{d}\theta^2} + mgL(1 - \cos\theta) \tag{5.13}$$

can be rescaled and written as

$$\hat{H} = -\frac{\mathrm{d}^2}{\mathrm{d}\theta^2} - V\cos\theta, \tag{5.14}$$

where $V = 2m^2L^3g/\hbar^2$ is dimensionless and the zero of energy was chosen to eliminate the constant additive factor.

For a simple rigid rotor (such as a rigid molecule in a $^1\Sigma$ state) in a DC electric field, the Hamiltonian can be written as

$$\begin{aligned}\hat{H} &= \boldsymbol{N}^2 - \omega\cos\theta \\ &= -\left\{\frac{\partial^2}{\partial\theta^2} + \cot\theta\frac{\partial}{\partial\theta} + \frac{1}{\sin^2\theta}\frac{\partial^2}{\partial\phi}\right\} - \omega\cos\theta,\end{aligned} \tag{5.15}$$

where $\omega = Ed/B_e$ is a dimensionless constant. If the rotor is prepared in an eigenstate of $\boldsymbol{N}_Z$, the Hamiltonian is

$$\hat{H} = -\frac{\mathrm{d}^2}{\mathrm{d}\theta^2} - \cot\theta\frac{\mathrm{d}}{\mathrm{d}\theta} + \frac{M_N^2}{\sin^2\theta} - \omega\cos\theta. \tag{5.16}$$

Compare Eq. (5.14) and Eq. (5.16). The difference between the two equations is the middle two terms in Eq. (5.16). This suggests that a planar pendulum can be used as a good model for understanding the behavior of the rigid rotor in an electric field. Therefore, as an important step to understand molecular rotations in electric fields, let us take a closer look at the eigenstates of a planar pendulum.

First we note the following:

- In the limit of vanishing potential $V \to 0$, Hamiltonian (5.14) describes a quantum particle on a ring.
- For finite V, the eigenstates of Hamiltonian (5.14) with energies $\epsilon < V$ are the bound states in the cosine potential. These states represent oscillations between $\theta = \theta_{\mathrm{min}}$ and $\theta = \theta_{\mathrm{max}}$ with various amplitudes.
- In the limit of very large V and $\epsilon \ll V$, the pendulum librates in a very small range of θ near $\theta = 0$. The cosine term in the interaction potential can be expanded as $\cos\theta \approx 1 - \theta^2/2$, and the planar pendulum becomes a one-dimensional harmonic oscillator.

The Schrödinger equation for the planar pendulum

$$\left[-\frac{\mathrm{d}^2}{\mathrm{d}\theta^2} - V\cos\theta\right]\psi(\theta) = \epsilon\psi(\theta) \tag{5.17}$$

is deceptively simple. By the following change of variables

$$\theta = 2z,$$
$$\lambda = 4\epsilon,$$
$$q = -2V,$$

it can be reduced to the angular Mathieu differential equation [42]

$$\left[\frac{\mathrm{d}^2}{\mathrm{d}z^2} + \lambda - 2q\cos(2z)\right]\psi(z) = 0. \tag{5.18}$$

The solutions of this equation – the Mathieu functions – are known to be "among the most difficult special functions used in physics" [42] due to the parametric dependence of the corresponding eigenvalues on the continuous parameter q entering the Mathieu equation. That said, the problem of finding the eigenstates of a planar pendulum can be reduced to the diagonalization of a simple matrix [58].

Note that depending on the values of λ and q, Eq. (5.18) permits both periodic and aperiodic solutions $\psi(z)$. The physical solutions for a planar pendulum must be periodic, with period π or 2π. These are the solutions we will be seeking.

The general prescription for calculating the Mathieu functions is described in Ref. [58]. Note that Eq. (5.18) is a differential equation with a periodic potential. We discussed such equations in Section 4.1 as a prelude to the Floquet theory. As we did there, to find the periodic solutions of Eq. (5.18), we can use the Floquet's theorem to write

$$\psi(z) = \mathrm{e}^{i\nu z}p(z), \tag{5.19}$$

where the functions $p(z)$ have period π, i.e. the same period as $\cos 2z$. The characteristic exponent ν is a function of both λ and q. We are interested only in solutions, for which $\nu(\lambda, q) = n$, where n is an integer. In this case, the functions $\psi(z)$ have period π for even n and 2π for odd n and we can use the Fourier expansion

$$\psi(z) = \sum_{k=-\infty}^{\infty} c_k\, \mathrm{e}^{ikz}. \tag{5.20}$$

The substitution of expansion (5.20) into the Schrödinger equation leads to

$$\sum_{k=-\infty}^{\infty} [(-k^2 + \lambda)\mathrm{e}^{ikz} - q(\mathrm{e}^{iz(k+2)} + \mathrm{e}^{iz(k-2)})]c_k = 0. \tag{5.21}$$

Multiplying on the left by $\mathrm{e}^{ik'z}$ and integrating over z, we obtain the recursive equation for the coefficients c_k

$$(k^2 - \lambda)c_k + q(c_{k-2} + c_{k+2}) = 0. \tag{5.22}$$

It is evident that the eigenvalues λ and the coefficients c_k can be calculated by diagonalizing the following matrix [58]:

$$\begin{pmatrix} \cdot & \cdot & \cdot & \cdot & \cdot & \cdot & \cdot \\ \cdot & (-2)^2 & 0 & q & 0 & 0 & \cdot \\ \cdot & 0 & (-1)^2 & 0 & q & 0 & \cdot \\ \cdot & q & 0 & 0^2 & 0 & q & \cdot \\ \cdot & 0 & q & 0 & (+1)^2 & 0 & \cdot \\ \cdot & 0 & 0 & q & 0 & (+2)^2 & \cdot \\ \cdot & \cdot & \cdot & \cdot & \cdot & \cdot & \cdot \end{pmatrix}. \tag{5.23}$$

The eigenvectors of this matrix are the column vectors of the coefficients c_k that give the Mathieu functions by Eq. (5.20).

Figure 5.3a depicts the allowed energies as well as a few selected wave functions for a planar pendulum with $V = 11.25$. This value of V is chosen to compare the results with those presented in Ref. [59]. The solutions with $\epsilon < V$ are the bound states in the cosine potential, representing the libration of the pendulum in a limited range of angles. At energies greater that V, the pendulum is allowed to make full 360-degree turns. Note that, when in low energy states, the pendulum likes to stay near the minimum of the potential at $\theta = 0$, but in high energy states, the probability density is the largest near the maximum of the interaction potential.

5.1.3 Pendular States of Molecules

For a rigid rotor in an electric field, we need to consider the effect of the middle two terms in Eq. (5.16). Let us rewrite Eq. (5.16) in a more suggestive form

$$\hat{H} = -\left\{ \frac{\mathrm{d}^2}{\mathrm{d}\theta^2} + \cot\theta \frac{\mathrm{d}}{\mathrm{d}\theta} \right\} + \left\{ \frac{M_N^2}{\sin^2\theta} - \omega\cos\theta \right\}. \tag{5.24}$$

Notice that the electric field magnitude appears only in the last term through the constant ω. The term in the first curly braces can be thought of as the kinetic energy and in the second – as an effective potential. As in the case of a quantum pendulum, the eigenstates of the Hamiltonian (5.16) at energies below ω are the bound states in the potential given by the expression in the second curly braces. At low electric fields (meaning shallow cosine potential), when the potential does not support any bound states, the molecule is completely disoriented, sampling all possible angles. As the electric field increases, the potential becomes deeper, binding the molecule to librate in a limited range of θ. The positive term $M_N^2/\sin^2\theta$ makes the potential shallower, and thus the molecule less oriented. This is in agreement with our earlier observation following Eq. (5.5).

The eigenstates of the Hamiltonian (5.24) can be calculated using the basis of spherical harmonics, as was done in Section 2.4.1, or the Fourier basis (5.20). Figure 5.3 compares the energies and a few selected eigenstates of a rigid rotor

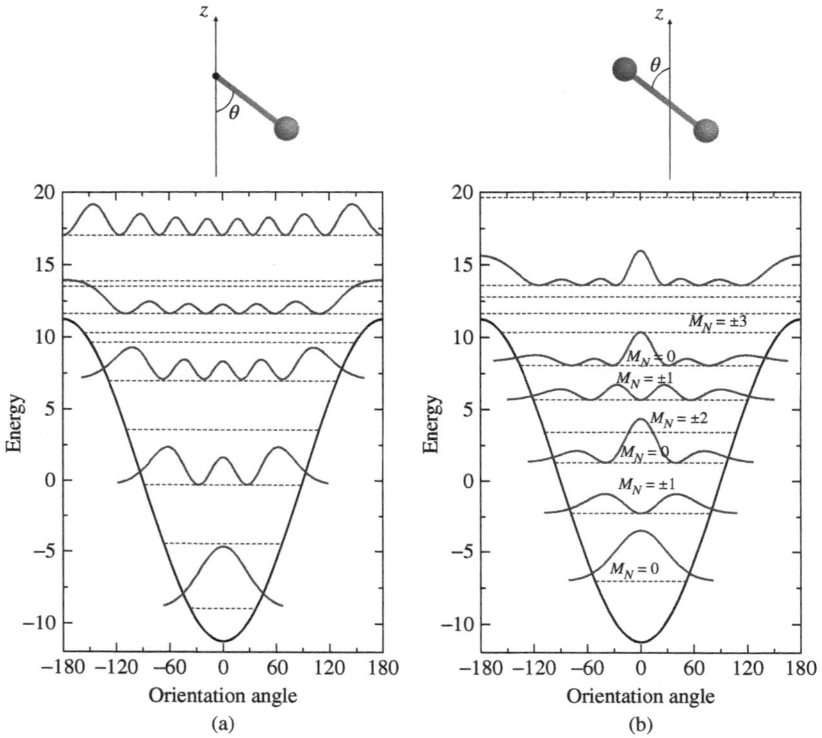

Figure 5.3 The eigenstates of a quantum pendulum (a) and a rigid rotor in a DC field (b). The vertical lines show the energies and the curves – the square of the corresponding wave functions plotted as functions of the orientation angle θ. The wave functions are not normalized and scaled for better visibility. The bound states of the rigid rotor are labeled by the quantum number M_N. The bound states and wave functions are calculated for $V = \omega = 11.25$. The energy is in the units of $\hbar^2/2mL^2$ for the planar pendulum and in the units of B_e for the rigid rotor.

in an electric field with those of a quantum pendulum. Note that the eigenstates of the rotor can be labeled by M_N, which remains a good quantum number if the electric field is directed along the quantization Z-axis. The states with $\pm M_N$ are degenerate.

The expectation value of the orientation angle cosine for a particular eigenstate can be calculated from the (measurable) ω-dependence of the corresponding eigenenergy. Using the Hellman–Feynman theorem, we can write [59]:

$$\langle \cos\theta \rangle = -\left\langle \frac{\partial}{\partial \omega} \hat{H} \right\rangle = -\frac{\mathrm{d}}{\mathrm{d}\omega} \epsilon. \tag{5.25}$$

The minimum and maximum angles of the oscillation for a bound state can then be estimated as

$$-\theta_{\min} = \theta_{\max} = \arccos(-\epsilon/\omega). \tag{5.26}$$

At high electric field, when ω is large, the potential in Eq. (5.16) is very deep and the rotor librates in a small range of θ around zero. For small θ, the Hamiltonian (5.16) can be written as

$$\hat{H} = -\left\{\frac{\partial^2}{\partial\theta^2} + \frac{1}{\theta}\frac{\partial}{\partial\theta}\right\} + \left\{\frac{M_N^2}{\theta^2} + \frac{1}{2}\omega\theta^2 - \omega\right\}. \tag{5.27}$$

If the last constant term is removed by changing the zero of energy, this Hamiltonian becomes exactly the same as the radial part of the Hamiltonian for the three-dimensional harmonic oscillator written in cylindrical polar coordinates. The eigenstates are [43]

$$\epsilon_{M_N} = -\omega + (M_N + 2n + 1)\sqrt{2\omega}, \quad \text{with} \quad n = 0, 1, 2, \ldots \tag{5.28}$$

The eigenenergies of a rigid rotor in a strong electric field are thus equidistant, with the energy separation $\sqrt{8\omega}$. This can be observed by examining the high-field limit of the energy levels in Figure 2.3.

Applying Eq. (5.25) to the deeply bound states (5.28), we see that for pendular states near the bottom of the cosine potential in a strong electric field,

$$\langle\cos\theta\rangle = 1 - (M_N + 2n + 1)\frac{1}{\sqrt{2\omega}}. \tag{5.29}$$

The maximum degree of orientation attainable for a rigid rotor in a DC electric field is thus

$$\langle\cos\theta\rangle = 1 - \frac{1}{\sqrt{2\omega}}. \tag{5.30}$$

The libration range for this state is approximately

$$\Delta\theta = \theta_{\max} - \theta_{\min} = 2\arccos\left(1 - \frac{2}{\sqrt{2\omega}}\right), \tag{5.31}$$

approaching zero as ω increases.

Pendular states can be observed spectroscopically, see, for example, Refs. [49, 52, 60].

5.1.4 Alignment of Molecules by Intense Laser Fields

Consider a dipole placed in an oscillating electric field polarized along the z-axis

$$\boldsymbol{E}(t) = \xi\hat{Z}\cos\omega t. \tag{5.32}$$

If the electric field is strong enough, the dipole follows the field vector and oscillates back and forth along the polarization axis of the field, as discussed in Section 4.5.2. On average, the dipole is as likely to point in the positive Z-direction as in negative Z-direction so it is *aligned* along the polarization axis.

Molecules can be aligned by intense off-resonant laser fields. As I pointed out in Chapter 4, the effect of laser fields on molecular energy levels can be made much stronger than the effect of DC fields. It is no surprise, therefore, that the literature abounds with examples of laser-field alignment of molecules, small and large.[2] In order to understand the quantum mechanics of laser-field alignment of molecules, consider a rigid rotor placed in an off-resonant laser field.

The potential energy of a rigid rotor in an off-resonant laser field is given by Eq. (4.134). We assume that the field is linearly polarized along the Z-axis. If the first term in Eq. (4.134) is eliminated by shifting the zero of energy, the Hamiltonian for a rigid rotor in an off-resonant laser field can be written as

$$\hat{H} = N^2 - \gamma \cos^2\theta, \tag{5.33}$$

where $\gamma = \xi^2 \Delta\alpha / 4B_e = \xi^2(\alpha_{\parallel} - \alpha_{\perp})/4B_e$ is a dimensionless constant, and B_e is the rotational constant of the molecule. This Hamiltonian is similar to Eq. (5.16), except that the second term contains the *square* of the orientation angle cosine. Squaring a cosine potential converts a potential with a single well in the angle interval between 0 and 2π to a double-well potential.

> Orienting molecules is thus equivalent to trapping them in a single-well potential, while aligning molecules corresponds to trapping them in a double-well potential.

Squaring the cosine also changes the parity of the potential energy operator. While $\cos\theta \propto Y_{10}$ has odd parity, $\cos^2\theta$ does not change under parity inversion so the second term in Eq. (5.33) couples states of the *same* parity. The matrix elements of this operator can be evaluated by writing

$$\langle NM_N | \cos^2\theta | N'M'_N \rangle = \frac{4\pi}{3} \langle NM_N | Y_{10} Y_{10} | N'M'_N \rangle, \tag{5.34}$$

and inserting the identity operator

$$\hat{1} = \sum_{N''} \sum_{M''_N} |N''M''_N\rangle \langle N''M''_N| \tag{5.35}$$

between the spherical harmonics $Y_{10}Y_{10}$. This yields

$$\langle NM_N | \cos^2\theta | N'M'_N \rangle = \frac{4\pi}{3} \sum_{N''} \sum_{M''_N} \langle NM_N | Y_{10} | N''M''_N \rangle \langle N''M''_N | Y_{10} | N'M'_N \rangle, \tag{5.36}$$

2 Some of the representative references include the work published in [61–78].

which allows us to use Eq. (2.37) to write

$$\begin{aligned}
&\langle NM_N|\cos^2\theta|N'M'_N\rangle \\
&= \sum_{N''}\sum_{M''_N}(-1)^{M_N+M''_N}\sqrt{(2N+1)(2N'+1)(2N''+1)^2} \\
&\times\begin{pmatrix} N & 1 & N'' \\ -M_N & 0 & M''_N \end{pmatrix}\begin{pmatrix} N & 1 & N'' \\ 0 & 0 & 0 \end{pmatrix}\begin{pmatrix} N'' & 1 & N' \\ -M''_N & 0 & M'_N \end{pmatrix}\begin{pmatrix} N'' & 1 & N' \\ 0 & 0 & 0 \end{pmatrix}.
\end{aligned} \tag{5.37}$$

Since the $3j$-symbols with vanishing projections are nonzero only when the sum of the three angular momenta is an even number (see Appendix B), the matrix elements (5.37) are nonzero only when both $N + N'' + 1$ and $N'' + N' + 1$ are even, so $N + N'$ must be even. In addition, the $3j$-symbols impose the selection rules $M_N = M''_N$ and $M''_N = M'_N$, which reduces to $M_N = M'_N$, illustrating that M_N is a good quantum number and can be used as a label of the eigenstates of aligned molecules. We have assumed that the quantization axis is directed along the polarization axis of the laser field.

Eq. (5.37) can be written in a more useful explicit form

$$\begin{aligned}
&\langle NM_N|\cos^2\theta|N'M_N\rangle = \sqrt{(2N+1)(2N'+1)(2N+3)^2} \\
&\times\begin{pmatrix} N & 1 & N+1 \\ -M_N & 0 & M_N \end{pmatrix}\begin{pmatrix} N & 1 & N+1 \\ 0 & 0 & 0 \end{pmatrix}\begin{pmatrix} N+1 & 1 & N' \\ -M_N & 0 & M_N \end{pmatrix}\begin{pmatrix} N+1 & 1 & N' \\ 0 & 0 & 0 \end{pmatrix} \\
&+\sqrt{(2N+1)(2N'+1)(2N-1)^2} \\
&\times\begin{pmatrix} N & 1 & N-1 \\ -M_N & 0 & M_N \end{pmatrix}\begin{pmatrix} N & 1 & N-1 \\ 0 & 0 & 0 \end{pmatrix}\begin{pmatrix} N-1 & 1 & N' \\ -M_N & 0 & M_N \end{pmatrix}\begin{pmatrix} N-1 & 1 & N' \\ 0 & 0 & 0 \end{pmatrix},
\end{aligned} \tag{5.38}$$

which shows that the matrix elements (5.37) vanish unless $N' = N$ or $N' = N \pm 2$. Using Eq. (5.4), we can write

$$\langle NM_N|\cos^2\theta|NM_N\rangle = \left[\frac{(N+M_N+1)(N-M_N+1)}{(2N+3)(2N+1)}\right] + \left[\frac{(N+M_N)(N-M_N)}{(2N+1)(2N-1)}\right], \tag{5.39}$$

$$\begin{aligned}
&\langle NM_N|\cos^2\theta|N+2,M_N\rangle \\
&= \left[\frac{(N+M_N+1)(N-M_N+1)(N+M_N+2)(N-M_N+2)}{(2N+1)(2N+3)(2N+3)(2N+5)}\right]^{1/2},
\end{aligned} \tag{5.40}$$

$$\begin{aligned}
&\langle NM_N|\cos^2\theta|N-2,M_N\rangle \\
&= \left[\frac{(N-M_N-1)(N+M_N-1)(N-M_N)(N+M_N)}{(2N-1)^2(2N-3)(2N+1)}\right]^{1/2}.
\end{aligned} \tag{5.41}$$

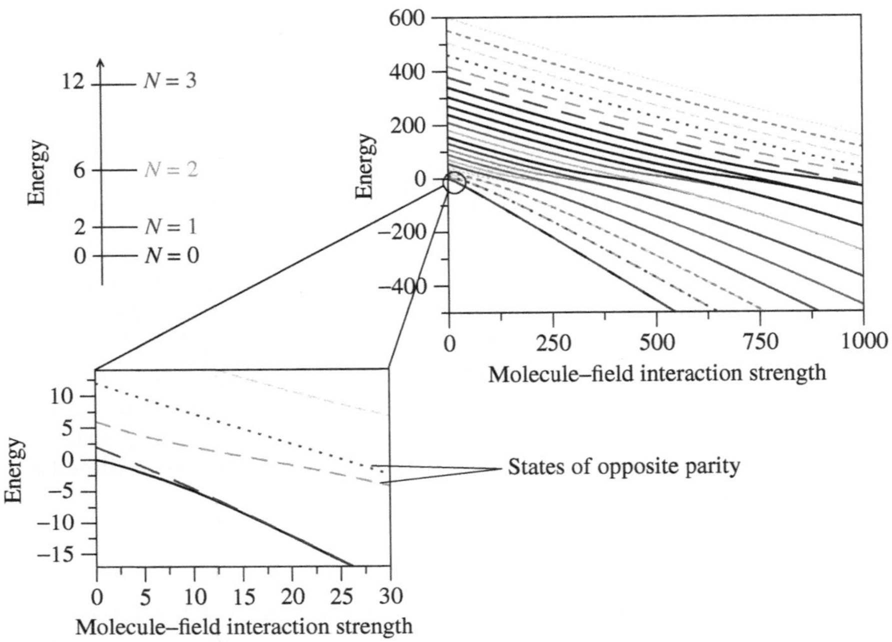

Figure 5.4 Energy levels of a rigid rotor in an off-resonant laser field. The energy is in units of the rotational constant B_e and the molecule–field interaction strength is in units of $\gamma = \xi^2 \Delta\alpha / 4B_e$. The figure illustrates that the interaction with the laser field brings the molecular states of opposite parity together.

Figure 5.4 shows 25 lowest energy levels of a rigid rotor in an off-resonant laser field computed by diagonalizing the Hamiltonian (5.33) in the basis of $|NM_N\rangle$ states with $M_N = 0$. The figure illustrates two important observations:

- At high fields, all energy levels become high-field seeking.
- The interaction with the laser field, miraculously, brings the molecular states of opposite parity (even and odd N) together. At high fields, the molecular states form nearly degenerate doublets of opposite parity.

The wave functions of the rigid rotor in a strong laser field are shown in Figure 5.5. The figure illustrates that the low-energy states are the bound states in the $\gamma\cos^2\theta$ potential, which has a double-well shape. The doublets of opposite parity states shown in Figure 5.4 arise due to the symmetry of the field-induced potential. The splitting between the energy levels in the doublets is determined by the probability of tunneling between the wells so these states are often referred to as the tunneling doublets. As the field intensity increases, the potential becomes deeper, binding more levels and bringing more of the opposite parity states to form nearly degenerate doublets.

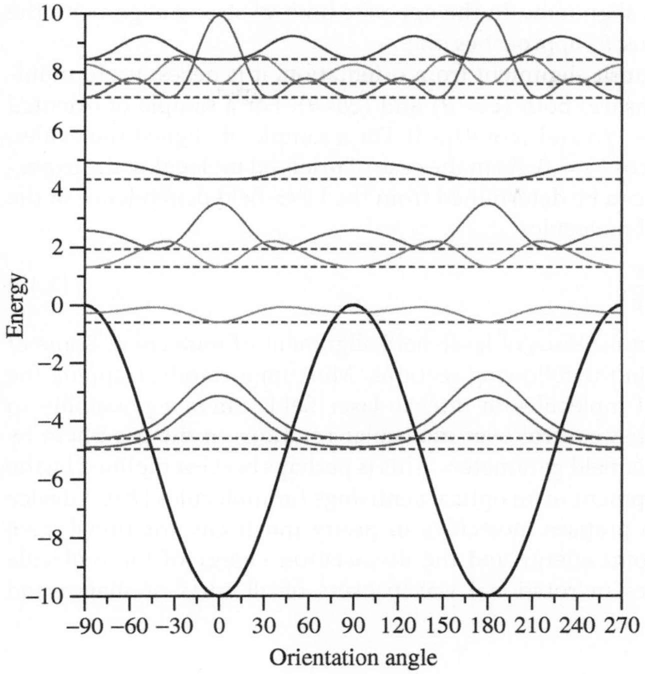

Figure 5.5 The eigenstates of a rigid rotor in an AC electric field. The horizontal lines show the energies and the curves – the square of the corresponding wave functions plotted as functions of the orientation angle θ. The wave functions are not normalized and enhanced for better visibility. The bound states of the rigid rotor are labeled by the quantum number M_N. The bound states and wave functions are calculated for $\gamma = 10$. The energy is in the units of B_e.

The degree of alignment of the molecular axis can be quantified by the expectation value of $\cos^2\theta$. If molecules are prepared in an angular momentum state with a given N and in a uniform (isotropic) distribution of states $|NM_N\rangle$ with different M_N, the expectation value is

$$\frac{1}{2N+1}\sum_{M_N}\langle NM_N|\cos^2\theta|NM_N\rangle$$
$$= \frac{1}{3}(2N+3)\begin{pmatrix} N & 1 & N+1 \\ 0 & 0 & 0 \end{pmatrix}^2 + \frac{1}{3}(2N-1)\begin{pmatrix} N-1 & 1 & N \\ 0 & 0 & 0 \end{pmatrix}^2$$
$$= \frac{1}{3}, \tag{5.42}$$

where we have used the orthonormality property of the $3j$-symbols and Eq. (5.4). The expectation value $\langle\cos^2\theta\rangle = 1/3$ thus corresponds to the complete absence of alignment. In the opposite limit of strong alignment, the expectation value $\langle\cos^2\theta\rangle$ approaches one.

In order to distinguish alignment from orientation, it is necessary to monitor (calculate or measure) both $\langle\cos^2\theta\rangle$ and $\langle\cos\theta\rangle$. For a sample of oriented molecules, $\langle\cos^2\theta\rangle > 1/3$ and $\langle\cos\theta\rangle > 0$. For a sample of aligned molecules, $\langle\cos^2\theta\rangle > 1/3$, but $\langle\cos\theta\rangle = 0$. As in the case of oriented molecules, the expectation value $\langle\cos^2\theta\rangle$ can be determined from the laser-field dependence of the energy of the aligned molecules

$$\langle\cos^2\theta\rangle = -\frac{d\epsilon}{d\gamma}. \tag{5.43}$$

There are many applications of laser-field alignment of molecules. Some of them are discussed in the following sections. Most importantly, trapping the rotational motion of molecules in intense laser fields offers a possibility to achieve unprecedented control over molecular rotations in the gas phase by manipulating the laser field parameters. This is perhaps best exemplified by the experimental development of an optical centrifuge for molecules [79], a device that can be used to prepare molecules in pretty much any rotational state between the zero point energy and the dissociation energy of the molecule or prepare molecules in rotational wavepackets of all sorts of shapes and forms [80–89].

5.2 Molecular Centrifuge

Consider a molecule aligned by a strong off-resonant laser field. If the polarization axis of the field is rotated in the laboratory frame, the molecular axis follows the field, undergoing controlled rotational motion. Accelerating the rotation of the field can then be used to spin the molecule up. This is the basic operational principle of an optical centrifuge for molecules.

As originally proposed by Karczmarek et al. [79], an optical centrifuge can be created by superimposing two counter-rotating circularly polarized fields. For example, the superposition of two fields

$$\begin{aligned} \boldsymbol{E}_1 &= \frac{\xi}{2}\{\hat{X}\cos[\omega t + \phi_L(t)] + \hat{Y}\sin[\omega t + \phi_L(t)]\}, \\ \boldsymbol{E}_2 &= \frac{\xi}{2}\{\hat{X}\cos[\omega t - \phi_L(t)] - \hat{Y}\sin[\omega t - \phi_L(t)]\}, \end{aligned} \tag{5.44}$$

yields the electric field

$$\boldsymbol{E} = \boldsymbol{E}_1 + \boldsymbol{E}_2 = \xi\cos\omega t[\hat{X}\cos\phi_L(t) + \hat{Y}\sin\phi_L(t)], \tag{5.45}$$

Whose polarization axis rotates in the space-fixed XY-plane with the frequency $\Omega = \dot{\phi}_L$. If $\phi_L = \beta t^2/2$, the polarization axis rotates with constant acceleration

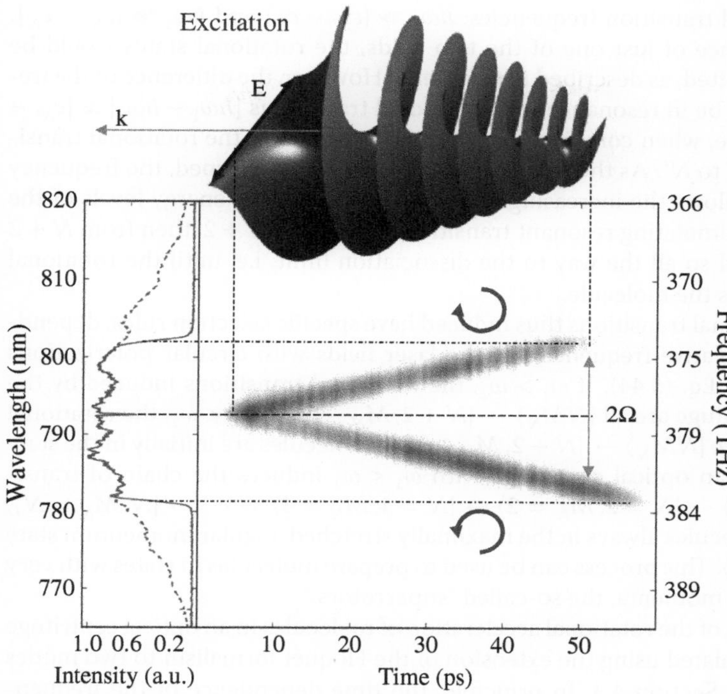

Figure 5.6 An optical centrifuge for molecules. The spinning electric field is created by splitting a laser pulse at the center of its spectrum and applying an opposite frequency chirp to the two halves. *Source*: Reproduced with permission from Korobenko et al. 2014 [80]. © 2014, the American Physical Society.

$\beta = \dot{\Omega}$. The time-dependent phase $\phi_L = \beta t^2/2$ can be engineered by linearly chirping the frequencies of the two fields $\omega_1 = \omega + \beta t/2$ and $\omega_2 = \omega - \beta t/2$. In practice, this can be done by splitting a laser pulse at the center of its spectrum and applying an opposite frequency chirp to the two halves, as illustrated in Figure 5.6, borrowed from the recent work of Korobenko et al. [80]. While the magnitude and time-dependence of ϕ_L control the acceleration of the field polarization vector, the sign of ϕ_L determines the direction of the rotation: negative ϕ_L leads to clockwise rotation, while positive ϕ_L to counterclockwise. The rotation is about the axis of the laser beam propagation. Applying the molecular centrifuge to an ensemble of molecules can thus be used to make all molecules to rotate synchronously, either clockwise or counterclockwise.

To develop a quantum mechanical picture of the rotational acceleration stimulated by an optical centrifuge, consider a molecule with dipole moment $\boldsymbol{d}$ placed in two sumperimposed laser fields (5.44), with frequencies $\omega_1 = \omega + \beta t/2$ and $\omega_2 = \omega - \beta t/2$. The frequencies of both fields are far-detuned from the rotational transition frequencies: $\hbar\omega_1 \gg |\epsilon_{N'} - \epsilon_N|$ and $\hbar\omega_2 \gg |\epsilon_{N'} - \epsilon_N|$. In the presence of just one of the two fields, the rotational states would be AC-Stark-shifted, as described in Chapter 4. However, the difference of the frequencies can be in resonance with rotational transitions $|\hbar\omega_1 - \hbar\omega_2| \approx |\epsilon_{N'} - \epsilon_N|$. Therefore, when combined, the two fields can drive the rotational transitions from N to N'. As the frequencies of the fields are chirped, the frequency difference follows the increasing energy gap between the energy levels of the rigid rotor, stimulating resonant transitions from N to $N+2$, then from $N+2$ to $N+4$, and so all the way to the dissociation limit, i.e. until the rotational energy breaks the molecule.

The rotational transitions thus induced have specific selection rules, depending on the relative frequencies of the laser fields with circular polarizations as defined in Eq. (5.44). If $\omega_1 > \omega_2$, the rotational transitions induced by the optical centrifuge are all $|NM_N\rangle \rightarrow |N+2, M_N - 2\rangle$. If $\omega_1 < \omega_2$, the rotational transitions are $|NM_N\rangle \rightarrow |N+2, M_N+2\rangle$. If molecules are initially in the state with $N = 0$, an optical centrifuge with $\omega_1 < \omega_2$ induces the chain of transitions $|N=0\rangle \rightarrow |N=2, M_N=2\rangle \rightarrow |N=4, M_N=4\rangle \rightarrow \cdots \rightarrow |N, M_N = N\rangle$, keeping molecules always in the maximally stretched angular momentum state $|N, M_N = N\rangle$. This process can be used to prepare molecules in states with very high angular momenta, the so-called "superrotors."

The theory of the rotational acceleration of molecules in an optical centrifuge can be formulated using the extension of the Floquet formalism to two modes described in Section 4.3. In principle, the time dependence of the frequencies ω_1 and ω_2 perturbs the periodicity of the Hamiltonian. However, as can be seen from Figure 5.6, the fields oscillate much faster than the timescale of the frequency chirping. Therefore, we can assume that ω_1 and ω_2 are time independent and apply the Floquet analysis as described in Section 4.3.

To obtain the Floquet matrix for a molecule in two superimposed laser fields (5.44), we first note that

$$\hat{X}\cos\omega t \pm \hat{Y}\sin\omega t = \mathrm{Re}[(\hat{X} \mp i\hat{Y})\mathrm{e}^{i\omega t}] \tag{5.46}$$

and rewrite Eqs. (5.44) as

$$\boldsymbol{E}_1 = \frac{\xi}{4}(\boldsymbol{\epsilon}_- \mathrm{e}^{i\omega_1 t} + \boldsymbol{\epsilon}_+ \mathrm{e}^{-i\omega_1 t}), \tag{5.47}$$

$$\boldsymbol{E}_2 = \frac{\xi}{4}(\boldsymbol{\epsilon}_+ \mathrm{e}^{i\omega_2 t} + \boldsymbol{\epsilon}_- \mathrm{e}^{-i\omega_2 t}), \tag{5.48}$$

where $\boldsymbol{\epsilon}_\pm = \hat{x} \pm i\hat{y}$. The interaction of the molecular dipole $\boldsymbol{d}$ with the electric fields can be written as

$$-\boldsymbol{d}\cdot\boldsymbol{E}_1 = -\frac{d\xi}{4}\sqrt{\frac{8\pi}{3}}[Y_{1,-1}(\theta,\phi)\mathrm{e}^{i\omega_1 t} - Y_{1,+1}(\theta,\phi)\mathrm{e}^{-i\omega_1 t}], \tag{5.49}$$

$$-\boldsymbol{d}\cdot\boldsymbol{E}_2 = \frac{d\xi}{4}\sqrt{\frac{8\pi}{3}}[Y_{1,+1}(\theta,\phi)\mathrm{e}^{i\omega_2 t} - Y_{1,-1}(\theta,\phi)\mathrm{e}^{-i\omega_2 t}], \tag{5.50}$$

where d is the magnitude of $\boldsymbol{d}$ and

$$Y_{1\pm 1}(\theta,\phi) = \mp\sqrt{\frac{3}{8\pi}}\frac{d_x \pm i d_y}{d} \tag{5.51}$$

are the spherical harmonics [9].

The matrix in Eq. (4.46) encompassed two electronic states α and β of the molecule. For our present purposes, it is necessary to specify the angular part of the molecular states explicitely. I will denote the rotational states of the ground electronic state α as $|\alpha, JM\rangle$ and the rotational states of the excited state β as $|\beta, JM\rangle$. The energies of the corresponding molecular states are denoted as $\epsilon_{\alpha,J}$ and $\epsilon_{\beta J}$. In order to preserve generality, I will label the states by the quantum numbers of the total angular momentum J and the projection M of $\boldsymbol{J}$ on the z-axis. For a molecule in a $^1\Sigma$ state, $J = N$.

By the Wigner–Eckart theorem, the matrix elements

$$\langle \alpha, JM|Y_{1,1}|\beta, J'M'\rangle \propto \begin{pmatrix} J & 1 & J' \\ -M & 1 & M' \end{pmatrix}. \tag{5.52}$$

The 3j-symbol in this equation vanishes unless $J' = J$ or $J' = J \pm 1$ and unless $M' = M + 1$. The complex conjugate of the matrix elements (5.52) is

$$\langle \beta, J'M'|Y_{1,-1}|\alpha, JM\rangle \propto \begin{pmatrix} J' & 1 & J \\ -M' & -1 & M \end{pmatrix}, \tag{5.53}$$

which vanishes unless $M' = M - 1$.

Consider now the Floquet matrix in the basis of states $|\alpha, JM, n_1 n_2\rangle$, where n_1 and n_2 label the Fourier components corresponding, in the quantum picture, to the number of photons with ω_1 and ω_2 (see Section 4.3). First note that the state $|\alpha, JM, 00\rangle$ is resonant with the states $|\alpha, J+2, M'', 1, -1\rangle$. Here, we assume that $\epsilon_{\alpha,J+2} - \epsilon_{\alpha J} = \hbar(\omega_1 - \omega_2) > 0$. Eq. (5.52) prohibits direct couplings between the states $|\alpha, JM, 00\rangle$ and $|\alpha, J+2, M'', 1, -1\rangle$. These states are coupled via an intermediate state $|\beta, J'M', n_1' n_2'\rangle$. As shown in Sections 4.2 and 4.3, the off-diagonal elements of the Floquet matrix couple the Fourier components n_1 only with $n_1' = n_1$ and $n_1 \pm 1$ and the Fourier components n_2 only with $n_2' = n_2$ and $n_2' = n_2 \pm 1$. It is clear, therefore, that the state $|\beta, J'M', n_1' n_2'\rangle$ must correspond to $n_1' = 1$ and $n_2' = 0$. The matrix elements

$$\langle \alpha, JM, 00|\hat{H}|\beta, J'M', 1, 0\rangle \tag{5.54}$$

are mediated by the first term in Eq. (5.48) so they are nonzero only when $M' = M - 1$. The matrix elements

$$\langle \beta, J', M' = M - 1, 1, 0|\hat{H}|\alpha, J+2, M'', 1, -1\rangle \tag{5.55}$$

are mediated by the second term in Eq. (5.49) so they are nonzero only when $M'' = M - 2$.

As shown in Section 4.2, the amplitude for a transition from molecular state γ to molecular state γ' induced by a time-dependent field is

$$U_{\alpha,\beta}(t,t_0) = \sum_n \langle \gamma' n | e^{-i\hat{H}_F(t-t_0)/\hbar} | \gamma 0 \rangle e^{in\omega t}, \tag{5.56}$$

where $\hat{H}_F$ is the Floquet Hamiltonian and n labels the Fourier components that correspond to different photon-number states in the quantum picture (see Chapter 4). For a molecule in a bi-chromatic laser field, the analogous expression is [29]

$$U_{\alpha,\beta}(t,t_0) = \sum_{n_1} \sum_{n_2} \langle \gamma' n_1 n_2 | e^{-i\hat{H}_F(t-t_0)/\hbar} | \gamma 00 \rangle e^{i(n_1\omega_1 + n_2\omega_2)t}. \tag{5.57}$$

Using a spectral resolution of the identity operator, the amplitude for transitions from state $|\alpha, JM, 00\rangle$ to state $|\alpha, J+2, M'', 1, -1\rangle$ can be written in terms of the matrix elements (5.54) and (5.55). It is the interplay of the selection rules for the "up" (from α to β) and "down" (from β to α) transitions that determines the selection rules for the two-photon transitions

$$|\alpha, JM, 00\rangle \rightarrow |\alpha, J+2, M-2, 1-1\rangle.$$

It is easy to see from this analysis that, if the polarizations of the fields (5.44) are reversed, both the "up" and "down" transitions must increase the value of M by one, leading to the selection rule $M \rightarrow M'' = M + 2$.

5.3 Orienting Molecules Matters – Which Side Chemistry

The macroscopic effects of chemistry can be easily observed: we have all witnessed explosions in a chemistry lab! The microscopic details of individual molecular interactions (which ultimately lead to those explosions) are, on the other hand, much harder to probe. Despite all the wonderful developments in molecular spectroscopy and imaging techniques, understanding dynamics of molecular reactions at the microscopic, two-molecule level is still akin to untangling a skein of yarn.

And this is where orienting molecules in a lab frame can help.[3] Imagine an experiment where molecules are held oriented in space and projectiles (atoms or other molecules) are collided with the different sides of the molecules. Exactly this idea was pursued in the pioneering work by Bernstein and coworkers [90, 91] and Brooks et al. [92, 93]. These authors examined the reactions of symmetric top molecules CH_3I and CF_3I. Similarly to diatomic

3 Parts of the following discussion in this section are based on Ref. [1] that contains a more comprehensive list of references.

molecules in a Π electronic state discussed in Section 2.4.5, symmetric tops possess nearly degenerate doublets of opposite parity states – exactly what is needed to orient molecules effectively with relatively low electric fields. These experiments showed that the reactions of K with CH_3I [94] and Rb with CH_3I [95, 96] are more facile if the atoms approach the iodine side of H_3C-I. At the same time, it was demonstrated that the reaction of F_3C-I with K is faster when potassium interacts with the CF_3 side of the molecule [97]. The authors of Ref. [94] describe their experiment as follows: "... oriented CH_3I molecules are bombarded with K atoms to determine the extent chemical reactivity varies over the surface of the molecule." This degree of information about microscopic chemical interactions – afforded by orienting molecules with external fields – is extremely valuable for the development of accurate theories of chemical reactions and understanding chemistry in general. More recently, similar idea was explored in the experiments with Π-state diatomic molecules by ter Meulen and coworkers [98, 99]. In this work, OH radicals were prepared in a single fine-structure energy level of a single rotational state by hexapole state selection and oriented by a static electric field. The experiment was designed to allow the authors to orient the molecules either with their –O or with their –H end pointing toward the collision partner. The results showed that the probability of rotational excitation in collisions of OH molecules with Ar atoms differs by as much as a factor of 8, depending on which way the molecule points.

As should be clear from Chapter 2 and conclusions earlier in this chapter, it is much harder to orient diatomic molecules in a $^1\Sigma$ electronic state. Orienting such molecules requires mixing their rotational states, which are separated by a much bigger energy gap than the opposite parity states in symmetric tops or in Π-state molecules. For the rigid rotor, the energy gap between adjacent rotational energy levels increases with angular momentum J as $\epsilon_{J+1} - \epsilon_J \propto J$. This implies that it is much easier to orient molecules in low angular momentum states, i.e. molecules in low-temperature ensembles. Friedrich and Herschbach were the first to orient a diatomic molecule (ICl) in a $^1\Sigma$ electronic state [47, 48]. This work was quickly followed by an experiment exploring the effects of orientation of ICl on inelastic collisions with Ar [50]. This experiment showed that an external electrostatic field suppresses rotational transitions by increasing the energy gap between the rotational energy levels. Significant effects of the electric fields were observed for the lowest three rotational states in a field with a strength of 100 kV cm^{-1} at a collision energy of 40 meV. That one needs such a strong electric field for influencing the dynamics of elementary reactions again illustrates how difficult it is to achieve control over the microscopic behavior of molecules.

A few years later came the experiments investigating the influence of the molecular orientation on the velocity and angular distributions of the chemical reaction products, taking stereochemistry up to yet another level of detail. A few examples of such measurements include the reactions of K with CH_3Br [100], K with ICl [101], Li with HF [102], and K with C_6H_5I [103]. Reference [104] reviews this work.

It was also demonstrated that selecting molecules in specific states by hexapole electric fields can be used to alter the overall reactivity of molecules, the product recoil and the distributions of products of chemical reactions [105]. These experiments have truly laid the foundation for controlled chemistry. A decade later, these experiments would be taken to another level of precision and control after the development of techniques for cooling, slowing, and accelerating molecules. This will be discussed in Chapter 6.

Orienting molecules, from small to big [106, 107], could also be used as a tool to learn about the intrinsic properties of molecules. A wonderful example is provided by the work of Gijsbertsen et al. [108] in a paper titled "Direct Determination of the Sign of the NO Dipole Moment." In this work, the authors analyzed the distribution of ions produced by breaking diatomic molecules with extreme UV light after orienting them in space by an external DC electric field. This measurement showed that the distribution of charges inside the molecule NO corresponds to N^-O^+, thus making possible, for the first time, to determine the *direction*, and not just the magnitude, of the permanent dipole moment of a molecule.

5.4 Conclusion

The literature abounds with beautiful examples of experiments demonstrating various effects of molecular orientation and alignment on molecular interactions. The purpose of this chapter was to introduce the concepts that are key in the work on molecular orientation and alignment and offer a glimpse of the beautiful science made possible by controlling the rotational motion of molecules by electric fields. This chapter should by no means be treated as a complete review of experimental or theoretical work. The reader is invited to consult our recent article [1] for an extensive list of relevant references to year 2013.

Exercises

5.1 Label the curves in Figure 5.2 that correspond to the lowest-energy and highest-energy states of the three states shown.

5.2 Compute the energy levels of the structureless rigid rotor in a DC electric field by diagonalizing the matrix of the corresponding Hamiltonian in the Fourier basis (5.20).
This way of solving the rigid-rotor-in-a-field problem is a little more tedious than in the basis of spherical harmonics. However, it illustrates the similarity with a quantum pendulum.

5.3 Rewrite Eq. (5.22) in the form of an eigenvalue equation.

5.4 Derive the analog of Eq. (5.57) for the transition amplitude of the rotational transitions in an optical centrifuge.

6
External Field Traps

If one extends the rules of two-dimensional focusing to three dimensions, one possesses all ingredients for particle trapping.

—Paul [109], 1989 Physics Nobel Laureate

Since electric and magnetic fields – by virtue of the Zeeman and Stark effects – change the potential energy of molecules, gradients of fields can be used to alter the translational motion of molecules in specific quantum states. If the translational energy of molecules is smaller than the potential energy change due to the field gradients, the molecular motion can be confined, either along one axis, or in two dimensions, or in three dimensions. Confining molecules in two dimensions can be used for *focusing* of molecular beams. Confining molecules in two dimensions can also be used to trap them in a storage ring, analogous to circular particle accelerators for charged particles. A major thrust of current research is to trap molecules in three dimensions. This can be done with an external field configuration providing a three-dimensional minimum of potential energy for molecules. External field traps can be designed using static or time-varying fields, or a combination of both.

It is notoriously more difficult to trap neutral atoms or molecules than ions in an external field trap. This is mainly because the trapping of neutral particles is based either on the second-order Stark effect or the Zeeman effect. The trap depth – a quantity defined as the difference in potential energy between the point of the energy minimum inside the trap and at the edge of the trap – is very small for neutral particles. The trap depth puts the limit on the maximum kinetic energy of molecules to be trapped. Every experiment aiming at trapping neutral molecules must therefore start with reducing the translational energy (a process we typically call "cooling") to below the trap depth or synthesizing molecules with low translational energy in the laboratory frame. The latter can be done by associating ultracold atoms, already in a trap.

Molecules in Electromagnetic Fields: From Ultracold Physics to Controlled Chemistry, First Edition. Roman V. Krems.

In order to set the ground for the discussion of external field traps for molecules, I begin this chapter with a brief summary of the classic experiments on deflection and focusing of molecular beams.

6.1 Deflection and Focusing of Molecular Beams

As can be seen from Figures 2.3–2.5 to 3.1-3.2, when molecules are placed in strong electric or magnetic fields, the Stark or Zeeman shifts of the molecular energy levels are, at best, on the order of 1 cm^{-1}. Thus, a typical molecule prepared in a low-field-seeking state loses about 1 cm^{-1} of kinetic energy – give or take – when climbing a field gradient from zero to the maximum field that can be created in the laboratory. That is not a lot of energy. The Boltzmann constant in the units of cm^{-1} is $k_B = 0.695$ cm^{-1} K^{-1} (an important number to remember) so 1 cm^{-1} is equivalent to $1/0.695 = 1.44$ K. This shows that the translational motion of molecules at room temperature (~300 K) is unlikely to be affected by an applied electric or magnetic field. Unless the molecules are made to travel along an extended path and the external field is applied to deflect the motion from the straight path.

Exactly that was done in the experiment by Gerlach and Stern [110] illustrated in Figure 6.1. Their experiment is the earliest and, probably, the most famous example of manipulation of the translational motion of neutral particles (silver atoms) by an external (magnetic) field. At about the same time (a little earlier, actually), Kallman and Reiche [111] proposed an experiment to deflect a beam of polar molecules using an inhomogeneous *electric* field.[1] They argued that such a measurement could demonstrate whether a dipole moment is an inherent property of an isolated molecule or whether molecules

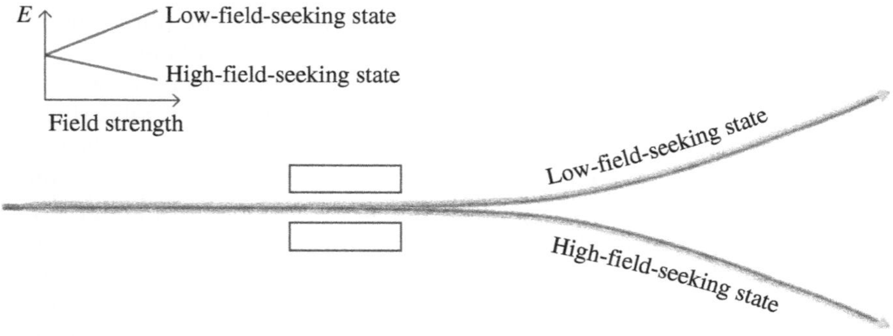

Figure 6.1 The low-field-seeking and high-field-seeking states in the Stern–Gerlach experiment.

1 Parts of the following discussion in this section as well as in Sections 6.2, 6.4, and 6.5 are reproduced with permission from Ref. [1]. Consult Ref. [1] for a more comprehensive list of references.

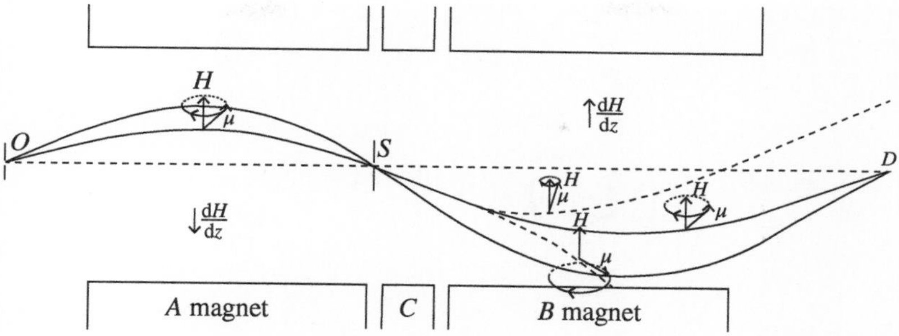

Figure 6.2 Paths of molecules. The two solid curves indicate the paths of two molecules having different moments and velocities and whose moments are not changed during passage through the apparatus. This is indicated by the small gyroscopes drawn on one of these paths, in which the projection of the magnetic moment along the field remains fixed. The two dotted curves in the region of the *B* magnet indicate the paths of two molecules the projection of whose nuclear magnetic moments along the field has been changed in the region of the *C* magnet. This is indicated by means of the two gyroscopes drawn on the dotted curves, for one of which the projection of the magnetic moment along the field has been increased, and for the other of which the projection has been decreased.
Source: Reproduced with permission from Rabi et al. 1939 [114]. © 1939, the American Physical Society.

become polarized only when they interact with other molecules. As discussed in Ref. [112], the work of Stern and Gerlach on magnetic-field deflection was actually inspired by the theoretical paper of Kallman and Reiche on electric-field deflection. The goal of their proposal was to determine whether a dipole moment is an inherent property of individual molecules or if it arises only in the bulk due to intermolecular interactions. The first experiment demonstrating the deflection of a beam of molecules by an electric field gradient was reported by Wrede [113] who was a student of Stern [112].

Deflection of molecular beams by gradients of fields was put to good use by later researchers. In particular, Rabi et al. [114] proposed a technique, illustrated in Figure 6.2, for accurate measurements of the magnetic dipole moments of atoms and molecules. The method involves passing a molecular beam through two regions of inhomogeneous magnetic field (magnet A and magnet B) with opposite gradients. The magnets force the molecules in a particular Zeeman state to follow a sigmoid path leading to a detector at point D. The first magnet selects the molecules in the desired state; the second – directs the molecules in this state onto a detector. As the beam travels between the magnets, it is exposed to another perturbation, such as an rf field, which transfers some of the molecules to a different state. If transferred to a different state, molecules, while passing through the second magnet, get deflected away from the detector. The trajectories of such molecules are shown in Figure 6.2 by dashed curves. Magnetic deflection of molecular beams has later become a useful method for state analysis and selection [115].

(a)
(b)
(c)

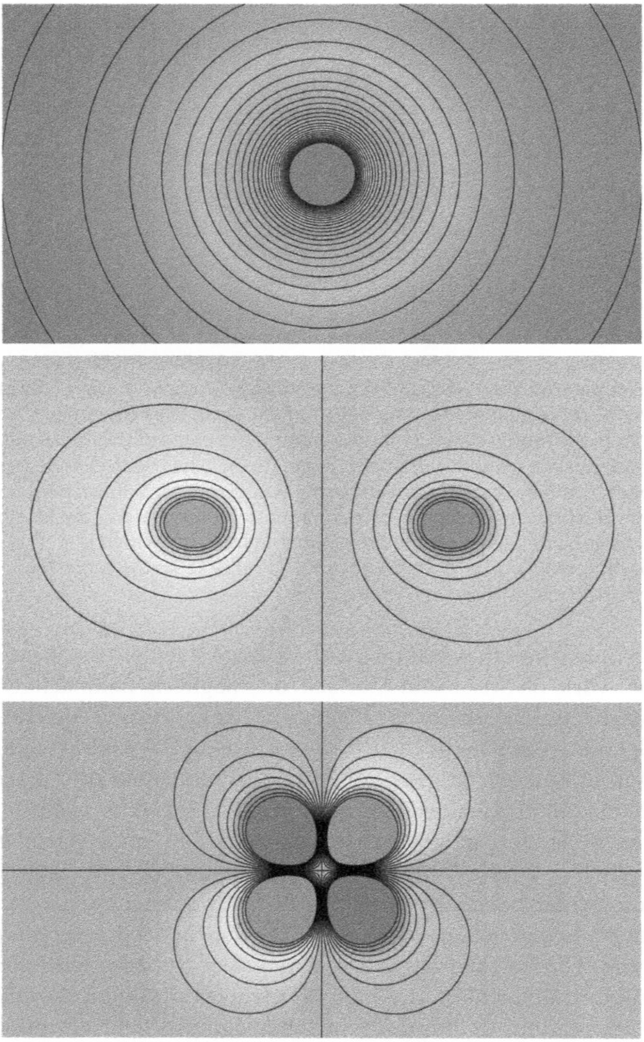

Figure 6.3 The electric field potential generated by a monopole (a), dipole (b), and quadrupole (c).

A beam of polar molecules can be *focused* when passed through an electric quadrupole. To see this, it is useful to examine the electric field potential of a quadrupole shown in Figure 6.3. It is easy to see that the electric field produced by four charges of alternating sign placed in the corners of a square vanishes in the middle of the square. If a molecule is prepared in a low-field-seeking state, it will be forced toward the middle of the square when passing through the quadrupole. Four charged rods arranged in a quadrupole configuration can be used to focus an ensemble of molecules. This was most famously exploited in a landmark experiment by Townes and coworkers who used an electric quadrupole to focus a state-selected beam of ammonia molecules to a microwave cavity. This experiment resulted in the invention of the maser [116, 117].

If bent, the quadrupole rods can be used to separate slow molecules, which follow the path of the minimum of the electric field potential, from fast molecules, which cannot make the curve. The quadrupole thus becomes a velocity selector [118] and a *guide* for slow molecules [119–122]. Bent quadrupole guides have been recently used to build complex devices, such as the centrifugal decelerator [123] illustrated in Figure 6.4. Such devices are modern extensions of the Townes' electric quadrupole setup, designed to achieve extreme control over the translational motion of molecules. More on this is in Section 6.2.

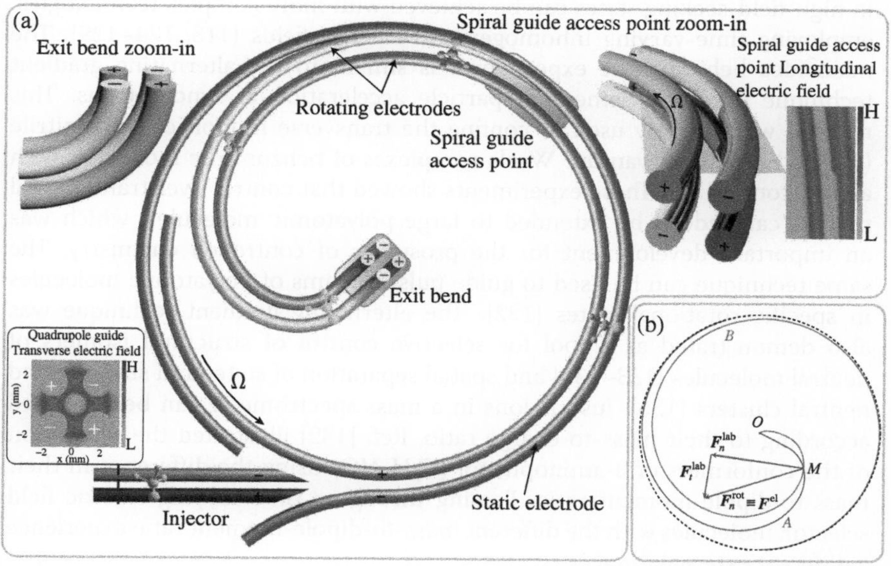

Figure 6.4 An illustration of a device called "centrifuge decelerator" for the production of slow molecules. The molecules are injected into the space between four electrodes bent into a spiral shape. The spiral is then rotated to decelerate molecules moving toward the center by the inertial force. The electrodes are bent upward at the end of the spiral, thus guiding only the slowest molecules (with translational temperature less than 1 K) toward the exit. *Source*: Reproduced with permission from Chervenkov et al. 2014 [123]. © 2014, American Physical Society.

When a molecule is placed in a field, the energy of its quantum states goes either up or down. In the former case, we talk about low-field-seeking states; in the latter – about high-field-seeking states. See Figure 6.1 for an illustration.

Molecules in high-field-seeking states are forced toward regions of strong field so the simple quadrupole shown in Figure 6.3 does not focus molecules in a high-field-seeking state. This is unfortunate because the ground state of any molecule is high-field seeking (see Figure 2.3 and the discussion in Chapter 2). Even worse, polyatomic molecules have no low-field-seeking states at reasonable magnitudes of electric fields! This can be understood using Figure 2.3 as an example. As can be seen in this figure, *all* Stark states of a $^1\Sigma$ molecule become *high-field seeking* at high fields. This is a general phenomenon, which can also be observed in other figures of Chapter 2. The crossover of the Stark states from low-field-seeking to high-field seeking is determined by the magnitude of the rotational constant, which determines the separation between states of opposite parity at zero electric field. Polyatomic molecules have a very small rotational constant, and consequently a large density of rotational states. This makes all Stark states to become high-field seeking at very low field strengths.

Focusing (and Stark deceleration discussed in the next section) of molecules in high-field-seeking states can be achieved with more complex technologies, employing time-varying inhomogeneous electric fields [118, 124–129]. The main idea behind these experiments is similar to the alternating gradient technique used for elementary particle acceleration in synchrotrons. This method was recently used to confine the transverse motion of benzonitrile (C_7H_5N) [130] and van der Waals complexes of benzonitrile molecules with argon atoms [131]. These experiments showed that control over translational motion can indeed be extended to large polyatomic molecules, which was an important development for the prospects of controlled chemistry. The same technique can be used to guide pulsed beams of polyatomic molecules in specific rotational states [132]. The alternating gradient technique was also demonstrated as a tool for selective control of structural isomers of neutral molecules [133–135] and spatial separation of state- and size-selected neutral clusters [136]. Just as ions in a mass spectrometer can be separated according to their mass-to-charge ratio, Ref. [132] illustrated the separation of the conformers of 3-aminophenol (C_6H_7NO) using the difference in their mass-to-dipole-moment ratio. Passing through a time-varying electric field selector, molecules with the different mass-to-dipole-moment ratio experience a different external force.

The AC Stark effect can also be used to deflect molecular beams by applying laser field with a gradient of intensity [137]. As discussed in Chapter 4, the laser field does not discriminate much between polar and nonpolar molecules. This makes the use of laser fields for the deflection of molecular beams hugely advantageous: it can be applied to both polar and nonpolar species. The deflection of nonpolar molecules I_2 and CS_2 by laser fields was demonstrated in an experiment by Sakai et al. [138].

Laser fields are easy to shape and tune. This can be exploited to build laser-field "lenses", which could disperse or focus molecular beams. For example, Zhao et al. showed that nanosecond IR laser pulses can be tailored to provide a cylindrical lens for molecules [139]. In later experiments, a similar principle was used to demonstrate a laser-field "molecular prism" that could separate a mixture of benzene and nitric oxide [140], or a mixture of benzene and carbon disulfide [141]. This is akin to chromatography, where the separation of molecules is achieved due to the difference in the way they interact with laser field pulses.

Since laser fields can be used to control the motion of molecules and the dynamics of the molecular wavepackets depends on the mass of the atoms in the molecule, one can also use laser fields to separate different isotopes of the same molecular species. This was done by Averbukh et al. who proposed and demonstrated spatial separation of molecular isotopes using nonadiabatic excitation of vibrational wavepackets [142, 143]. Molecular motion can also be manipulated with microwave fields, as demonstrated in Refs. [144, 145].

Interesting effects can be obtained by using a combination of laser fields with static electric or magnetic fields. For example, one can align paramagnetic molecules by laser fields and then pass them through a Stern–Gerlach magnet [146]. Aligning molecules by femto-second pulses also has a significant effect on their deflection in inhomogeneous laser fields [147]. Such control of molecular motion is very important for matter wave optics with molecules [146, 148–151]. One can imagine the development in a near future of coherent sources of molecules, which can then be controlled by means of shaped laser fields for applications such as deposition of molecules on surfaces with extremely high precision. The experiments discussed in this section have also formed the basis for the discussion of the possibility of slowing and trapping molecules by optical dipole forces [152].

6.2 Electric (and Magnetic) Slowing of Molecular Beams

In the previous section, we discussed the experiments on manipulating the *transverse* motion of molecular beams. As we mentioned, these experiments go back to the work of Stern and Gerlach, Kallmann and Reiche and Wrede in the early to mid 1920s. In stark contrast, control over the *longitudinal* motion of molecular beams was achieved only very recently. Why the nearly 80 years gap? Applying gradients of fields along the trajectory of a molecular beam can, in principle, slow molecules down, or accelerate them. The problem is that molecules in a beam produced by an expansion from a pressurized chamber move way too fast, often as fast as bullets. Since a strong electric field ($\approx$100 kV cm^{-1}) changes the energy of typical polar molecules by about

1 cm^{-1} – which is a very small amount compared to the energy of a bullet – climbing up or sliding down an electric field gradient produced by a single pair of charged electrodes hardly affects the longitudinal motion of a molecular beam. However, if subjected to multiple electric field deceleration or acceleration stages, molecules in a beam may noticeably change their longitudinal motion.

According to the authors of Chapter 14 of [153], electric field deceleration of neutral molecules was probably first attempted by John King at the Massachusetts Institute of Technology. The goal of the experiment was to slow a beam of ammonia in order to reduce the linewidth of a maser. A few years later (1966), Lennard Wharton and coworkers at the University of Chicago proposed the design of a molecular *accelerator*, a machine that would produce controllable molecular beams for chemical reaction dynamics experiments [118]. Unfortunately, both of these experiments were discontinued [118], as the focus of the research field shifted elsewhere (see Chapter 14 of [153] for more details). The ideas were revived in 1999 (>30 years later!) by Meijer and coworkers who demonstrated that the longitudinal motion of molecules can be controlled by the "Stark decelerator." In what can now be called a landmark experiment, they slowed a beam of metastable CO molecules from 225 to 98 m s^{-1} [154]. Working in parallel, Gould and coworkers [155] presented a proof-of-principle experiment demonstrating the slowing of a beam of Cs atoms.

The main principle of the Stark deceleration of a molecular beam is illustrated by Figure 6.5. In a Stark decelerator, an array of electrodes creates an inhomogeneous electric field with periodic minima and maxima of the field strength. Polar molecules traveling in this electric field potential lose kinetic energy when climbing the potential hills and gain kinetic energy when going down the hills. To prevent kinetic energy gain, the electric field is temporally modulated so that the molecules most often travel against the gradient of the electric field potential. Obviously, the most efficient deceleration can be achieved if the field is suddenly switched off when the molecules reach the maximum of the potential hill and switched back on when the molecules are at the minimum. This idea, while simple, is challenging to implement due to many technical difficulties. The spacing of the electrodes, the time variation of the electric field, and the initial beam velocity all affect the dynamics of molecules in a Stark decelerator. The spread of molecules in a beam is also a cause of problems, since some of the molecules lagging behind or traveling too fast may actually experience velocity gain during the deceleration stages. A lot of work had gone into simulations of molecular dynamics inside Stark decelerators [157–161]. With the help of analytical and numerical models of Stark deceleration and acceleration, the molecular beams were finally "tamed" [162]. The group of Meijer and other research groups demonstrated Stark deceleration for beams of metastable CO ($a^3\Pi$) [154], ND_3 [163],

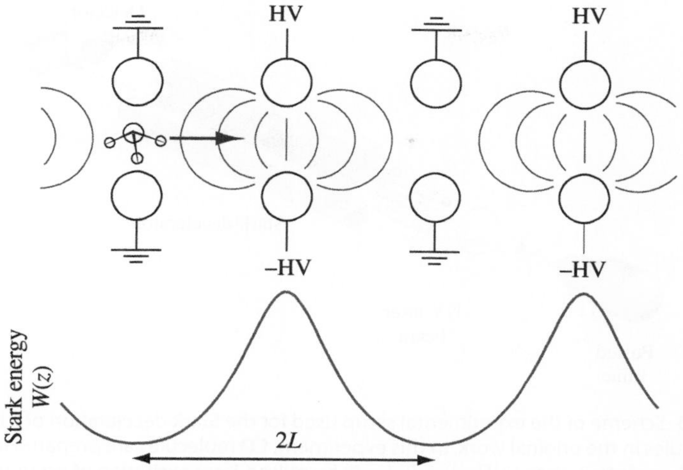

Figure 6.5 Schematic diagram of a Stark decelerator. The Stark energy of an ND_3 molecule in a low-field-seeking quantum state is shown as a function of position *z* along the molecular beam axis. The Stark energy has a period of 2*L*. *Source*: Reproduced with permission from Heiner et al. 2006 [156]. © 2006, the Royal Society of Chemistry.

OH [164–166], OD [167], NH ($a^1\Delta$) [168], H_2CO [169], SO_2 [170], LiH [171], CaF [172]. The experiments for slowing beams of CH_3F [173], CH [174], and SrF [175] were also proposed. The analysis of molecular motion in Stark decelerators [157–161] allowed the development of time sequences for field variations, leading to macroscopic confining potentials moving, and slowing, in sync with molecules in a beam [176, 177].

The original Stark decelerator consisted of 63 electric field stages combined in a structure 71 cm long [154], shown in Figure 6.6. Thanks to the many simulations and experiments, detailed understanding of the deceleration dynamics allowed the development of methods for controlling the longitudinal motion of molecular beams on much smaller length scales and it is currently possible to design powerful molecular beam decelerators of much smaller size. For example, Marian et al. [178] demonstrated a Stark decelerator made of wires 11 cm long (compare this to the 11 m design from 1966 [118]!). In another experiment, Meek et al. [179–182] developed a molecular chip – a decelerator about 50 mm long – and using it brought molecules to standstill! Reflection [183] and focusing [184] of molecules in such micro-devices were demonstrated soon thereafter.

The Stark deceleration experiments are usually state-selective and decelerate molecules in a particular low-field-seeking state. Extending the technique to molecules in high-field-seeking states has proven difficult, primarily because high-field seekers are attracted to the electrodes, disturbing the transverse

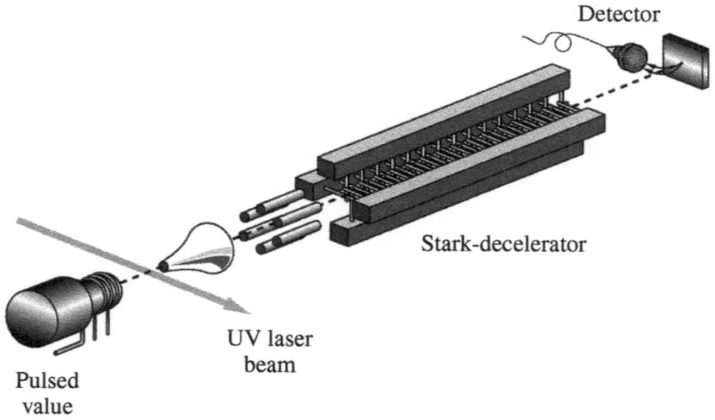

Figure 6.6 Scheme of the experimental setup used for the Stark deceleration of a beam of CO molecules in the original work. In this experiment, CO molecules are prepared in a single, metastable quantum state ($a^3\Pi_1$, $v = 0, J = 0$) by pulsed-laser excitation of ground-state CO molecules. The beam of metastable CO molecules is slowed down on passage through a series of 63 pulsed electric field stages. The time-of-flight distribution of the metastable CO molecules over the 54 cm distance from laser preparation to detection is measured via recording of the amount of electrons emitted from a gold surface when the metastable CO molecules impinge on it. *Source*: Adapted with permission from Bethlem et al. 1999 [154]. © 1999, the American Physical Society.

stability of the beam. To overcome this problem, it was proposed to use alternating gradient focusers [118, 126, 129], leading to the demonstration of deceleration of metastable CO [124], YbF [125], OH [127] and benzonitrile (C_7H_5N) [130] in high-field-seeking states. This is particularly important for heavy polyatomic molecules, which, as discussed in Section 6.1, do not have low-field-seeking rotational states. The main idea behind this technique is similar to the main operation principle of ion traps (or AC electric traps, as discussed in Section 6.4).

The Stark deceleration method, as demonstrated by Meijer and coworkers, is applicable only to polar molecules, because its operational principle is based on the DC Stark effect described in Chapter 2. And what about atoms and non-polar species? To make the technique more general, Merkt and coworkers demonstrated an alternative method based on exciting molecules to a Rydberg state, which responds to an electric field due to an enormous dipole moment produced by the Rydberg electron and the ionic core. It was shown that a beam of H_2 molecules can be effectively slowed using this technique [185, 186].

Recently, the idea of Stark deceleration has been successfully generalized to build *Zeeman* decelerators. Imagine the same picture as in Figure 6.5 but with Zeeman energy instead of Stark energy. The problem is magnetic fields are generally more difficult to tune than electric fields. Extending the main operation principle of the Stark decelerator to deceleration of radicals with switching magnetic fields has met with a number of challenges. The technical difficulties were overcome (almost simultaneously) in two different laboratories. The experiments demonstrated the slowing of beams of neutral H and D atoms [187–189], metastable Ne atoms [190, 191], and oxygen molecules [192, 193]. One should expect much progress in the design of Zeeman accelerators in a near future. As the dynamics of molecules in decelerators can be readily simulated by classical trajectory studies [194], it may be possible to design sophisticated decelerators with field sequences determined by machine learning algorithms. Already, there is a variety of new designs of the Zeeman decelerators being proposed and demonstrated, including one where molecules are trapped in a moving three-dimensional magnetic trap, which is slowed down [195–198].

The DC fields are generally harder to manipulate than laser fields. This raises the question, given the success of Stark and Zeeman deceleration, can one design a similar deceleration technique, where traveling molecules ride the potential energy hills generated by optical fields? In 2000, Friedrich proposed the idea of slowing the translational motion of molecules by gradients of an optical dipole force [199], a so-called "optical scoop." The group of Peter Barker, now at University College London, has demonstrated that molecules can be trapped in a moving periodic potential of a laser beam [200, 201]. Molecules thus trapped can be accelerated or decelerated in a process preserving narrow velocity spreads. This experiment can be thought of as a demonstration of an optical analog of a Stark decelerator [202, 203]. When it comes to manipulating molecules with AC fields, one is, of course, not limited to laser fields. As discussed in Section 4.7, microwave fields may have a very strong impact on the dynamics of molecules. This prompted Enomoto and Momose to propose a microwave field decelerator [204]. Their idea was experimentally realized by Merz et al. [205]. In another interesting development, Ahmad et al. [206] proposed to decelerate molecules by coherent pulse trains. Given the obvious advantages that AC fields offer when it comes to manipulating the translational motion of molecules, there is much to expect in this research area in the near future.

6.3 Earnshaw's Theorem

Many of the difficulties in the research field aimed at slowing and trapping molecules are due to the Earnshaw's theorem. This theorem prohibits the trapping of high-field-seeking molecules in a three-dimensional static field, which effectively means that it is impossible to trap a neutral molecule in the absolute ground state in a static field trap.

The most concise way to formulate the theorem is by the answer to the following question: Is it possible to assemble a collection of stationary charges – such as the ones shown in Figure 6.3 – in a stable configuration without any external forces? The Earnshaw's theorem says no.

What is important to us here is the consequence of this theorem that it is impossible to engineer a static electric field potential to have a three-dimensional extremum (maximum or minimum) in free space.

Proven by Earnshaw in 1842 [118], this theorem follows directly from two of the Maxwell's equations (see Section 2.1):

$$\boldsymbol{E} = -\nabla\varphi - \frac{\partial \boldsymbol{A}}{\partial t}, \tag{6.1}$$

$$\nabla \cdot \boldsymbol{E} = 0. \tag{6.2}$$

The second equation is the Gauss law for the electric field in free space.

For a static field, the time derivative of the vector potential is zero. Taking the divergence of $\boldsymbol{E}$, we thus obtain for a static electric field in vacuum

$$\nabla \cdot \boldsymbol{E} = -\nabla^2\varphi = 0 \tag{6.3}$$

or

$$\frac{\partial^2\varphi}{\partial x^2} + \frac{\partial^2\varphi}{\partial y^2} + \frac{\partial^2\varphi}{\partial z^2} = 0. \tag{6.4}$$

Equation (6.4) immediately shows that the electric field potential φ cannot have a three-dimensional extremum in free space. If φ had a three-dimensional extremum, we would have

$$\frac{\partial\varphi}{\partial x} = \frac{\partial\varphi}{\partial y} = \frac{\partial\varphi}{\partial z} = 0, \tag{6.5}$$

and the second derivatives of φ with respect to each of x, y, and z would have to be of the same sign. If this were true, the sum of the three second derivatives (6.4) could not be zero.

Note that Eq. (6.3) does allow saddle points, where Eq. (6.5) is satisfied and where, for example,

$$\frac{\partial^2\varphi}{\partial x^2} + \frac{\partial^2\varphi}{\partial y^2} = -\frac{\partial^2\varphi}{\partial z^2}. \tag{6.6}$$

If both of the derivatives on the left-hand side of Eq. (6.4) are positive, this electric field potential has a minimum in the xy-plane and a maximum along the z-direction.

It would have been nice if the Earnshaw's theorem was not true. A positive charge placed in a minimum of an electric field potential is forced to stay there, just like a negative charge placed in a maximum of an electric field potential. An electric field configuration with an extremum in free space could be used to develop *electrostatic* traps for ions. Since this is not possible, ions are trapped by a combination of static and AC fields, which leads to undesirable effects such as micromotion [207]. As mentioned earlier, the Earnshaw's theorem also has unpleasant consequences for trapping neutral molecules.

The potential energy of molecules in an electric field depends on the character of the molecular state (low-field seeking vs high-field seeking) and the *magnitude* of the field. Due to the arguments above, the electric field magnitude

$$E = |\boldsymbol{E}| = \left[\left(\frac{\partial \varphi}{\partial x} \right)^2 + \left(\frac{\partial \varphi}{\partial y} \right)^2 + \left(\frac{\partial \varphi}{\partial z} \right)^2 \right]^{1/2} \tag{6.7}$$

cannot have a three-dimensional maximum. Therefore, it is impossible to trap molecules in high-field-seeking states by electrostatic forces alone. This is unfortunate because the ground state of any molecule is high-field seeking (see Figure 2.3). Furthermore, large polyatomic molecules have no low-field-seeking states, as discussed in Section 6.1. The Earnshaw's theorem thus prohibits electrostatic traps for molecules in the absolute ground state and for heavy polyatomic molecules in any state.

Fortunately, Eq. (6.3) does allow saddle points, where Eq. (6.5) is satisfied and the electric field magnitude (6.7) has a minimum. The potential energy of molecules in the *low-field-seeking* states is minimal at these points so the Earnshaw's theorem does not prohibit electrostatic traps for low-field-seeking molecules. An example of an electrostatic trap is the field produced by charged electrodes of a quadrupole [208]. If the electrodes are arranged in the particular configuration used for the radio-frequency ion trap [207], the electric field potential is [208]

$$\varphi(x, y, z) = \varphi_0(x^2 + y^2 - 2z^2)/2r_0^2, \tag{6.8}$$

which corresponds to the squared electric field magnitude

$$|\boldsymbol{E}|^2 = \varphi_0^2(x^2 + y^2 + 4z^2)/r_0^4, \tag{6.9}$$

increasing in all directions from the origin.

What about the magnetic field magnitude? Can static magnetic fields be arranged to have a maximum of the field magnitude in free space? The answer is no.

To prove this, we will use similar arguments as for the electric field potential and the other two Maxwell's equations:

$$\nabla \times \boldsymbol{B} = \frac{1}{c^2} \frac{\partial \boldsymbol{E}}{\partial t}, \tag{6.10}$$

$$\nabla \cdot \boldsymbol{B} = 0. \tag{6.11}$$

First, we note that using the vector identity [42]

$$\boldsymbol{a} \times (\boldsymbol{b} \times \boldsymbol{c}) = \boldsymbol{b}(\boldsymbol{a} \cdot \boldsymbol{c}) - \boldsymbol{c}(\boldsymbol{a} \cdot \boldsymbol{b}), \tag{6.12}$$

we can write

$$\nabla^2 \boldsymbol{B} = \nabla(\nabla \cdot \boldsymbol{B}) - \nabla \times (\nabla \times \boldsymbol{B}). \tag{6.13}$$

In the absence of electric currents, the right-hand side of Eq. (6.10) vanishes, which leads us to conclude that

$$\nabla^2 \boldsymbol{B} = 0. \tag{6.14}$$

Since this is a vector equation, it must hold for each vector component and we must conclude that in a space void of charges and currents

$$\nabla^2 B_x = \nabla^2 B_y = \nabla^2 B_z = 0. \tag{6.15}$$

Let us now examine the gradient and the Laplacian of the *magnitude* of the magnetic field. For the gradient, we obtain

$$\nabla B = \nabla(B_x^2 + B_y^2 + B_z^2)^{1/2} = \frac{1}{B}(B_x \nabla B_x + B_y \nabla B_y + B_z \nabla B_z). \tag{6.16}$$

For the Laplacian,

$$\nabla^2 B = \frac{1}{B}(|\nabla B_x|^2 + |\nabla B_y|^2 + |\nabla B_z|^2 + B_x \nabla^2 B_x + B_y \nabla^2 B_y + B_z \nabla^2 B_z)$$
$$- \frac{1}{B^3}(B_x \nabla B_x + B_y \nabla B_y + B_z \nabla B_z)^2. \tag{6.17}$$

At an extremum, the gradient is zero, so we have

$$\frac{\partial B}{\partial x} = \frac{\partial B}{\partial y} = \frac{\partial B}{\partial z} = 0 \tag{6.18}$$

and

$$\frac{\partial^2 B}{\partial x^2} + \frac{\partial^2 B}{\partial y^2} + \frac{\partial^2 B}{\partial z^2} = \frac{1}{B}(|\nabla B_x|^2 + |\nabla B_y|^2 + |\nabla B_z|^2) \geq 0, \tag{6.19}$$

where I have used Eq. (6.15). At a three-dimensional maximum, each of the second derivatives on the left-hand side of Eq. (6.19) must be negative. Since the right-hand side of Eq. (6.19) is necessarily positive, the magnetic field magnitude can have no maximum in free space. Just like the electric field magnitude, it is allowed to have a minimum.

6.4 Electric Traps

The Earnshaw's theorem implies that molecules in high-field-seeking states cannot be trapped in three dimensions by *electrostatic* forces. On the other hand, molecules in *low-field-seeking* states are trappable. In fact, low-field-seeking molecules can be trapped using the same setup of electrodes as the one used for trapping ions developed by Paul [109]. The setup of the Paul trap shown in Figure 6.7 consists of a pair of end-cap electrodes and a ring electrode. In an ion-trapping experiment, the voltage oscillates between the end-cap and ring electrodes with the typical frequency 100 kHz to 100 MHz. This creates a time-averaged potential that confines charged particles to the middle of the trap [207].

The static limit of the Paul trap can be used as an electrostatic trap for molecules. If a constant voltage is applied to the electrodes, they produce a static electric field whose magnitude vanishes in the center of the ring and

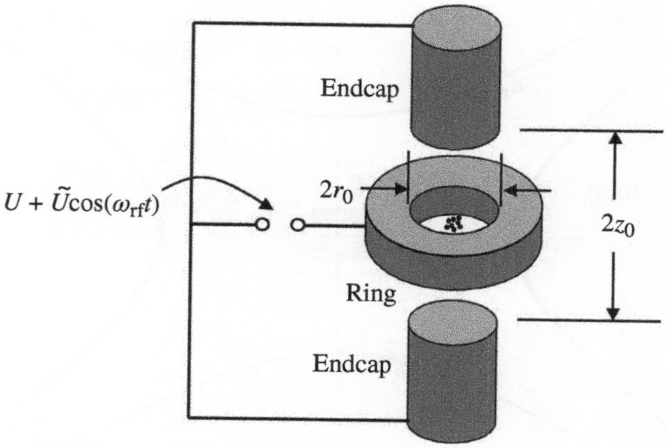

Figure 6.7 Schematic drawing of the electrodes for a cylindrically symmetric 3D rf trap. Typical dimensions are $r_0 \approx \sqrt{2}z_0 \approx 100$ μm to 1 cm, with $\tilde{U} = 100$–500 V, $|U| = 0$–50 V, and $\omega_{rf}/2\pi = 100$ kHz to 100 MHz. *Source*: Reproduced with permission from Leibfried et al. 2003 [207]. © 2003, the American Physical Society.

increases in all directions from the center. This is shown in Figure 6.8. The electric field potential inside the trap is [208]

$$\varphi(x, y, z) = \varphi_0(x^2 + y^2 - 2z^2)/2r_0^2, \tag{6.20}$$

where $\varphi_0/2$ is the magnitude of the potential applied to each electrode and r_0 is the radius of the ring electrode. This potential corresponds to the squared electric field magnitude

$$|\boldsymbol{E}|^2 = \varphi_0^2(x^2 + y^2 + 4z^2)/r_0^4. \tag{6.21}$$

Such field configuration could not be used to trap ions or high-field-seeking molecules as they would be attracted to the electrodes. However, low-field-seeking molecules placed in an electrostatic trap are forced toward the center of the ring.

The electrostatic trap for molecules was first realized by Bethlem et al. [159, 163]. It is much harder to trap neutral molecules than ions. Ion traps require a voltage of 100–500 V to generate a strongly confining potential. Typical depths of ion traps exceed hundreds of Kelvin. The trapping force of an electrostatic trap for molecules is due to the second-order Stark effect so it is much weaker. Bethlem et al. [163] applied 12 kV to the ring electrode, with the end-cap electrodes held at 0 and −400 V. The voltage difference of 12 kV generated the confining potential for ND_3 molecules with the trap depth of only 0.24 cm^{-1} (0.35 K). To produce molecules with the translational energy below the trap depth, Bethlem et al. used a 63-stage Stark decelerator with the electric field gradients reaching about 100 kV cm^{-1} at each stage. Thus, they were able to

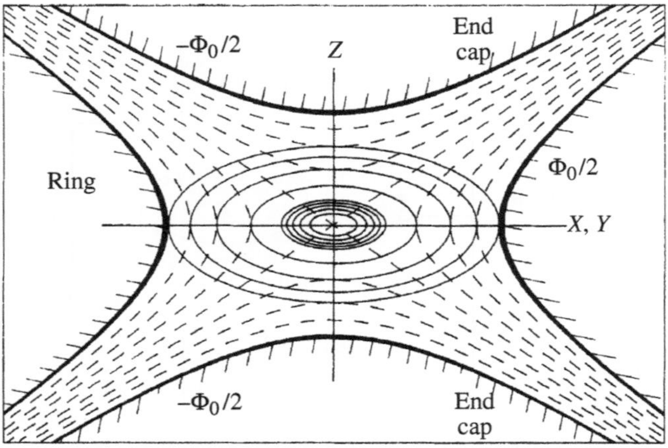

Figure 6.8 Electrostatic quadrupole trap geometry in cross section. The figure has rotational symmetry about the Z-axis. Heavy shaded curves: electrode surfaces, held at constant potentials $\pm\Phi_0/2$. The ring radius is R_0 and the end-cap half-spacing $R_0/\sqrt{2}$. Dashed curves: surfaces of constant Φ with values $\Phi = [-0.375(0.125)0.37514\Phi_0$. Full curves: surfaces of constant E^2 with values $E^2 = [0.02(0.02)0.10, 0.25(0.25)1.00]\Phi_0^2/R_0^2$. A particle whose electric polarizability is negative will have minimum potential energy at the origin, where $E^2 = 0$. *Source*: Reproduced with permission from Wing 1980 [208]. © 1980, the American Physical Society.

decelerate the molecules from the initial kinetic energy 57 cm^{-1} (mean velocity 260 m s^{-1}) to 0.13 cm^{-1} (mean velocity 13 m s^{-1}).

Trapping molecules in high-field-seeking states requires a *dynamical* trap. A stable trapping field configuration for high-field-seeking molecules (or atoms) can be generated – as proposed by Peik [209] – by superimposing an alternating hexapole field and a static homogeneous field. The rotationally symmetric hexapole field can be produced by four electrodes consisting of two end caps and two rings. This setup is similar to the Paul trap shown in Figure 6.7, except that it contains two rings separated by a gap instead of a single ring in the middle.

Trapping of ND_3 molecules in the lowest energy, high-field-seeking state was demonstrated in the experiment of van Veldhoven et al. [210]. The trapping field in this experiment was generated by two ring electrodes with an inner diameter of 10 mm separated by a 2.9 mm gap and two end-cap electrodes, with the closest separation of 9.1 mm. The molecules initially prepared in a low-field-seeking state were decelerated by a Stark decelerator to about 15 m s^{-1} and focused by a hexapole onto the area inside the AC trap. Once the molecules reached the inside of the trap, an alternating voltage was applied to the electrodes. The voltage was alternating by varying the potential between 5 and 11 kV on the first end cap, 7.5 and 1.65 kV on the first ring, −7.5 and

−1.65 kV on the second ring, and −5 and −11 kV on the second end cap, with the frequency of 1100 Hz. The (5, 7.5, −7.5, −5)-voltage creates an electric field that focuses high-field-seeking molecules in the axial direction (along the axis z passing through the end caps), while the (11, 1.65, −1.65, −11)-voltage configuration focuses high-field-seeking molecules in the radial direction (perpendicular to z, see Figure 6.7). The same field configuration could be used to trap low-field-seeking molecules. The low-field-seeking molecules would be focused in the radial direction by the (5, 7.5, −7.5, −5)-voltage and in the axial direction by the (11, 1.65, −1.65, −11)-voltage.

The work of Bethlem and coworkers built on the earlier work of Jongma et al. [211] and was followed by multiple experiments demonstrating the electrostatic trapping of OH [165], OD [167], metastable CO [212], and metastable NH [213]. A variety of electric traps with different geometries and operational principles were developed by subsequent authors. The literature abounds with examples of various trap designs. For example, Kleinert et al. [214] trapped ultracold NaCs molecules in a trap consisting of four thin wires forming the electrodes of a quadrupolar trapping field. The small dimensions of the wires allowed them to superimpose the trap onto a magneto-optical trap. This is useful for transferring molecules from an optical trap into a purely electrostatic trap. Trapping molecules in different kinds of trap may be necessary for some experiments, as the presence of certain fields (AC or DC, electric or magnetic) may sometimes be undesirable.

In another experiment, Rieger et al. [121] constructed an electrostatic trap consisting of five ring-shaped electrodes and two spherical electrodes at both ends. The authors demonstrated that the trap could be continuously loaded by molecules directed by a quadrupole guide. The continuous loading results in a steady-state population of trapped molecules, which is particularly useful for experiments probing cold and ultracold chemistry. For some applications, it may be useful to confine molecules in combined electric and magnetic fields, as was done, for example, in Ref. [215]. This experiment demonstrated the confinement of OH molecules in a *magneto-electrostatic* trap, combining the principles of a magnetic trap described in Section 6.5 with electrostatic trapping.

Trapping molecules in the ground state has been a particular focus of intense research, since molecules in the ground state are collisionally stable. The ground state of a molecule is necessarily high-field seeking. Schnell et al. [216, 217] demonstrated a linear AC trap for polar molecules in the ground state. A cylindrical AC trap was demonstrated by van Veldhoven et al. [210, 218] for confining molecules in their ground state.

The development of the Stark deceleration techniques was exploited for the design of a storage ring for neutral molecules [219], first realized by Crompvoets et al. [220]. This work has recently culminated in the demonstration of the molecular synchrotron [221–223]. The molecular synchrotron is a ring-shaped trap, in which packets of neutral molecules can be accelerated or decelerated with extremely high degree of control. Since multiple packets of molecules trapped in the synchrotron can be sent in the same or opposite directions and made to interact multiple times, the molecular synchrotron can be thought of as the ultimate molecule collider. It is expected to find a huge array of applications in the area of molecular collision dynamics [224].

6.5 Magnetic Traps

Historically, the first experiment on confining a *neutral* microscopic particle was done with magnetic fields [225]. This was a 1977 experiment by Kügler et al. who demonstrated the trapping of neutrons in a torroidal superconducting magnetic ring for up to 20 min. Eight years later, Migdall et al. reported the first magnetic trap for neutral atoms [226]. Another 13 years had to pass before magnetic trapping of a neutral molecule was reported by Weinstein et al. in the research group of Doyle and coworkers [227]. These time delays were due to the technical difficulties that had to be overcome in order to cool the translational motion of neutral atoms and molecules to energies below the trap depth offered by magnetic traps. While sources of "ultracold" slowly moving neutrons had been available since 1970s, it is only with the invention of laser cooling techniques [228] that the magnetic trapping of neutral atoms became possible. Unfortunately, laser cooling is difficult to extend to molecules.[2] In order to trap molecular radicals in a magnetic trap, Weinstein and coworkers had to develop a new method for cooling molecules and design powerful magnetic traps based on superconducting magnets.

Since Eq. (6.19) prohibits magnetic field maxima in free space, only atoms and molecules in *low-field-seeking* Zeeman states can be trapped in a static magnetic trap. The simplest possible design of a magnetic trap is illustrated in Figure 6.9. The trapping field is generated by a pair of parallel circular coils running current in the opposite directions (anti-Helmholtz configuration). The magnetic field thus generated is zero at the center of the area between the coils and increases linearly in all directions from the center. If the currents in the coils are run in the same direction (Helmoltz configuration), they produce a uniform magnetic field in the area between the coils. Now, the problem with magnetic traps is that one needs a lot of current to generate strong trapping fields.

In the pioneering experiment on magnetic trapping of sodium atoms, Migdall et al. [226] used coils with a mean radius of 2.7 cm and carrying currents of 1900 A. The heat produced by the currents was relieved with fans and wet

---

2 This statement, although common 10 years ago, must now be made with caution, due to the work from the group of David DeMille at Yale University, demonstrating that laser cooling of molecules is indeed feasible [229].

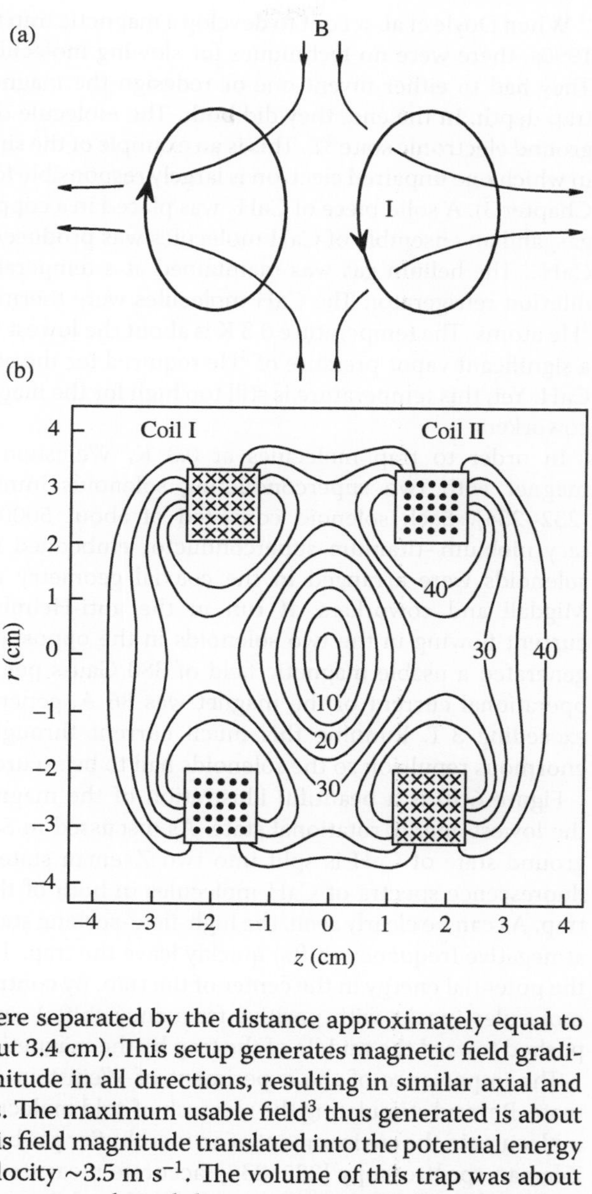

Figure 6.9 Magnetic trap for neutral atoms. (a) Spherical quadrupole trap with lines of the magnetic field. (b) Equipotentials of the trap (with field magnitudes indicated in millitesla) in a plane perpendicular to the coils. *Source*: Reproduced with permission from Phillips 1998 [230]. © 1998, the American Physical Society.

towels [231]. The coils were separated by the distance approximately equal to 1.25 of the coil radii (about 3.4 cm). This setup generates magnetic field gradients of about equal magnitude in all directions, resulting in similar axial and radial confinement forces. The maximum usable field[3] thus generated is about 0.025 T. For Na atoms, this field magnitude translated into the potential energy 0.017 K, equivalent to velocity $\sim$3.5 m s^{-1}. The volume of this trap was about 20 cm^3. A beam of Na atoms was slowed down to energies below 0.017 K by laser cooling [228].

---

3 The maximum usable field is the maximum field for which the equipotential surface does not pass through any part of the apparatus.

When Doyle et al. set out to develop a magnetic trap for molecules in the early 1990s, there were no techniques for slowing molecules to such low energies. They had to either invent one or redesign the magnetic trap to increase the trap depth. In the end, they did both. The molecule of choice was CaH in the ground electronic state $^2\Sigma$. This is an example of the simplest molecular radical, in which one unpaired electron is largely responsible for the Zeeman effect (see Chapter 3). A solid piece of CaH_2 was placed in a copper cell filled with helium gas, and an ensemble of CaH molecules was produced by the laser ablation of CaH_2. The helium gas was maintained at a temperature of about 0.3 K by a dilution refrigerator. The CaH molecules were thermalized by collisions with ^{3}He atoms. The temperature 0.3 K is about the lowest temperature allowing for a significant vapor pressure of ^{3}He required for the efficient thermalization of CaH. Yet, this temperature is still too high for the magnetic trap of Migdall and coworkers.

In order to trap molecules at 0.3 K, Weinstein and coworkers used a magnet with two superconducting solenoids immersed in liquid helium [232–234]. Each solenoid consisted of about 5000 turns of multifilamentary niobium–titanium superconductor embedded in a copper matrix. The solenoids were arranged in the coaxial geometry as in the experiment of Migdall and coworkers. If run in the anti-Helmholtz configuration with current flowing in the two solenoids in the opposite directions, this magnet generated a usable magnetic field of 380 Gauss per Ampere. The maximum operational current of the magnet was 86 A, generating the trapping fields exceeding 3 T. Running this much current through the solenoids induces enormous repulsion so the solenoids had to be secured by a titanium cask.

Figure 6.10 is a beautiful illustration of the magnetic trapping of CaH in the lowest energy rotational state. As discussed in Section 3.1, the rotational ground state of CaH is split into two Zeeman states. Figure 6.10 shows the fluorescence spectra of CaH molecules in both of these states in a magnetic trap. As can be clearly seen, the high-field-seeking states (represented by peaks at negative frequency shifts) quickly leave the trap. They feel the maximum of the potential energy in the center of the trap. By contrast, the low-field-seeking molecules (peaks with positive frequency shifts) remain in the trap and are pushed toward the middle of the trap by the magnetic field gradients.

The importance of the experiment of Weinstein et al. is hard to overestimate. Beyond stimulating the research of cold molecules, magnetic trapping of molecules in the environment of an inert buffer gas became a useful tool for the Zeeman spectroscopy [235, 236], accurate measurement of the radiative lifetimes of long-lived molecular levels [212], and the study of atomic and molecular collisions at temperatures near and below 1 K. In addition to CaH, the experiments demonstrated the trapping of CrH and MnH [237] as well as NH [238, 239] radicals. Long lifetimes of magnetically trapped molecules exceeding 20 s have been achieved by subsequent authors [239]. Vanhaecke et al. [240] trapped Cs_2 molecules in an exited, metastable $^3\Sigma_u^+$ state

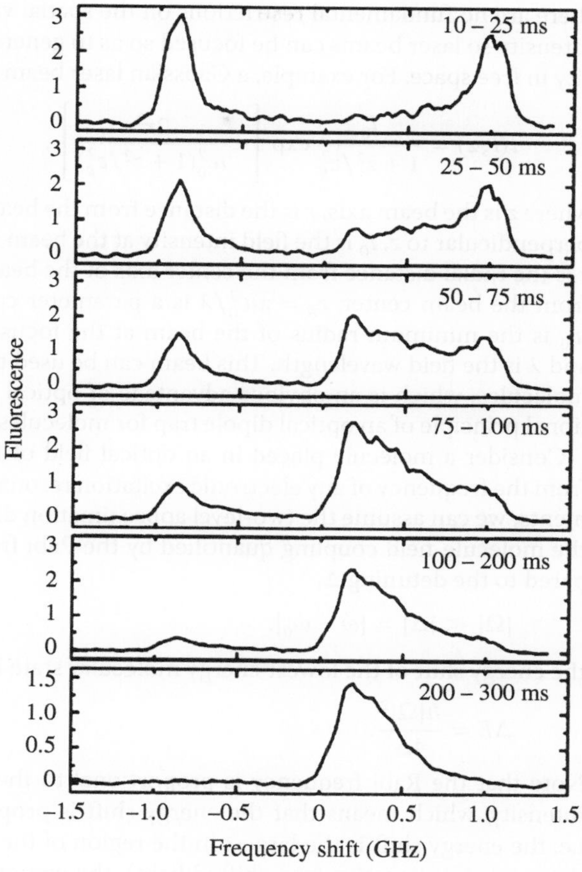

Figure 6.10 Time evolution of the CaH spectrum in a magnetic trap. These spectra reveal that CaH molecules in the high-field-seeking state (negative frequency shifts) quickly leave the trap. The trapped molecules in the low-field-seeking state (positive frequency shifts) are confined and compressed toward the center of the trap. *Source*: Reproduced with permission from Weinstein et al. 1998 [227]. © 1998, Nature Publishing Group.

in a quadrupole magnetic trap, holding the molecules for about one second. Hogan et al. [241] demonstrated the loading of magnetic atoms in a trap after Zeeman deceleration, a technique that can be likewise used for molecules. Riedel et al. [242] demonstrated that multiple packets of molecules from a pulsed molecular beam can be loaded in the same magnetic trap, thus resulting in accumulation of molecules and increased phase space density. In this experiment, the NH molecules were decelerated while in an electronically excited state. Once the molecules reached the center of the magnetic trap, they were optically pumped to another electronically excited state, which decays spontaneously to the ground state.

6.6 Optical Dipole Trap

The AC Stark effect can be used to trap molecules by laser fields. An optical dipole trap is simply a focused laser beam. Unlike in the case of static traps, there are no fundamental restrictions on the spatial variation of the laser field intensity so laser beams can be focused so as to generate a maximum of intensity in free space. For example, a Gaussian laser beam has the intensity profile

$$I(r,z) = \frac{I_0}{1 + z^2/z_R^2} \exp\left[-\frac{2r^2}{w_0^2(1 + z^2/z_R^2)}\right], \tag{6.22}$$

where z is the beam axis, r is the distance from the beam center in the direction perpendicular to z, I_0 is the field intensity at the beam center at the focus point, r is the radial distance from the center axis of the beam, z is the axial distance from the beam center, $z_R = \pi w_0^2/\lambda$ is a parameter called the Rayleigh range, w_0 is the minimum radius of the beam at the focus point (the beam waist), and λ is the field wavelength. This beam can be used to trap high-field-seeking molecules, which is an obvious advantage of optical dipole traps. The operational principle of an optical dipole trap for molecules is the same as for atoms.

Consider a molecule placed in an optical field of frequency ω far detuned from the frequency of any electronic excitation resonance. For qualitative arguments, we can assume the two-level approximation discussed in Section 4.4. If the molecule-field coupling quantified by the Rabi frequency Ω is weak compared to the detuning Δ,

$$|\Omega| \ll |\Delta| = |\omega - \omega_0|, \tag{6.23}$$

the energy shift of the lowest energy molecular state is given by

$$\Delta E = \frac{\hbar|\Omega|^2}{4\Delta}. \tag{6.24}$$

Note that the Rabi frequency is proportional to the square root of the field intensity, which means that the energy shift is proportional to the intensity, i.e. the energy shift is the largest in the region of the most intense field. If the detuning Δ is negative (red-shifted light), the energy shift is negative, i.e. the state is high-field seeking. If the detuning is positive (blue-shifted light), then the energy shift is positive, which makes this molecular state low-field seeking. Thus, it is possible – in principle – to convert a high-field-seeking state into a low-field-seeking state simply by changing the frequency of the applied field. In practice, molecules often possess a dense manifold of excited states that are coupled directly to the ground state. The response of the molecule to the laser field is determined by a combination of contributions from all these couplings, as follows from Eq. (4.94). This may make applying blue-detuned light problematic. However, it should always be possible to decrease the potential energy of a molecule in the ground electronic state by applying an optical field that is red detuned from all excited states.

The possibility of confining polarizable diatomic molecules by a laser field was proposed by Friedrich and Herschbach [36] and first demonstrated for Cs_2 molecules in the experiment of Takekoshi et al. [243]. The molecules – Cs dimers – were produced by the photoassociation of ultracold Cs atoms trapped in a magneto-optical trap and confined at the focus of a linearly polarized beam of a CO_2 laser having the wavelength 10.6 μm and power 17 W. The laser beam was focused to 64 μm, creating a dipole trap depth for Cs molecules of about 350 μK. In more recent years, optical trapping has become a widely used tool for trapping such species as KRb, LiCs, RbCs, as well as a broad range of homonuclear diatomic molecules, such as Cs_2, Sr_2, Rb_2, K_2 [153, 244–250].

As Eq. (6.23) suggests the depth of an optical trap depends on both Ω and Δ. While the detuning Δ appears in the denominator, there are practical limitations restricting the trap depth. The detuning must be large enough and the Rabi frequency small enough to prevent the loss of molecules due to light-induced electronic transitions. The typical depth of an optical dipole trap is below 1 microKelvin.

6.7 Microwave Trap

Microwave traps have not yet been realized but they are envisioned to be a wonderful analog of an optical dipole trap [251]. The operational principle of a microwave trap is the same as for an optical trap. Trapping molecules is based on the AC Stark shifts, this time induced by off-resonant microwave fields. The effect of an off-resonant microwave field on the rotational energy levels of molecules was discussed in Section 4.7. The trapping field can be produced by a confocal microwave cavity. The confocal shape of the mirrors focuses the field intensity in the center of the cavity, producing a Gaussian profile of the microwave field [251]. As in the case of optical traps, red-detuned microwave fields can be used to trap molecules in high-field-seeking states.

There are two potential advantages of a microwave trap over optical dipole traps. The first is the size of the trapping volume. Microwave cavities can be tens of centimeter in size. For example, the design of a microwave trap proposed in the original work by DeMille et al. [251] is based on a Fabry–Perot cavity with the separation between the mirrors equal to $L = 21.5$ cm, using a microwave field at frequency $\omega = 2\pi \times 15$ GHz, with wavelength 2 cm. This yields the beam waist size $\omega_0 = 2.7$ cm at the center of the cavity.

Even more important is the potentially large trap depth of microwave traps. Because the AC Stark effect induced by a microwave field is due to mixing of the rotational states and the lifetime of excited rotational states is very long, it is possible to apply fields with high intensity and very low detuning without triggering substantial losses of molecules. Assuming an input power of 2 kW (which, as mentioned in Ref. [251], is the power of microwave field commonly available from klystron-based satellite communication amplifiers) and perfect mode matching at the input of the field into the cavity, DeMille and coworkers estimated the peak electric field at the center of the cavity to be 28 kV cm^{-1}.

This much field can change the potential energy of molecules by more than 1 K! For example, the depth of the trap proposed by DeMille and coworkers would be 2.8 K for a highly polar molecule SrO, possessing the dipole moment 8.9 Debye.

6.8 Optical Lattices

At the focus of intense current research is the development of experiments for trapping molecules in optical lattices. These experiments are stimulated by the promise of synthesizing highly controllable quantum many-body systems made of molecules suspended by light. An optical lattice is a standing wave pattern of an electric field generated by the interference of two or more laser beams propagating in different directions. An atom or molecule placed in this field experiences a change of potential energy, which varies periodically in space. The electric field patterns – and hence the potential energy landscape experienced by the atom or molecule – can be controlled by varying the polarization, frequency, or direction of the superimposed oscillating fields. For certain cases, as discussed below, the optical lattice potentials can also be varied by changing the phase of the laser beams.

As a simple example, consider two counter-propagating plane waves of electric field, having the same polarization and frequency,

$$\boldsymbol{E}_1(z,t) = \xi \mathrm{Re}[\epsilon \mathrm{e}^{-i(\omega t - kz - \varphi_1)}] \tag{6.25}$$

$$\boldsymbol{E}_2(z,t) = \xi \mathrm{Re}[\epsilon \mathrm{e}^{-i(\omega t + kz - \varphi_2)}]. \tag{6.26}$$

The total field is the sum of the two [252]:

$$\boldsymbol{E}(z,t) = 2\xi \cos\left(kz + \frac{\varphi_1 - \varphi_2}{2}\right) \mathrm{Re}\left[\epsilon \mathrm{e}^{-i(\omega t - \frac{\varphi_1+\varphi_2}{2})}\right]. \tag{6.27}$$

The potential energy of a molecule placed in this field is given by Eq. (4.94) so it can be written as

$$U(z) \propto \xi^2 \cos^2\left(kz + \frac{\varphi_1 - \varphi_2}{2}\right). \tag{6.28}$$

The potential is thus a function of the spatial coordinate z exhibiting periodic minima or maxima separated by half the wavelength $\lambda/2$. The possibility of trapping atoms and molecules in a standing light wave was proposed in 1976 in a visionary article by Letokhov et al. [253].

While the lattice constant of the potential (6.27) can be varied by changing the frequency of the oscillating fields, the topology of the potential does not depend on the detuning of the laser frequency. This means that both red-detuned and blue-detuned fields can be used to form an optical lattice. Red-detuned fields will confine atoms or molecules at the bottom of the potential wells. Blue-detuned fields will confine atoms or molecules at the nodes of the potential. Note also, that the topology of the potential remains unchanged under the variation of the phases φ_1 and φ_2. Changing the phase difference merely shifts the potential. This means that the simple one-dimensional lattice potential (6.27) cannot be tuned by changing the phase of the laser beams. The potential does depend on the relative polarization of the counter-propagating beams, which can be used to form optical lattice potentials of various forms and shapes, especially in two and three dimensions.

The one-dimensional periodic potential (6.27) can be extended to two or three dimensions by superimposing more than two laser beams. The interference of multiple oscillating electric fields

$$E_i(r,t) = \xi_i \mathrm{Re}[\epsilon_i e^{-i(\omega t - k_i \cdot r - \varphi_i)}] \tag{6.29}$$

with different amplitudes ξ_i, propagation directions k_i and polarizations ϵ_i yield an electric field pattern leading to the dipole potential [252]

$$U(r) \propto \sum_{i=1} \xi_i^2 + \sum_{i \neq j} \xi_i \xi_j (\epsilon_i^* \cdot \epsilon_j) e^{i[(k_i - k_j)\cdot r + \varphi_i - \varphi_j]}. \tag{6.30}$$

If the fields have the same amplitude ξ and the same polarizations, this reduces to

$$U(r) \propto \xi^2 \left\{ N + 2 \sum_{i>j} e^{i[(k_i - k_j)\cdot r + \varphi_i - \varphi_j]} \right\}, \tag{6.31}$$

where N is the number of the superimposed laser beams.

Consider now a special case of *three* electric field waves, all propagating in the (x, y)-plane and all having the same polarizations and amplitudes. If the wave vectors k_2 and k_3 propagate symmetrically with respect to k_1

$$k_1 = k\hat{x} \tag{6.32}$$

$$k_2 = k(\hat{x}\cos\theta + \hat{y}\sin\theta) \tag{6.33}$$

$$k_3 = k(\hat{x}\cos\theta - \hat{y}\sin\theta), \tag{6.34}$$

then the dipole potential due to the superposition of these three fields is [252]

$$U(x,y) \propto \xi^2 \left[3 + 2\cos(2K_\perp y + \varphi_2 - \varphi_3) + 4\cos\left(Kx + \varphi_1 - \frac{\varphi_2 + \varphi_3}{2}\right) \times \cos\left(K_\perp y + \frac{\varphi_2 - \varphi_3}{2}\right)\right], \tag{6.35}$$

where $K_\perp = k\sin\theta$ and $K = k(1 - \cos\theta)$. This potential is periodic in both the x- and y-directions. The periodicity in the x-direction is $\lambda_x = \lambda/\sin\theta$. The periodicity in the y-direction is $\lambda_y = \lambda/(1 - \cos\theta)$. For $\theta = \pi/2$, this is a square lattice. For $\theta = \pi/3$ or $2\pi/3$, this is a hexagonal lattice (Figure 6.11). It is evident from Eq. (6.35) that changing the phase φ_1 shifts the potential along x but does not change the topology of the potential surface. Similarly, changing the

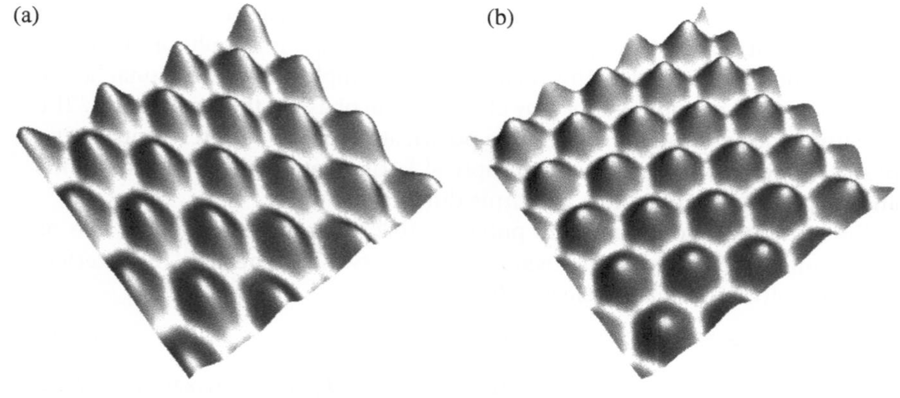

Figure 6.11 Two-dimensional optical lattice potential produced by interfering three oscillating electric fields propagating in the same plane along the directions given by Eq. (6.34). The potential can be controlled by changing the angle between the wave vectors of the interfering fields. The potential is shown for $\theta = \pi/2$ (a) and $\theta = 2\pi/3$ (b).

phases φ_2 or φ_3 shifts the potential along both x and y without changing the overall topology.[4]

In the case of four overlapping laser beams, the two-dimensional potential becomes phase-dependent. Consider, for example, two counter-propagating fields with $\boldsymbol{k}_1 = -\boldsymbol{k}_3 = k\hat{x}$ and $\boldsymbol{k}_2 = -\boldsymbol{k}_4 = k\hat{y}$. This leads to a dipole potential [252]

$$
\begin{aligned}
U(x,y) \propto \xi^2 \Bigg[& \cos^2\left(\boldsymbol{k}_1 \cdot \boldsymbol{r} + \frac{\varphi_1 - \varphi_3}{2}\right) + \cos^2\left(\boldsymbol{k}_2 \cdot \boldsymbol{r} + \frac{\varphi_2 - \varphi_4}{2}\right) \\
& + 2\cos\left(\frac{\varphi_1 + \varphi_3 - \varphi_2 - \varphi_4}{2}\right) \cos\left(\boldsymbol{k}_1 \cdot \boldsymbol{r} + \frac{\varphi_1 - \varphi_3}{2}\right) \\
& \times \cos\left(\boldsymbol{k}_2 \cdot \boldsymbol{r} + \frac{\varphi_2 - \varphi_4}{2}\right)\Bigg]. \qquad (6.36)
\end{aligned}
$$

The shape of this potential depends on the relative phase of the electric field waves. In particular, having fixed the phase differences $\varphi_1 - \varphi_3$ and $\varphi_2 - \varphi_4$, one can change the topology of the potential by varying the value of $\varphi_1 + \varphi_3 - \varphi_2 - \varphi_4$.

In three dimensions, an optical lattice potential can be produced by interfering a minimum of four electric field waves. The topology of the potential of four laser beams is phase independent. If more than four beams are employed to create an optical lattice, it is necessary to control the phases of the electric fields in order to preserve the topology of the potential [252].

4 The phases of the laser beams correspond to a specific choice of the origin of the spatial coordinate system and time. Since any such choice is arbitrary, it is always possible to choose the origin of time to eliminate the phase of one laser field.

Trapping molecules in an optical lattice is challenging, primarily because the trap depth of optical lattice potentials is very low, usually below 1 microkelvin.

6.9 Some Applications of External Field Traps

Table 6.1 lists most of the molecules than have been trapped in various experiments and the typical parameters of the traps. This table is not all inclusive. The field is developing very quickly and new traps of new kinds of molecules appear in the literature often. The purpose of Table 6.1 is to provide an overview of the state of the art in this research field.

Confining molecules in external field traps can be used as a tool for numerous applications. The following are some of the most apparent and prominent applications.

- First of all, external field traps are used to isolate molecules from their thermal environment. This is a necessary step in most cooling experiment aiming to produce molecular ensembles at temperatures below 1 K. Once molecules are trapped, they can be cooled to an ultracold temperature by a process involving dissipation of energy. Atoms are often cooled to ultracold temperatures[5] by evaporative cooling. Molecules can be cooled to ultracold temperatures either by evaporative cooling or by sympathetic cooling, i.e. by immersing molecules in a bath of ultracold atoms.[6] Cooling trapped molecular radicals to ultracold temperatures is expected to lead to radically new developments in science and technology [268]. More on this in subsequent chapters.
- External field traps confine atoms and molecules in specific Zeeman or Stark states, whether low-field seeking or high-field seeking. As clear from the previous chapters, these states are characterized by a well-defined projection of some angular momentum on the trapping field direction. By preselecting molecules in such quantum states, external field traps effectively orient either the electron spins or the molecular axes. As discussed in Chapter 5, this can

5 Ultracold temperature in reference to atomic and molecular gases is often thought of as temperature below 1 mK. More rigorously, the division between the cold and ultracold temperature regimes is based on the number of partial waves contributing to the scattering cross section of specific atoms or molecules under consideration [267]. Each partial wave is characterized by a well-defined angular momentum for the end-over-end rotational motion of the colliding particles. The lower the collision energy, the smaller the number of partial waves contributing to the scattering cross section. The ultracold temperature regime is often defined as temperatures, at which the scattering cross section is determined by a single partial wave contribution (see Chapter 8). For many atomic and molecular species, this happens at temperatures below $\approx 10^{-3}$ K.

6 More on the details and challenges of evaporative and/or sympathetic cooling of molecules in Chapter 11.

Table 6.1 Summary of the external field traps developed for neutral molecules as of 2013. Only selected representative references are given. See text for a more comprehensive list of references.

Trap type	Molecule	Trap depth (K)	Molecule number	Density (cm^{-3})	Representative references
Electrostatic	OH, OD, CO($a^3\Pi_1$), NH_3, ND_3	0.1–1.2	10^4–10^8	10^6–10^8	[121, 165, 167, 212, 254]
	NH($a^1\Delta$), CH_3F, CH_2O, CH_3Cl				[213, 255, 256]
Chip electrostatic	CO($a^3\Pi_1$)	0.03–0.07	10^3–10^4	10^7 (10/site)	[180]
Thin-wire electrostatic	NaCs	5×10^{-6}	10^4–10^5	10^5–10^6	[214]
AC electric	$^{15}ND_3$	5×10^{-3}	10^3–10^4	$\sim 10^5$	[210, 216–218]
Magneto-electrostatic	OH	0.4	10^5	3×10^3	[215]
Magnetic	OH	0.2	10^5	10^6	215, 257, 258
Magnetic	CaH, NH	~1	10^8–10^{10}	10^7–10^{10}	227, 238, 242
Magnetic	CrH, MnH	1	10^5	10^6	[237]
Magnetic	Cs_2	4×10^{-6}	10^5	$<10^8$	[240]
Magnetic	KRb	0.5	~20–30	10^4	[259]
Optical dipole trap	Alkali dimers, Sr_2	10^{-6}	10^5	10^{12}	[243–247, 260]
					[248–253, 261–263]
Optical lattice	KRb, Cs_2, Sr_2	10^{-6}	10^4	(0.1–1)/site	249, 250, [264–266]

Source: Reproduced with permission from Lemeshko et al. 2013 [1]. © 2013, Taylor & Francis.

be used for a variety of applications such as controlled chemistry [269] or novel spectroscopy of large oriented molecules [270–272].

- Holding molecules in an external field trap for a long time provides long interrogation time for precision spectroscopy experiments [273–279] and accurate measurements of the radiative lifetimes, for example, of vibrationally excited states [212, 280]. The latter is especially important for free radicals, for which there are few alternative methods. Coupled with the cooling experiments, improvements in the precision of the spectroscopic measurements may lead to far-reaching applications. It is expected that experiments with ultracold molecules will provide a new, much improved, limit on the value of the electric dipole moment of the electron [281–284] and the time variation, if any, of the fundamental constants [285–288].
- Trapping molecules in optical lattices offers an opportunity to study condensed-matter physics in previously unimaginable ways. The quantum theory of the solid state is based on lattice models. These models often lead to simple Hamiltonians. However, the dimension of the Hilbert spaces required to diagonalize these Hamiltonians is often prohibitively large so accurate solutions of the Schrödinger equation cannot be obtained except for a few simple cases in one dimension. By interrogating atoms and molecules trapped in optical lattices, it is possible to study exact quantum phenomena described by the lattice models. Moreover, the trapping potential of optical lattices is highly tunable, allowing for the possibility of creating lattice systems with various geometries and the possibility of studying the eigenstates of the lattice models as functions of the Hamiltonian parameters. In this way, atoms and molecules trapped in optical lattices can be used to study quantum phase transitions. Compared to ultracold atoms, the added bonus of molecules is their polarity, which enables long-range interactions between particles on a lattice, thereby allowing the design of a wider class of lattice models.
- Trapping molecules also provides stationary targets for molecular scattering experiments.

In conclusion of this chapter, it is important to point out that the experiments on trapping molecules are still at a relatively early stage of development and much more is expected to happen in the next few years, leading to discoveries and new applications not yet foreseen.

Exercises

6.1 Calculate the average velocity (in meter per second) of NH_3 molecules in a gas at room temperature, at 1 °K and at 10^{-6} °K. Compare this velocity with the speed of a bullet fired from a typical gun.

6.2 Prove that spin-1/2 particles can be trapped by a static magnetic field only if they are in a low-field-seeking state.

6.3 Consider a gas of oxygen molecules $^{16}O_2$ made of two ^{16}O atoms. If the molecules are trapped in a static magnetic field and are in a thermal equilibrium, what is the lowest energy state present in the trap? How does the answer change for $^{17}O_2$ molecules made of two ^{17}O atoms? Is the answer different for $^{17}O^{16}O$?

6.4 Which of the following molecules cannot be confined in a magnetic trap and why? (O_2, C_2, N_2, CH_3, CaF).

7

Molecules in Superimposed Fields

We have seen that

- Magnetic fields induce interactions that couple states of the same parity[1];
- Electric fields couple states of the opposite parity[2];
- Off-resonant laser fields couple rotational states of the same parity.[3] These couplings bring pairs of the opposite parity states into closely spaced tunneling doublets.[4]

These features of the molecule–field interactions suggest interesting possibilities for controlling molecules with *combined* fields. In this chapter, I will show that magnetic fields can bring molecular states of opposite parity into degeneracy, thus allowing for first-order Stark effect in diatomic molecules, even at very low electric fields. We will also discuss the enhanced orientation of molecules by DC electric fields in the presence of an off-resonant laser field and the effect of combined fields on interactions between molecules.

7.1 Effects of Combined DC Electric and Magnetic Fields

7.1.1 Linear Stark Effect at Low Fields

In Section 2.3, we argued that the Stark shifts of molecular energy levels in diatomic molecules are always quadratic at low electric fields and become linear at high electric fields. We argued that there are no truly degenerate states of different parity in diatomic molecules. We were wrong!

Consider a molecule in a $^2\Sigma$ electronic state. In such molecules, the electron spin is coupled to the molecular axis by the spin–rotation interaction,

1 Chapter 3.
2 Chapter 2.
3 Section 4.6.
4 Section 5.1.4.

which is typically very weak. If a $^2\Sigma$ molecule is placed in a magnetic field, the magnetic-field-induced interaction quickly overtakes the spin–rotation interaction so the Zeeman spectrum of the molecule consists of doublets of the spin-up and spin-down states, with "up" and "down" referring to the orientation of the electron spin with respect to the magnetic field axis. At the same time, because neither the magnetic-field-induced interaction nor the spin–rotation interaction mixes different rotational states, each Zeeman doublet is associated with a particular quantum number of the rotational angular momentum N.

The Zeeman shifts of the rotational energy levels of $^2\Sigma$ molecules were discussed in Section 3.5.1. Take a close look at Figure 3.1. That figure shows that the low-field-seeking state (state with energy going upward) stemming at zero magnetic field from the $N = 0$ rotational state crosses with several high-field-seeking Zeeman states (states with energy going downward) stemming from the $N = 1$ rotational state. The parity of the rotational states in a $^2\Sigma$ molecule is determined by the factor $(-1)^N$. The intersecting Zeeman states thus have the opposite parity. The crossings between these states are real, because parity is conserved for molecules in a magnetic field and, consequently, there are no interactions coupling the intersecting states.

If the molecule is now subjected to an electric field, different parity states interact and the crossings become avoided. This is illustrated in Figure 7.1, where two Zeeman states of the opposite parity are labeled β and γ. These states are coupled directly by the electric-field-induced interaction, as can be inferred from Eq. (2.45). This means that the energy separation between β and γ at the avoided crossing increases with the strength of the electric field. The avoided crossings analogous to the ones shown in Figure 7.1 were recently imaged in the experiments by Cahn et al. [290].

The close examination of Figure 7.1 leads us to the following observations:

- The positions of the crossings between the opposite parity states depend on the rotational constant of the molecule and the strength of the spin–rotation interaction. Molecules with smaller rotational constants exhibit the crossings at lower magnetic fields.
- At the particular strengths of the magnetic field corresponding to the crossings, molecules will exhibit the linear Stark effect, even in the limit of very weak fields. The energy level structure of molecules at these magnetic fields must be very sensitive to small variations in the electric field.
- By applying a DC electric field and tuning the magnetic field strength adiabatically, it is possible to convert molecules from the magnetic-low-field-seeking state (state β at low magnetic fields) to the magnetic high-field-seeking state (state β at high magnetic fields). This corresponds to changing the orientation of the molecule's magnetic moment with respect to the magnetic field axis.

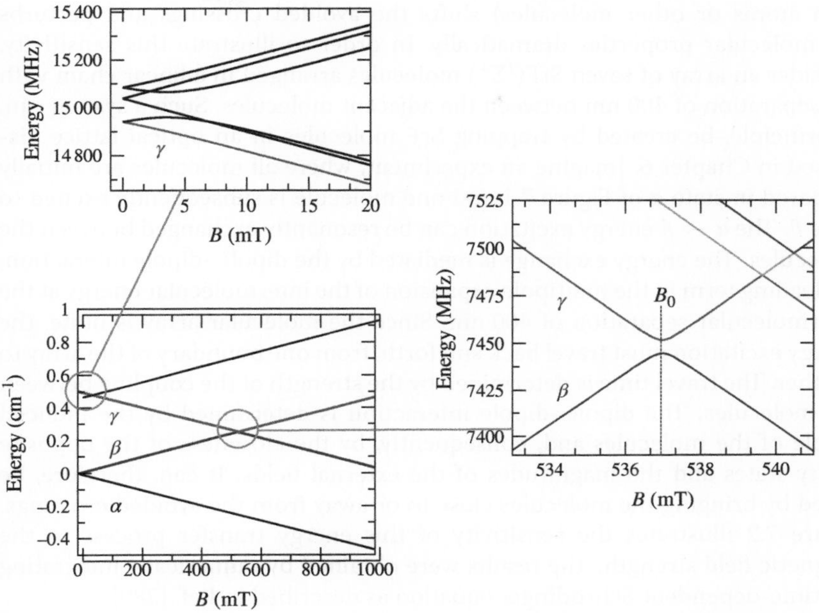

Figure 7.1 Energy levels of the SrF($^2\Sigma^+$) molecule as functions of a magnetic field in the presence of an electric field of 1 kV cm^{-1}. The rotational constant of SrF is 0.251 cm^{-1}, the spin–rotation interaction constant γ_{SR} is 2.49 × 10^{-3} cm^{-1}, and the dipole moment is 3.47 D. States β and γ undergo an avoided crossing at the magnetic field value B_0. The value of B_0 varies with the electric field. *Source*: Adapted with permission from Rios et al. 2010 [289]. © 2010, the Institute of Physics.

- Molecules subject to electric and magnetic fields near the avoided crossings depicted in Figure 7.1 must be very sensitive to any perturbations that shift the molecular energy levels.
- The effect illustrated in Figure 7.1 is general for open-shell molecules that possess a magnetic moment largely decoupled from the rotational angular momentum of the molecular axis. For such molecules, it should always be possible to bring states of the opposite parity into degeneracy by applying a magnetic field. In particular, the avoided crossings of the type depicted in Figure 7.1 should be expected to occur in molecules in Σ electronic states with any spin multiplicity.

Molecules placed in superimposed electric and magnetic fields near the avoided crossings shown in Figure 7.1 are fragile: a small variation of the fields or a small change of the energy level structure (for example, due to collisions with atoms or other molecules) shifts the avoided crossings and perturbs the molecular properties dramatically. In order to illustrate this sensitivity, consider an array of seven $SrF(^2\Sigma^+)$ molecules arranged in a linear chain with the separation of 400 nm between the adjacent molecules. Such a system can, in principle, be created by trapping SrF molecules in an optical lattice discussed in Chapter 6. Imagine an experiment, where all molecules are initially prepared in state α of Figure 7.1 and one molecule is subsequently excited to state β. The $\alpha \rightarrow \beta$ energy excitation can be resonantly exchanged between the molecules. The energy exchange is mediated by the dipole–dipole interaction, the leading term in the multipole expansion of the intermolecular energy at the intermolecular separation of 400 nm. Since the molecular array is finite, the energy excitation must travel back and forth, from one boundary of the array to another. The travel time is determined by the strength of the coupling between the molecules. The dipole–dipole interaction is determined by the induced dipole of the molecules and, consequently, by the closeness of the opposite parity states and the magnitudes of the external fields. It can, therefore, be tuned by bringing the molecules close to or away from the avoided crossings. Figure 7.2 illustrates the sensitivity of this energy transfer process to the magnetic field strength. The results were obtained by numerically integrating the time-dependent Schrödinger equation as described in Ref. [289].

The fragility of molecules near the avoided crossings induced by superimposed electric and magnetic fields can be used for multiple applications. For example, bringing the molecular states of the opposite parity into degeneracy can be used to enhance the orientation of the molecular axis by a DC electric field [291]. The extreme sensitivity of the molecular energy level structure near the avoided crossings to variations of both the magnetic and electric fields can be used for precision imaging of electromagnetic fields [292]. The sensitivity to external perturbations can be used as a mechanism for controlling molecular collisions [293].

7.1.2 Imaging of Radio-Frequency Fields

An obvious application of the avoided crossings discussed in Section 7.1.1 is for sensitive detection of electromagnetic fields. Of particular interest is the possibility to image AC electric fields with very low frequencies. Low-frequency electromagnetic fields can be detected by measuring the transitions between hyperfine states of atoms [294]. The problem with using hyperfine transitions is that they are between states of the same parity. The hyperfine transitions are therefore electric-dipole forbidden and must be induced by the magnetic component of the electromagnetic field. The electric and magnetic field amplitudes of an electromagnetic field are related as $E = Bc$, where c is the speed of light. The magnetic-field-induced transitions are, therefore, much weaker than the electric-dipole-allowed transitions.

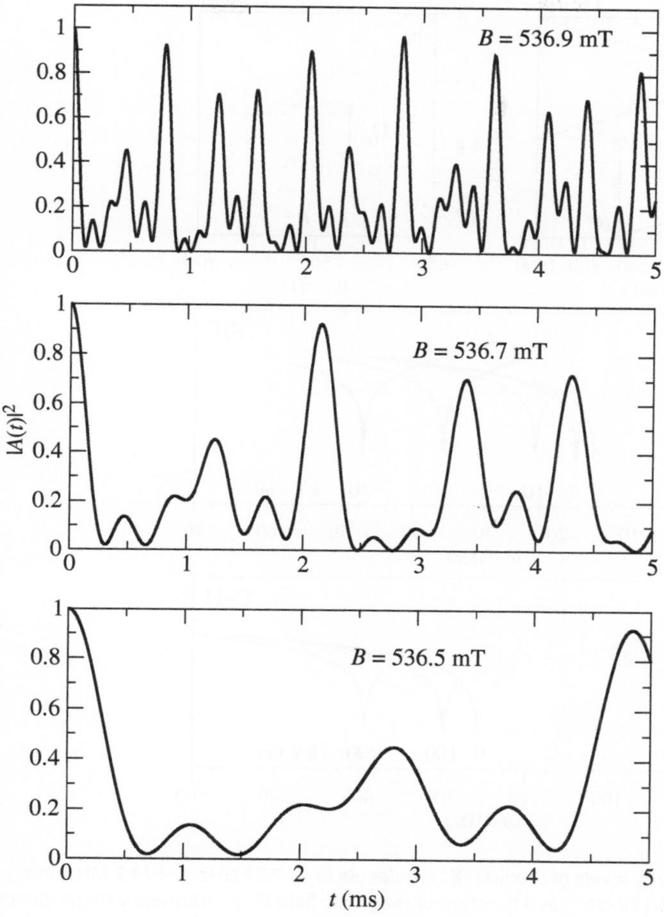

Figure 7.2 Time-dependent probability of the spin excitation to be localized on molecule four in a one-dimensional array of seven $SrF(^2\Sigma^+)$ molecules in an optical lattice with lattice spacing $a = 400$ nm for different magnetic fields near the avoided crossing shown in Figure 7.1. The electric field magnitude is 1 kV cm^{-1}. The value of B_0 varies with the electric field. *Source*: Adapted with permission from Perez-Rios et al. 2010 [289]. © 2010, the Institute of Physics.

The avoided crossings shown in Figure 7.1 provide closely spaced energy levels (β and γ), with each being a superposition of different parity states. The $\beta \rightarrow \gamma$ transitions are, therefore, electric-dipole-allowed. Moreover, the energy separation between these states can be tuned by varying the strength of the applied DC field. This seems ideal for imaging the electric field component of the radio-frequency fields [292].

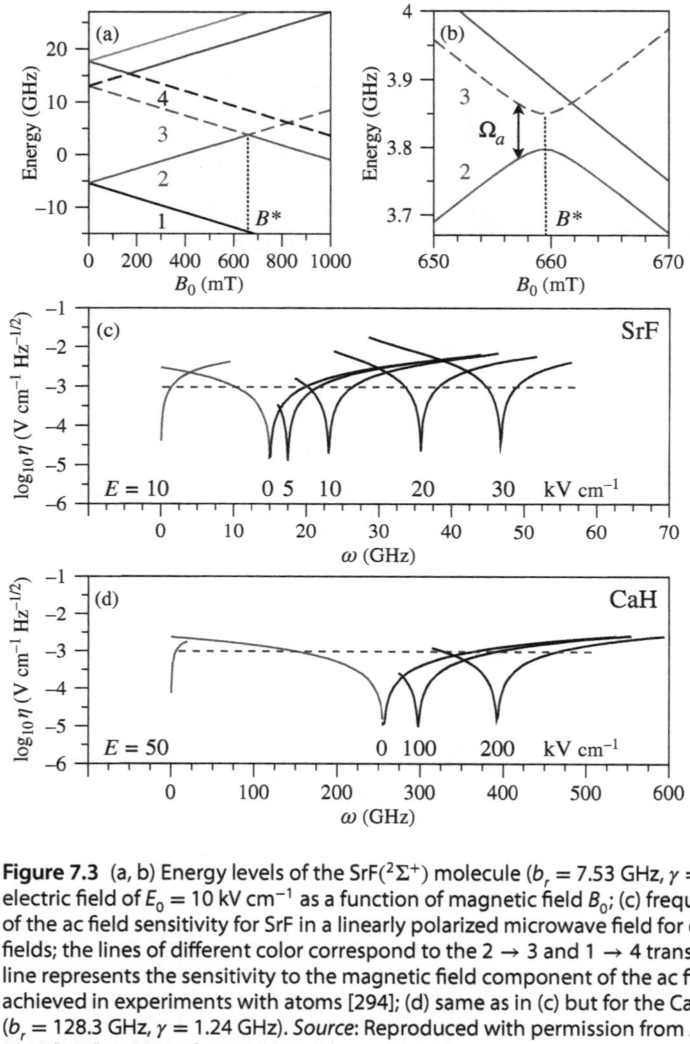

Figure 7.3 (a, b) Energy levels of the SrF($^2\Sigma^+$) molecule ($b_r = 7.53$ GHz, $\gamma = 74.7$ MHz) in an electric field of $E_0 = 10$ kV cm^{-1} as a function of magnetic field B_0; (c) frequency dependence of the ac field sensitivity for SrF in a linearly polarized microwave field for different electric fields; the lines of different color correspond to the 2 → 3 and 1 → 4 transitions. The dashed line represents the sensitivity to the magnetic field component of the ac field that can be achieved in experiments with atoms [294]; (d) same as in (c) but for the CaH($^2\Sigma^+$) molecule ($b_r = 128.3$ GHz, $\gamma = 1.24$ GHz). *Source*: Reproduced with permission from Alyabyshev et al. 2012 [292]. © 2012, the American Physical Society.

As an illustration of this point, I present in Figure 7.3 the predicted sensitivity of a measurement to the AC electric field defined as [294]

$$\eta[\mathrm{V\,cm^{-1}Hz^{-1/2}}] = \frac{2\sqrt{3}\hbar}{100\, d(B,E)\sqrt{nV_{\mathrm{eff}}}\sqrt{\tau}}, \tag{7.1}$$

where n is the density of molecules, $V_{\text{eff}} = 2\pi\sigma_{\text{eff}}^2\rho$ is the effective imaging volume, σ_{eff} is the dispersion of the spatial coordinate, ρ represents the $1/e$ radius of the molecular cloud, τ is the measurement time, and $d(B, E)$ is the transition dipole moment. The transition dipole moment depends on the strength of both the magnetic and electric fields and is dramatically enhanced near the avoided crossings between β and γ.

7.2 Effects of Combined DC and AC Electric Fields

7.2.1 Enhancement of Orientation by Laser Fields

As shown in Figure 5.2, it takes a lot of electric field strength to orient the axis of diatomic molecules effectively. This is mainly because orienting molecules requires mixing the opposite parity states, and the opposite parity states in the case of a structureless rigid rotor correspond to states of different N, often separated by large energy gaps. Using magnetic fields as discussed in Section 7.1.1 or off-resonant laser fields as discussed in Section 5.1.4 to bring the opposite parity states together can be used to make the orientation of molecules by DC electric fields easier [20, 295]. Here, I will illustrate the enhancement of the orientation by an off-resonant laser field.

Combining the discussion of Chapter 2 and Sections 4.6 and 5.1, we can write the general Hamiltonian for a rigid rotor in superimposed DC electric and off-resonant laser fields as

$$\hat{H} = N^2 - \omega \cos\theta_R - \gamma \cos^2\theta_L, \tag{7.2}$$

where θ_R is the angle between the vector of the DC field and the molecular axis and θ_L is the angle between the polarization axis of the laser field and the molecular axis. If the laser field is linearly polarized along the direction of the DC field, $\theta_R = \theta_L = \theta$. This is the case we will consider here. The extension to nonparallel fields is straightforward [20]. The Hamiltonian (7.2) can be diagonalized by expanding the wave function of the molecule in the eigenstates of the rigid rotor, as discussed in Section 5.1.

For a typical molecule with the permanent dipole moment 1 D, the polarizability anisotropy $\Delta\alpha = 10\ \text{Å}^3$ and the rotational constant 1 cm^{-1}, the parameters $\omega = Ed/B_e$ and $\gamma = \xi^2\Delta\alpha/4B_e$ can be in the range between zero and $\omega = 1.7$ for 100 kV cm^{-1} and $\gamma = 2$ for the laser field intensity $\xi^2/2 = 10^9$ W cm^{-2}. Figure 7.4 shows the expectation value $\langle\cos\theta\rangle$ for the lowest-energy state of the rigid rotor as a function of both ω and γ. It can be seen that applying an off-resonant laser field, even if of a moderate strength, leads to a significant improvement of the molecular axis orientation.

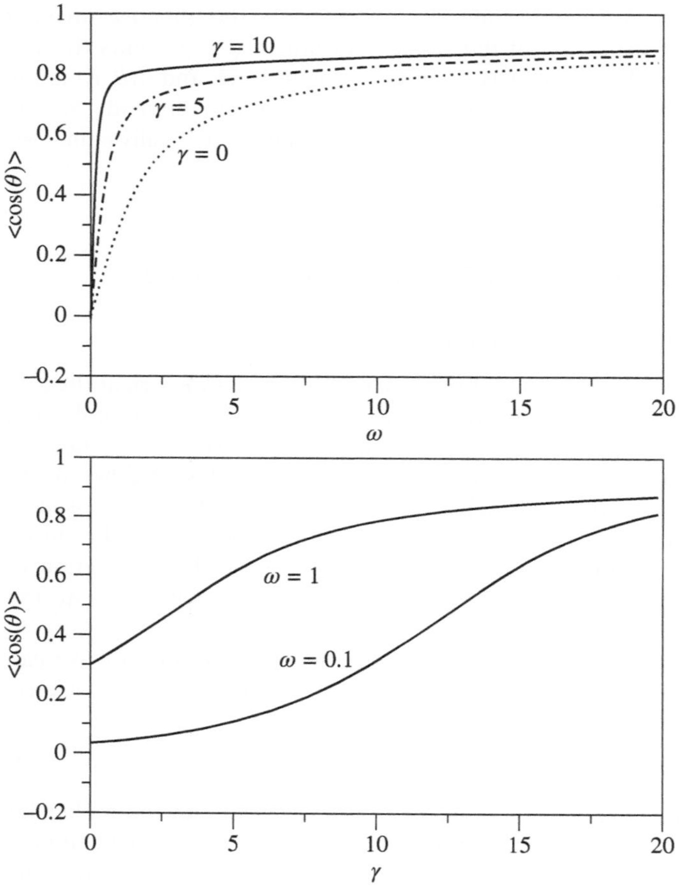

Figure 7.4 Enhancement of the molecular axis orientation by an off-resonant laser field. The last term in Eq. (7.2) brings the opposite parity states into closely spaced tunneling doublets, while the second last term orients the molecule with respect to the direction of the DC field by mixing the opposite parity states. The expectation value $\langle \cos\theta \rangle$ is shown for the ground state of the molecule.

7.2.2 Tug of War Between DC and Microwave Fields

Recent literature abounds with proposals of schemes for controlling interactions of molecules by a combination of DC and microwave electric fields.[5] The aim of that work is to engineer quantum many-body systems with ultracold molecules trapped in optical lattices and allowed to interact via

5 For representative work, see Refs. [296–305].

intermolecular dipole–dipole interactions. If the molecules are separated by a large distance where the intermolecular interactions are perturbatively small, the dipolar interactions can be tuned by tuning the energy level structure of the participating molecules with external fields.

That the dipole–dipole interactions must be sensitive to external electric fields is clear already from the basic considerations in Section 2.1. In the absence of electric fields, the permanent dipole moment of a molecule rotates freely, leading to zero net dipole moment in the space-fixed coordinate frame, and hence zero dipolar interactions. An electric field constrains the rotational motion by disturbing the isotropy of space and induces the dipolar interactions. A single DC electric field is sufficient to induce significant dipolar interactions between molecules in optical lattices. What microwave fields add is versatility.

A single field defines the quantization axis and hybridizes the molecular states with the same projection of the total angular momentum on the field axis. An additional field, be it static or oscillating, can be directed at an angle with respect to the first field. The second field can then hybridize states with different projections of total angular momentum, making the molecular structure more complex and tunable. In addition, the frequency of the microwave field can be chosen to be close to a resonance for the transition between some molecular states, enhancing their mixing and adding to the versatility of control.

As an illustrative example, consider a rigid rotor placed in a superposition of a DC electric and a single-frequency microwave field with circular polarization. The microwave field is assumed to be detuned from any of the rotational transitions, although it can be nearly resonant with some of the transitions. We will chose the z-axis to lie along the direction of the DC electric field and let the microwave field rotate in the xy-plane. The Hamiltonian of this system is

$$\hat{H} = B_e N^2 - \boldsymbol{d} \cdot \boldsymbol{E} - \xi \boldsymbol{d} \cdot (\hat{x} \cos \omega t + \hat{y} \sin \omega t), \tag{7.3}$$

which can be rewritten in an already more familiar form

$$\hat{H} = B_e N^2 - \mathrm{d}E \cos\theta - \frac{\xi}{2}(\boldsymbol{d} \cdot \hat{\epsilon}_{-} e^{i\omega t} + \boldsymbol{d} \cdot \hat{\epsilon}_{+} e^{-i\omega t}), \tag{7.4}$$

where $\hat{\epsilon}_{\pm} = \hat{x} \pm i\hat{y}$ and we have used the identity

$$\hat{x} \cos \omega t + \hat{y} \sin \omega t = \mathrm{Re}[(\hat{x} - i\hat{y})e^{i\omega t}]. \tag{7.5}$$

Note that by shifting the time variable from t to $t + \pi/2\omega$, we can make the transformation

$$\hat{x} \cos \omega t + \hat{y} \sin \omega t \Rightarrow i^2(\hat{x} \sin \omega t + \hat{y} \cos \omega t) = -i\, \mathrm{Im}\, [(\hat{x} + i\hat{y})e^{i\omega t}], \tag{7.6}$$

which would bring the last term in Eq. (7.4) in line with Eq. (4.168). However, it is easier to work with matrix elements that are real so we will use Eq. (7.4).

As a side note, I would like to point out that if a Hamiltonian is invariant under time reversal, it is always possible to choose a phase convention such that the matrix elements of the Hamiltonian are all real.

For the practical purpose of evaluating the matrix elements of the Hamiltonian, it is convenient to write the molecule–field interactions in terms of the spherical harmonics. Given that

$$\cos\theta = \sqrt{\frac{4\pi}{3}} Y_{1,0}(\theta, \phi = 0) \tag{7.7}$$

and that

$$\boldsymbol{d} \cdot \boldsymbol{\epsilon}_{\pm} = \mp d \sqrt{\frac{8\pi}{3}} Y_{1,\pm 1}(\theta, \phi), \tag{7.8}$$

the Hamiltonian (7.4) assumes the following form:

$$\begin{aligned} \hat{H} = B_e N^2 - \mathrm{d}E \sqrt{\frac{4\pi}{3}} Y_{1,0}(\theta, \phi = 0) \\ - \frac{\mathrm{d}\xi}{2} \sqrt{\frac{8\pi}{3}} [Y_{1,-1}(\theta, \phi) \mathrm{e}^{i\omega t} - Y_{1,+1}(\theta, \phi) \mathrm{e}^{-i\omega t}]. \end{aligned} \tag{7.9}$$

Note that the angles θ and ϕ are the conventional spherical polar angles. Eq. (7.9) will permit us to use Eq. (D.23) from Appendix D for the integrals over products of three spherical harmonics in order to evaluate the matrix elements of the Hamiltonian.

In order to find the eigenstates of the Hamiltonian (7.9), we can use the Floquet formalism described in Section 4.2. The time-independent part of the Hamiltonian (7.9) can be diagonalized using the basis of the eigenstates of the field-free molecule. Therefore, we can use the tensor product of the eigenstates of the rigid-rotor $|NM_N\rangle$ and the Fourier component basis $|n\rangle$ of Section 4.2 as a basis for the Floquet Hamiltonian. The matrix elements of the Floquet Hamiltonian in the basis $|NM_N n\rangle$ are

$$\begin{aligned} \langle NM_N n | \hat{H}_{\mathrm{F}} | N' M'_N n' \rangle = \delta_{nn'} \left\{ B_e N(N+1) \delta_{NN'} + \hbar\omega \delta_{NN'} \delta_{M_N M'_N} \right. \\ \left. - \mathrm{d}E \sqrt{\frac{4\pi}{3}} \langle NM_N | Y_{1,0} | N' M'_N \rangle \right\} \\ - \frac{\mathrm{d}\xi}{2} \sqrt{\frac{8\pi}{3}} [\delta_{n'-n,1} \langle NM_N | Y_{1,-1} | N' M'_N \rangle \\ - \delta_{n'-n,-1} \langle NM_N | Y_{1,+1} | N' M'_N \rangle]. \end{aligned} \tag{7.10}$$

The eigenstates of this Hamiltonian represent the molecule in the presence of a DC field of strength E directed along the z-axis and a microwave field of amplitude ξ propagating along the z-axis and having circular polarization.

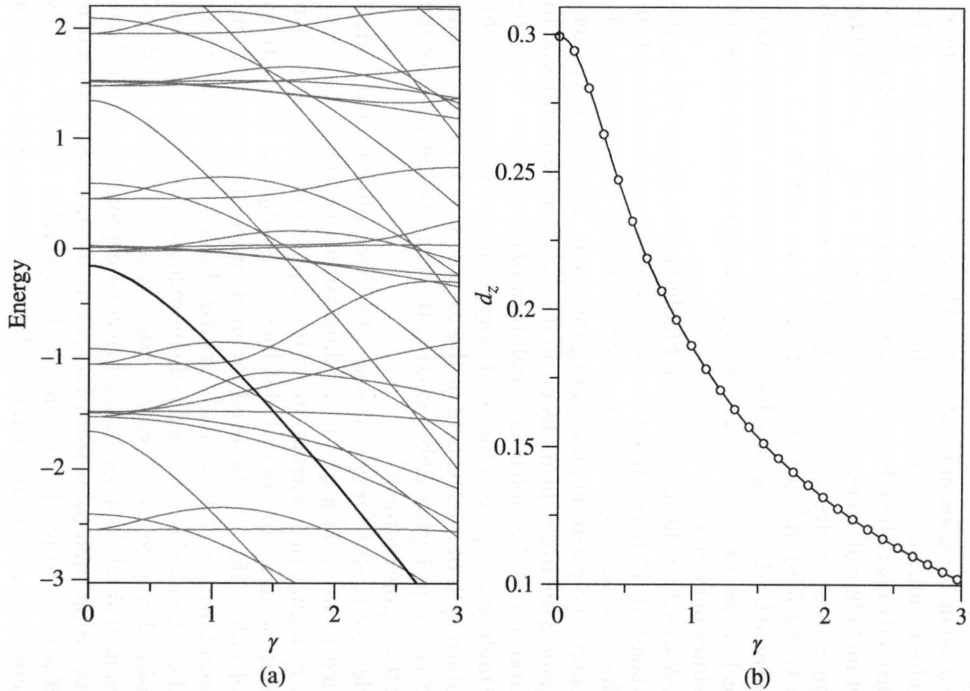

Figure 7.5 (a) The energy of the Floquet states for a rigid rotor in a superposition of a DC electric field directed along the z-axis and a circularly polarized microwave field rotating in the xy-plane. The strength of the DC field is $E = 0.1B_e/d$. The intensity of the microwave field is represented by the parameter $\gamma = d\xi/2B_e$. The frequency of the microwave field is $\omega = 1.5B_e/\hbar$. (b) The modification of the z-component of the molecule's dipole moment by the microwave field.

The eigenstates of the Floquet Hamiltonian for a polar rigid rotor in a DC field with strength $E = 0.1B_e/d$ and a microwave field with frequency $\omega = 1.5B_e/\hbar$ are shown in Figure 7.5a as functions of the microwave field intensity. The Floquet matrix has infinite size so there is an infinite series of eigenvalues. The eigenvalues form a periodic pattern consisting of bunches of states separated by multiples of $\hbar\omega$. The reader may observe some deviations from the periodicity. These occur because, in practice, the Floquet basis $|NM_N n\rangle$ must be restricted to a finite number of states, and the states at the boundaries of the matrix are not properly described. For the calculation presented in Figure 7.5, I used a basis set including all states with $N = 0$–7 and $n = -5, \ldots, +5$. While this basis is large enough to describe the shifts of the molecular states with $N = 0$ and $N = 1$, the field-induced shifts of the states with $N = 7$ are not accurately represented by the energy-level diagram in Figure 7.5. This is usually not a problem as long as the Floquet basis is large enough to encapsulate the dynamics of all states of interest.

Consider now a specific field-dressed state highlighted in Figure 7.5 by the bold curve. This state correlates with the lowest energy state of the molecule in the absence of the microwave field. As discussed in Section 5.1.1, the DC field forces the molecular dipole to orient itself along the z-axis. The microwave field rotating in the xy-field polarizes the molecule in the direction perpendicular to the z-axis, effectively tilting the molecular dipole toward the xy-plane. Thus, the dynamics of the molecular dipole in the laboratory frame can be controlled by tuning either the strength of the DC field or the parameters of the microwave field. This is illustrated in Figure 7.5b depicting the expectation value of the z-component of the dipole moment.

If the molecules are fixed to reside in sites of a three-dimensional optical lattice or confined to move in a quasi-two-dimensional geometry by a pair of laser beams, changing the orientation of the dipole moment, as shown in Figure 7.5, can be used to control the long-range dipolar interactions between the molecules. This can be used for many interesting applications. For example, a combination of a circularly polarized microwave field and an electric DC field can be used to tilt the dipole moment of molecules confined in a two-dimensional trap to the so-called magic angle, at which the dipolar interactions vanish [302]. As a result, one can engineer long-range repulsive interactions, preventing molecules from incursions to short intermolecular interactions and leading to the formation of self-assembled dipolar crystals [301]. For a variety of interesting applications of the microwave field control of molecular dipoles, the reader is referred to the recent literature [296–305].

8

Molecular Collisions in External Fields

There are "good" collisions and then there are "bad" collisions...

The development of quantum theory of molecular collisions in electromagnetic fields has been motivated largely by the experiments on cooling molecules to ultracold temperatures, typically defined as temperatures below 0.001 K. In order to cool a molecular ensemble to temperatures so low, it is necessary to break its contact with the thermal environment and dissipate its energy. The former is achieved by confining molecules in external field traps, as discussed in Chapter 6. The latter can be achieved by removing the fastest-moving molecules from the trap, while allowing the remaining molecules to reequilibrate their kinetic energy. This return to thermal equilibrium occurs through momentum exchange in molecular collisions. Collisions are thus vital for the cooling experiments.

However, molecular collisions may also induce undesirable processes, such as the transfer of molecules from trappable states to untrappable states. The success of the cooling experiments, where energy dissipation is achieved by evaporation, relies on the prevalence of "good" collisions – that is collisions leading to thermal equilibrium – over "bad" collisions, which remove molecules from the trap. It is clear that the good-to-bad ratio of collisions depends on the structure of molecules. So the question is, which molecules are good for cooling and which are bad? To answer this question, it is necessary to understand the mechanisms of the "bad" collisions, hence the need for theoretical models of molecular collisions in external field traps.

When molecules are trapped, their kinetic energy is necessarily smaller than the potential energy change due to the trapping field gradients. One should expect therefore that the trapping fields must modify the collision dynamics of trapped molecules to a greater extent. Under these conditions, the most accurate theoretical treatment of the molecular collision problem must include the interactions of molecules with electromagnetic fields explicitly.

Molecules in Electromagnetic Fields: From Ultracold Physics to Controlled Chemistry, First Edition. Roman V. Krems.

The purpose of this chapter is to describe the theory of molecular collisions in external fields, at the level of detail sufficient for the preparation of computer codes for quantum calculations of the scattering cross sections and collision event rates. We will consider only the time-independent Hamiltonians. This will allow us to formulate the theory of molecular collisions in DC electric and magnetic fields as well as in off-resonant laser and microwave fields. The treatment of the off-resonant AC fields will rely on the Floquet formalism discussed in Chapter 4, which will allow us to formulate the time-dependent scattering problems with time-independent Hamiltonians.

8.1 Coupled-Channel Theory of Molecular Collisions

8.1.1 A Very General Formulation

Consider two particles, A and B, described by the time-independent Hamiltonians $\hat{H}_A$ and $\hat{H}_B$. We will label the eigenvalues and eigenstates of $\hat{H}_A$ by ϵ_α and $|\alpha\rangle$,

$$\hat{H}_A|\alpha\rangle = \epsilon_\alpha|\alpha\rangle, \tag{8.1}$$

and those of $\hat{H}_B$ by ϵ_β and $|\beta\rangle$,

$$\hat{H}_B|\beta\rangle = \epsilon_\beta|\beta\rangle. \tag{8.2}$$

The Hamiltonians $\hat{H}_A$ and $\hat{H}_B$ will include the interactions of the particles with the electromagnetic fields.

The *total* Hamiltonian of the two particles is

$$\hat{H}_{\text{tot}} = -\frac{\hbar^2}{2m_A}\Delta_A - \frac{\hbar^2}{2m_B}\Delta_B + U + \hat{H}_A + \hat{H}_B, \tag{8.3}$$

where the first two terms are the operators for the kinetic energy of the particles in some space-fixed coordinate frame and U is the potential energy due to the interaction between A and B. We have denoted the total mass of A by m_A and the total mass of B by m_B. The interaction potential U is generally a complicated function of the *relative* coordinates, such as the separation between the centers-of-mass of the particles and the angles describing the orientation of A with respect to B. The potential U vanishes when the particles are separated.

We will neglect the effect of external fields on the translational motion of the particles during a collision. This is an approximation that is accurate for the strengths of fields that can be generated in the laboratory experiments. The effects of external fields on collisions will thus be entirely through the modification of the eigenspectra and eigenstates of $\hat{H}_A$ and $\hat{H}_B$. In rare cases, it may be necessary to include the interactions of the translational states of the particles with external fields. See, for example, Ref. [306]. These cases are quite exotic and they can usually be treated by a straightforward extension of the formulation of the scattering problem described in this chapter.

The collision of A and B is governed by the Hamiltonian

$$\hat{H} = -\frac{\hbar^2}{2\mu}\Delta_R + U + \hat{H}_A + \hat{H}_B, \tag{8.4}$$

where the first term is the kinetic energy of the *relative* motion, with μ being the reduced mass $\mu = m_A m_B/(m_A + m_B)$. Equation (8.4) is obtained from Eq. (8.3) by separating the center-of-mass motion of the collision complex and fixing the origin of the space-fixed coordinate frame at the center of mass of A and B. The center-of-mass motion of the A–B pair has no consequence for most collision problems. It is important to include it explicitly into consideration when computing the collision observables that depend on the interference of different collision events. For example, schemes for coherent control of molecular collisions based on preparing the collision partners in coherent superpositions with desired properties must include the description of the center-of-mass motion explicitly [307].

We will assume that before the collisions, A and B reside in specific eigenstates with energies ϵ_α and ϵ_β, and each collision event is independent so the center-of-mass motion can be disregarded. In this case, all we need to do is solve the Schrödinger equation with the Hamiltonian (8.4).

The separation between the center of mass of A and the center of mass of B is described by a vector $\boldsymbol{R}$, which is free to assume any orientation in the space-fixed coordinate frame. The kinetic energy term in Eq. (8.4) can be written in spherical polar coordinates as

$$-\frac{\hbar^2}{2\mu}\Delta_R = -\frac{\hbar^2}{2\mu}R^{-2}\frac{\partial}{\partial R}R^2\frac{\partial}{\partial R} + \frac{\hbar^2 \boldsymbol{l}^2(\theta,\phi)}{2\mu R^2}, \tag{8.5}$$

where R is the magnitude of $\boldsymbol{R}$ and $\boldsymbol{l}$ is the angular momentum describing the rotation of $\boldsymbol{R}$ in the space-fixed coordinate frame. The angular momentum operator $\boldsymbol{l}$ depends on the polar angle θ and the azimuth angle ϕ that specify the orientation of $\boldsymbol{R}$ in the space-fixed frame. The eigenstates of $\boldsymbol{l}^2$ are the spherical harmonics labeled by the quantum numbers l and m_l:

$$\boldsymbol{l}^2|lm_l\rangle = l(l+1)|lm_l\rangle \tag{8.6}$$

with m_l being the eigenvalue of the z-component of $\boldsymbol{l}$.

The goal of the scattering calculations is to find the eigenstates $|\Psi\rangle$ of the Hamiltonian (8.4)

$$\hat{H}|\Psi\rangle = E|\Psi\rangle \tag{8.7}$$

corresponding to a particular value of the total energy E. In the simplest possible approach, we can represent each of these eigenstates by a basis set expansion in terms of the direct products of the eigenstates of $\hat{H}_A$, $\hat{H}_B$, $\boldsymbol{l}^2$, and l_z,

$$|\Psi\rangle = R^{-1}\sum_\alpha\sum_\beta\sum_l\sum_{m_l} F_{\alpha\beta l m_l}(R)|\alpha\beta l m_l\rangle, \tag{8.8}$$

where $|\alpha\beta l m_l\rangle = |\alpha\rangle \otimes |\beta\rangle \otimes |l m_l\rangle$, and the sum – in principle – extends over all eigenstates of $\hat{H}_A$, all eigenstates of $\hat{H}_B$, and an infinite number of $|l m_l\rangle$ states with $l = 0, \ldots, \infty$ and $m_l = -l, \ldots, +l$. In practice, the sum in Eq. (8.8) will be truncated to a certain number of $|\alpha\beta l m_l\rangle$ states and convergence must be sought with respect to the number of terms included in the sum.

The expansion coefficients in Eq. (8.8) are the functions of R, the only variable that is not covered by the basis states. The factor R^{-1} in front of the sum plays an important role of converting the mix of first and second derivatives in the Hamiltonian (8.4) into a single second derivative operator (see Eq. (1.11)):

$$R^{-2}\frac{\partial}{\partial R}R^2\frac{\partial}{\partial R}\frac{F(R)}{R} = R^{-1}\frac{\partial^2}{\partial R^2}F(R). \tag{8.9}$$

If we now substitute the expansion (8.8) into the Schrödinger equation (8.7) with the Hamiltonian (8.4), we will obtain

$$\sum_{\alpha}\sum_{\beta}\sum_{l}\sum_{m_l}\left[\frac{\mathrm{d}^2}{\mathrm{d}R^2} - \frac{l(l+1)}{R^2} + \frac{2\mu}{\hbar^2}(E - \epsilon_\alpha - \epsilon_\beta)\right] F_{\alpha\beta l m_l}(R)|\alpha\beta l m_l\rangle$$
$$= \frac{2\mu}{\hbar^2}\sum_{\alpha}\sum_{\beta}\sum_{l}\sum_{m_l} U|\alpha\beta l m_l\rangle F_{\alpha\beta l m_l}(R), \tag{8.10}$$

where I have multiplied both sides by the factor $-R\ 2\mu/\hbar^2$ and replaced $\boldsymbol{l}^2|l m_l\rangle$, $\hat{H}_A|\alpha\rangle$ and $\hat{H}_B|\beta\rangle$ with the right-hand side of Eqs. (8.1), (8.2), and (8.6). The next step is to multiply both sides of Eq. (8.10) by $\langle\alpha'\beta' l' m_l'|$ from the left, to obtain

$$\left[\frac{\mathrm{d}^2}{\mathrm{d}R^2} - \frac{l(l+1)}{R^2} + k_{\alpha\beta}^2\right] F_{\alpha\beta l m_l}(R)$$
$$= \frac{2\mu}{\hbar^2}\sum_{\alpha'\beta' l' m_l'} \langle\alpha\beta l m_l|U|\alpha'\beta' l' m_l'\rangle F_{\alpha'\beta' l' m_l'}(R), \tag{8.11}$$

where

$$k_{\alpha\beta}^2 = \frac{2\mu}{\hbar^2}(E - \epsilon_\alpha - \epsilon_\beta), \tag{8.12}$$

and I have swapped the primed and unprimed quantum numbers for convenience. We have assumed that the states $|\alpha\beta l m_l\rangle$ are orthonormal, as any eigenstates of Hermitian operators.

Equation (8.11) is a set of coupled, second-order differential equations for the expansion coefficients $F_{\alpha\beta l m_l}(R)$. The states $|\alpha\beta l m_l\rangle$ define the *collision channels*. It can be seen that the interaction potential U couples different collision channels, thereby inducing transitions between different quantum states $|\alpha\rangle$ and $|\beta\rangle$ and between different angular momentum states $|l m_l\rangle$. The eigenstate of the total Hamiltonian in Eq. (8.8) can also be viewed as a sum of contributions corresponding to different states $|lm_l\rangle$,

$$|\Psi\rangle = \sum_l \sum_{m_l} G_{lm_l}(R)|lm_l\rangle, \tag{8.13}$$

where

$$G_{lm_l} = R^{-1} \sum_\alpha \sum_\beta F_{\alpha\beta lm_l}(R)|\alpha\beta\rangle. \tag{8.14}$$

The total eigenstate of the two-particle system can thus be viewed as a superposition of different states $|lm_l\rangle$ describing the end-over-end rotation of the collision complex *A*–*B*. The states $|lm_l\rangle$ are called the ***partal waves***, and the expansion (8.13) is called the ***partial wave expansion***. The number of partial waves required for an accurate representation of a scattering state $|\Psi\rangle$ depends on

- the reduced mass of the system;
- the kinetic energy of the relative motion;
- the steepness of the inter-particle interaction U at large distances R;
- the dependence of the interaction potential U on the relative particle orientation angles.

We will return to the discussion of the partial wave decomposition later.

Given that the states $|lm_l\rangle$, $|\alpha\rangle$, and $|\beta\rangle$ are known, finding the solutions of the Schrödinger equation is reduced to finding the expansion coefficients $F_{lm_l\alpha\beta}$ in Eq. (8.8). These coefficients are functions of R found by solving the system of the differential equations defined in Eq. (8.11) subject to the boundary conditions discussed in the next section. A numerical algorithm for solving these equations is described in Section 8.1.6.

8.1.2 Boundary Conditions

The next important question is, what are the boundary conditions for the solutions of Eq. (8.11)? Since the differential equations are second order, we need two boundary conditions for each function $F_{\alpha\beta lm_l}(R)$. The first one is simple. As the particles A and B collide, they hit a repulsive wall at short inter-particle separations.[1] Since the coefficients $F(R)$ describe the R-dependence of the

---

1 This is not true for the so-called insertion chemical reactions. An example of an insertion chemical reaction is the reaction of an alkali metal atom with an alkali metal dimer. In this reaction, the incoming atom inserts itself in between the atoms in the molecule, leading to the formation of an intermediate reaction complex characterized by a strongly attractive potential well. Here, we consider nonreactive scattering, in which the interacting molecules do not change their chemical identity. Collisions of nonreactive molecules are generally characterized by strongly repulsive short-range interactions, which prevent chemical reactions.

two-particle wave function and hence the probability density amplitudes in R-space, we must conclude that

$$F_{\alpha\beta l m_l}(R \to 0) \to 0, \tag{8.15}$$

for all values of α, β, l, and m_l. This boundary condition simply reflects the fact that the particles cannot be very close together. As discussed in Section 8.1.6, this is an important starting point for the numerical algorithms used for solving the system of equations (8.11).

The second boundary condition is obtained from the solutions of Eq. (8.11) at $R \to \infty$. The inter-particle interaction potential U and the centrifugal term $l(l+1)/R^2$ vanish as R increases so for sufficiently large R the coupled equations (8.11) become uncoupled, i.e. they become a set of independent equations, each of the following form:

$$\left[\frac{\mathrm{d}^2}{\mathrm{d}R^2} + k_{\alpha\beta}^2\right] F_{\alpha\beta l m_l}(R \to \infty) = 0.$$

The general solution of each of these equations can be written as

$$F_{\alpha\beta l m_l}(R \to \infty) = a_- \mathrm{e}^{-i(k_{\alpha\beta}R - \phi_-)} + a_+ \mathrm{e}^{i(k_{\alpha\beta}R - \phi_+)}, \tag{8.16}$$

where the $\pm$ signs correspond to the incoming and outgoing waves. The question is, what are the coefficients $a_\pm$ and the phases $\phi_\pm$?

To determine the phases, return to the Hamiltonian (8.4) and consider its eigenstates at large distances R, where U is zero but the centrifugal term is not. This is always possible for systems with inter-particle interactions U decaying faster than $\propto 1/R^2$. At these distances, the Schrödinger equation for the collision complex in the state $|\alpha\beta l m_l\rangle$ can be written as

$$\left[\frac{\mathrm{d}^2}{\mathrm{d}R^2} + 2\frac{1}{R}\frac{\mathrm{d}}{\mathrm{d}R} - \frac{l(l+1)}{R^2} + k_{\alpha\beta}^2\right] G_{\alpha\beta l m_l}(R \to \infty) = 0. \tag{8.17}$$

Here, we have returned to the original form of the radial part of the kinetic energy operator

$$\frac{1}{R^2}\frac{\mathrm{d}}{\mathrm{d}R}R^2\frac{\mathrm{d}}{\mathrm{d}R} = \frac{\mathrm{d}^2}{\mathrm{d}R^2} + \frac{2}{R}\frac{\mathrm{d}}{\mathrm{d}R}. \tag{8.18}$$

Note that

$$G_{\alpha\beta l m_l} = F_{\alpha\beta l m_l}/R. \tag{8.19}$$

If all terms in Eq. (8.17) are divided by $k_{\alpha\beta}^2$, it becomes

$$\left[\frac{\mathrm{d}^2}{\mathrm{d}x^2} + \frac{2}{x}\frac{\mathrm{d}}{\mathrm{d}x} + \left(1 - \frac{l(l+1)}{x^2}\right)\right] y(x) = 0 \tag{8.20}$$

with $x = k_{\alpha\beta}R$ and $y = G_{\alpha\beta l m_l}$. If multiplied by x^2, this equation becomes the well-known spherical Bessel equation [42]:

$$\left[x^2\frac{\mathrm{d}^2}{\mathrm{d}x^2} + 2x\frac{\mathrm{d}}{\mathrm{d}x} + (x^2 - l(l+1))\right] y(x) = 0. \tag{8.21}$$

The solutions $y(x)$ of the spherical Bessel equation are the spherical Bessel functions of the first and second kind, $j_l(x)$ and $n_l(x)$. Note that the solutions depend parametrically on the quantum number of the angular momentum l. The solutions of the Bessel equation can also be written in terms of the spherical Hankel functions, defined as [42]

$$h_l^{(1)}(x) = j_l(x) + in_l(x),$$
$$h_l^{(2)}(x) = j_l(x) - in_l(x). \tag{8.22}$$

At large values of the argument, these functions behave as [42]

$$h_l^{(1)}(x \to \infty) \sim -i\frac{\mathrm{e}^{i(x-\pi l/2)}}{x}$$
$$h_l^{(2)}(x \to \infty) \sim i\frac{\mathrm{e}^{-i(x-\pi l/2)}}{x}. \tag{8.23}$$

Comparing these solutions to Eq. (8.16), we see that the phases in Eq. (8.16) must be equal to $\phi_- = \pi l/2 + \pi/2$ and $\phi_+ = \pi l/2 + \pi/2$. Since the overall phase of $|\Psi\rangle$ is undetermined, we can multiply all eigenstates of the full Hamiltonian by $\mathrm{e}^{-i\pi/2}$, which will lead to $\phi_- = \pi l/2$ and $\phi_+ = \pi l/2 + \pi$. This settles the question about the phases.

In order to determine the coefficients $a_\pm$ in Eq. (8.16), we note that the term with the negative exponent is an incoming wave, whereas the one with the positive exponent is an outgoing wave. Let us now assume that before the collision the particles are prepared in a quantum state with specific quantum numbers that I will label by $\tilde{\alpha}$, $\tilde{\beta}$, $\tilde{l}$, and $\tilde{m}_l$, with a tilde over the quantum number.

The state $|\tilde{\alpha}\tilde{\beta}\tilde{l}\tilde{m}_l\rangle$ is the *incoming* collision channel.

Since – by our assumption – there is only one incoming collision channel, it is clear then that Eq. (8.16) must be of the form

$$F_{\alpha\beta l m_l}(R \to \infty) = \delta_{\alpha,\tilde{\alpha}}\delta_{\beta,\tilde{\beta}}\delta_{l,\tilde{l}}\delta_{m_l,\tilde{m}_l}\ \mathrm{e}^{-i(k_{\alpha\beta}R-\pi l/2)}$$
$$- P_{\tilde{\alpha}\tilde{\beta}\tilde{l}\tilde{m}_l;\alpha\beta l m_l}\ \mathrm{e}^{i(k_{\alpha\beta}R-\pi l/2)}, \tag{8.24}$$

where $P_{\tilde{\alpha}\tilde{\beta}\tilde{l}\tilde{m}_l;\alpha\beta l m_l}$ is the probability amplitude for the transition $|\tilde{\alpha}\tilde{\beta}\tilde{l}\tilde{m}_l\rangle \to |\alpha\beta l m_l\rangle$ induced by the interaction potential U and the minus sign in front of the second term explicitly accounts for the π phase difference between $h^{(1)}$ and $h^{(2)}$.

Using Eq. (8.24), we can write the eigenstates (8.8) of the total Hamiltonian at large R as follows:

$$|\Psi_{\tilde{\alpha}\tilde{\beta}\tilde{l}\tilde{m}_l}\rangle = R^{-1}\sum_\alpha\sum_\beta\sum_l\sum_{m_l}\{\delta_{\alpha,\tilde{\alpha}}\delta_{\beta,\tilde{\beta}}\delta_{l,\tilde{l}}\delta_{m_l,\tilde{m}_l}\mathrm{e}^{-i(k_{\alpha\beta}R-\pi l/2)}$$
$$- P_{\tilde{\alpha}\tilde{\beta}\tilde{l}\tilde{m}_l;\alpha\beta l m_l}\ \mathrm{e}^{i(k_{\alpha\beta}R-\pi l/2)}\}|\alpha\beta l m_l\rangle. \tag{8.25}$$

These eigenstates correspond to well-defined values of $\tilde{\alpha}$, $\tilde{\beta}$, $\tilde{l}$, and $\tilde{m}_l$ so we included these quantum numbers as subscripts on the left-hand side. Note that the states $|\Psi_{\tilde{\alpha}\tilde{\beta}\tilde{l}\tilde{m}_l}\rangle$ with different values of $\tilde{l}$ and $\tilde{m}_l$ are the eigenstates of the full Hamiltonian (8.3) with the same eigenvalue E.

What Eqs. (8.24) and (8.25) are saying is that if we prepare our particles A and B in specific quantum states $|\tilde{\alpha}\rangle$ and $|\tilde{\beta}\rangle$ and ensure that the collision complex rotates as stipulated by a specific eigenstate $|\tilde{l}\tilde{m}_l\rangle$ of $\boldsymbol{l}^2$ and its z-component, then the total state must be written as

$$\text{incoming wave}_{(\tilde{\alpha}\tilde{\beta}\tilde{l}\tilde{m}_l)}$$
$$+\sum_{\alpha}\sum_{\beta}\sum_{l}\sum_{m_l}\text{some amplitude}_{(\alpha\beta l m_l)} \times \text{outgoing wave}_{(\alpha\beta l m_l)}.$$

The amplitudes of the outgoing waves depend on the amplitudes for the transition probabilities from the incoming collision channel $(\tilde{\alpha}\tilde{\beta}\tilde{l}\tilde{m}_l)$ to the specific outgoing collision channels $(\alpha\beta l m_l)$. Now, it is generally not realistic to prepare the collision complex in a quantum state with specific $\tilde{l}$ and $\tilde{m}_l$. We will deal with this issue in Section 8.1.3.

8.1.3 Scattering Amplitude

In experiments, it is often possible to prepare particles in well-defined quantum states $|\alpha\rangle$ and $|\beta\rangle$ but hard to prepare the collision complex in well-defined states of the orbital angular momentum $|lm_l\rangle$. So, with rare exceptions, there is usually no such thing as particles prepared in a single collision channel $|\tilde{\alpha}\tilde{\beta}\tilde{l}\tilde{m}_l\rangle$.

Imagine an experiment, where all we know is that we prepare A in state $|\tilde{\alpha}\rangle$ and B in state $|\tilde{\beta}\rangle$. In this case, we should write the quantum state that represents the colliding particles as a coherent superposition of states $|\Psi_{\tilde{\alpha}\tilde{\beta}\tilde{l}\tilde{m}_l}\rangle$ (defined by Eq. (8.25)) with different values of $\tilde{l}$ and $\tilde{m}_l$ (remember, these states are all degenerate):

$$|\Psi_{\tilde{\alpha}\tilde{\beta}}\rangle = \sum_{\tilde{l}}\sum_{\tilde{m}_l} A_{\tilde{\alpha}\tilde{\beta}\tilde{l}\tilde{m}_l}|\Psi_{\tilde{\alpha}\tilde{\beta}\tilde{l}\tilde{m}_l}\rangle. \quad (8.26)$$

Our next step is to determine the expansion coefficients $A_{\tilde{\alpha}\tilde{\beta}\tilde{l}\tilde{m}_l}$.

Before the collision, the relative motion of the particles is a plane wave with an *arbitrary* propagation direction[2] so the full state of the entire system can be written as

$$|\Psi_{\tilde{\alpha}\tilde{\beta}}\rangle = \exp[i\boldsymbol{k}_{\tilde{\alpha}\tilde{\beta}} \cdot \boldsymbol{R}]|\tilde{\alpha}\tilde{\beta}\rangle + |\psi^{\text{scattered}}\rangle. \quad (8.27)$$

2 Most standard treatments of molecular collisions in the absence of external fields will choose the z-axis to lie along the propagation direction of the incoming collision flux. This simplifies the final equations. For problems in external fields, we do not have this luxury since the quantization z-axis is better determined by the external field direction. We must assume a more general condition that the initial collision flux is directed at an arbitrary angle with respect to the z-axis.

The plane wave can be expanded in spherical harmonics as [8]

$$\exp[i\boldsymbol{k}\cdot\boldsymbol{R}] = \frac{i2\pi}{kR}\sum_l\sum_{m_l} i^l Y^*_{lm_l}(\boldsymbol{k}/k)$$
$$\times\{\exp[-i(kR-\pi l/2)] - \exp[i(kR-\pi l/2)]\}Y_{lm_l}(\boldsymbol{R}/R), \tag{8.28}$$

where $\boldsymbol{k}/k$ is the unit vector specifying the direction of the plane wave propagation before the collision and $\boldsymbol{R}/R$ is the unit vector along the direction of the inter-particle axis. What we are saying by Eq. (8.28) is that we have prepared the collision complex in a coherent superposition of different states $|lm_l\rangle = Y_{lm_l}(\boldsymbol{R}/R)$.

Since the scattered part in Eq. (8.27) does not contain any incoming waves, we can obtain the coefficients $A_{\tilde{\alpha}\tilde{\beta}l\tilde{m}_l}$ by comparing the coefficients in front of the incoming waves $e^{-i(k_{\tilde{\alpha}\tilde{\beta}}R-\pi l/2)}$ in Eqs. (8.26) and (8.27). This comparison yields

$$A_{\alpha\beta lm_l} = \frac{i2\pi}{k_{\alpha\beta}} i^l Y^*_{lm_l}(\boldsymbol{k}_{\alpha\beta}/k_{\alpha\beta}), \tag{8.29}$$

where I have dropped the tildas to simplify the notation. Effectively, this is the amplitude of the fraction of the plane wave that goes into a collision channel with particular values of l and m_l in a single collision event.

Remember that R is a *relative* separation of A and B. The full wave function of the A–B system will contain terms $\propto e^{ikR}$ and terms $\propto e^{-ikR}$. As mentioned in Section 8.1.2, the terms with $\propto e^{-ikR}$ describe the *incoming* part of the quantum state. All of the incoming part of the full quantum state comes from the plane wave (8.28). The terms with $\propto e^{ikR}$ describe the *outgoing* part of the quantum state. The outgoing part of the quantum state will consist of two components: the part coming from the plane wave (as is clear from Eq. (8.28), the plane wave is a superposition of waves going in and out); and the part coming from the scattering, i.e. the incoming part converted to outgoing waves. Our goal now is to determine the amplitude of the scattered part of the total quantum state.

The scattered part is the difference between the outgoing part of the full state and the outgoing part of the plane wave before the collision:

$$|\psi^{\text{scattered}}\rangle \Rightarrow \left[|\Psi_{\tilde{\alpha}\tilde{\beta}}\rangle - \exp[i\boldsymbol{k}_{\tilde{\alpha}\tilde{\beta}}\cdot\boldsymbol{R}]|\tilde{\alpha}\tilde{\beta}\rangle\right]_{\text{outgoing}}. \tag{8.30}$$

To obtain the outgoing part of the full state $|\Psi_{\tilde{\alpha}\tilde{\beta}}\rangle$, we should insert Eqs. (8.25) and (8.29) into Eq. (8.26) and retain only the terms with the positive exponents. This yields

$$|\Psi_{\tilde{\alpha}\tilde{\beta}}\rangle_{\text{outgoing}} = -\frac{1}{R}\sum_l\sum_{m_l}\frac{i2\pi}{k_{\tilde{\alpha}\tilde{\beta}}} i^l Y^*_{lm_l}(\boldsymbol{k}_{\tilde{\alpha}\tilde{\beta}}/k_{\tilde{\alpha}\tilde{\beta}})$$
$$\times\sum_\alpha\sum_\beta\sum_{l'}\sum_{m'_l} P_{\tilde{\alpha}\tilde{\beta}lm_l;\alpha\beta l'm'_l}|\alpha\beta l'm'_l\rangle e^{i(k_{\alpha\beta}R-\pi l'/2)}, \tag{8.31}$$

which should be interpreted as the amplitude of a state $|lm_l\rangle$ going into a collision × the amplitude of the $(\tilde{\alpha}\tilde{\beta}lm_l) \to (\alpha\beta l'm_l')$ transition × the outgoing state $|\alpha\beta l'm_l'\rangle$ with the corresponding radial part. For the outgoing part of the plane wave, we have from Eq. (8.28)

$$\left[\exp[i\boldsymbol{k}_{\tilde{\alpha}\tilde{\beta}} \cdot \boldsymbol{R}]|\tilde{\alpha}\tilde{\beta}\rangle\right]_{\text{outgoing}} = -\frac{i2\pi}{k_{\tilde{\alpha}\tilde{\beta}}R} \sum_l \sum_{m_l} i^l Y^*_{lm_l}(\boldsymbol{k}_{\tilde{\alpha}\tilde{\beta}}/k_{\tilde{\alpha}\tilde{\beta}}) \times |\tilde{\alpha}\tilde{\beta}lm_l\rangle e^{i(k_{\tilde{\alpha}\tilde{\beta}}R-\pi l/2)}. \quad (8.32)$$

The difference between the two is

$$|\psi^{\text{scattered}}_{\tilde{\alpha}\tilde{\beta}}\rangle = \frac{1}{R} \sum_l \sum_{m_l} \frac{i2\pi}{k_{\tilde{\alpha}\tilde{\beta}}} i^l Y^*_{lm_l}(\boldsymbol{k}_{\tilde{\alpha}\tilde{\beta}}/k_{\tilde{\alpha}\tilde{\beta}}) \times \sum_\alpha \sum_\beta \sum_{l'} \sum_{m_l'} (\delta_{\tilde{\alpha},\alpha}\delta_{\tilde{\beta},\beta}\delta_{ll'}\delta_{m_lm_l'} - P_{\tilde{\alpha}\tilde{\beta}lm_l;\alpha\beta l'm_l'})|\alpha\beta l'm_l'\rangle e^{i(k_{\alpha\beta}R-\pi l'/2)}. \quad (8.33)$$

If we look at the last equation closely, we can see that it has the following form:

$$|\psi^{\text{scattered}}_{\tilde{\alpha}\tilde{\beta}}\rangle = \sum_\alpha \sum_\beta \frac{e^{ik_{\alpha\beta}R}}{R} |\alpha\beta\rangle p_{\tilde{\alpha}\tilde{\beta};\alpha\beta}. \quad (8.34)$$

In other words, it is

$$|\psi^{\text{scattered}}_{\tilde{\alpha}\tilde{\beta}}\rangle = \sum_\alpha \sum_\beta \text{outgoing spherical wave}_{(\alpha\beta)} \times |\alpha\beta\rangle \times \text{some coefficients}_{(\tilde{\alpha}\tilde{\beta};\alpha\beta)}.$$

The coefficients in front of the spherical waves in this sum are

$$p_{\tilde{\alpha}\tilde{\beta};\alpha\beta} = \frac{2\pi}{k_{\tilde{\alpha}\tilde{\beta}}} f_{\tilde{\alpha}\tilde{\beta};\alpha\beta} \quad (8.35)$$

with

$$f_{\tilde{\alpha}\tilde{\beta};\alpha\beta} = \sum_{l'} \sum_{m_l'} \sum_l \sum_{m_l} i^{(l-l'+1)} Y^*_{lm_l}(\boldsymbol{k}_{\tilde{\alpha}\tilde{\beta}}/k_{\tilde{\alpha}\tilde{\beta}}) Y_{l'm_l'}(\boldsymbol{R}/R) \times (\delta_{\tilde{\alpha},\alpha}\delta_{\tilde{\beta},\beta}\delta_{ll'}\delta_{m_lm_l'} - P_{\tilde{\alpha}\tilde{\beta}lm_l;\alpha\beta l'm_l'}). \quad (8.36)$$

Equation (8.34) shows that the coefficients $p_{\tilde{\alpha}\tilde{\beta};\alpha\beta}$ are the amplitudes of the spherical waves carrying particles in states α and β produced by the scattering of a plane wave carrying particles in states $\tilde{\alpha}$ and $\tilde{\beta}$. The coefficients $f_{\tilde{\alpha}\tilde{\beta};\alpha\beta}$ are called the *scattering amplitudes*.

Note that $2\pi/k_{\tilde{\alpha}\tilde{\beta}}$ is the de Broglie wavelength of the colliding particles. The absolute square of $p_{\tilde{\alpha}\tilde{\beta};\alpha\beta}$ gives the probability of scattering to produce outgoing waves carrying particles in states α and β. We can see that this probability can be viewed as the square of the de Broglie wavelength of the colliding particles times the absolute square of the scattering amplitude. The quantity $|f_{\tilde{\alpha}\tilde{\beta};\alpha\beta}|^2$ can thus be thought of as the fraction of the scattering target area (square of the de Broglie wavelength) that produces states α and β upon scattering.

8.1.4 Scattering Cross Section

When dealing with a gas of molecules, it is impossible to detect the outcome of a single collision. The observables in a scattering experiment represent quantities reflecting a myriad of collisions. Therefore, in order to define the observables for a scattering experiment, we need to operate with a flux of particles, rather than a single pair of particles.

Consider a flux of particles moving in a particular direction toward a scattering center.

The *differential scattering cross section* is defined as the ratio of the number of particles scattered in a particular direction to the number of incoming particles per unit time.

Given the wave function ψ, the probability current vector for particles of mass μ can be calculated as following [308]:

$$j = \mathrm{Re}\left(\psi^* \frac{\hbar}{i\mu}\nabla\psi\right). \tag{8.37}$$

For the incoming collision flux, the wave function is the plane wave

$$\psi = \exp[i\boldsymbol{k}_{\tilde{\alpha}\tilde{\beta}} \cdot \boldsymbol{R}] \tag{8.38}$$

so that

$$j^{\tilde{\alpha}\tilde{\beta}}_{\text{incoming}} = \frac{\hbar}{\mu}\boldsymbol{k}_{\tilde{\alpha}\tilde{\beta}}. \tag{8.39}$$

The state (8.38) is a wave of unit density. Equation (8.39) shows that this wave represents a stream of particles of density 1 moving with uniform velocity along the vector $\boldsymbol{k}_{\tilde{\alpha}\tilde{\beta}}$.

For the scattered collision flux carrying particles in state $|\alpha\beta\rangle$, the radial part of the wave function is

$$\psi = p_{\tilde{\alpha}\tilde{\beta};\alpha\beta}\frac{e^{ik_{\alpha\beta}R}}{R}. \tag{8.40}$$

Note that the scattering amplitude depends on both the direction of the incident flux $\boldsymbol{k}_{\tilde{\alpha}\tilde{\beta}}/k_{\tilde{\alpha}\tilde{\beta}} \equiv (\theta_{\text{i}}, \phi_{\text{i}})$ and the direction of the outgoing flux $\boldsymbol{R}/R \equiv (\theta, \phi)$. The wave (8.40) represents a stream of scattered particles of density $|p_{\alpha;\alpha\beta}|^2/R^2$ moving away from the scattering center in the positive direction of $\boldsymbol{R}$. The current density is [308]

$$j^{\alpha\beta}_{\text{scattered}} = |p_{\tilde{\alpha}\tilde{\beta};\alpha\beta}|^2\frac{\hbar k_{\alpha\beta}}{\mu R^2}. \tag{8.41}$$

If we multiply the scattered flux by $R^2 \sin\theta \, d\theta \, d\phi = R^2 \, d\Omega$,

$$j^{\alpha\beta}_{\text{scattered}} R^2 \, d\Omega = |p_{\tilde{\alpha}\tilde{\beta};\alpha\beta}|^2 \frac{\hbar k_{\alpha\beta}}{\mu} \, d\Omega, \tag{8.42}$$

we obtain the number of particles that pass per unit time through a spherical surface element of large radius, which subtends the solid angle $d\Omega$ in the direction of $\boldsymbol{R}$ [308].

The differential scattering cross section can then be defined as

$$\frac{d\sigma}{d\Omega} = \frac{|j^{\alpha\beta}_{\text{scattered}}|R^2}{|j^{\tilde{\alpha}\tilde{\beta}}_{\text{incoming}}|} = \frac{k_{\alpha\beta}}{k_{\tilde{\alpha}\tilde{\beta}}} \left(\frac{2\pi}{k_{\tilde{\alpha}\tilde{\beta}}}\right)^2 |f_{\tilde{\alpha}\tilde{\beta};\alpha\beta}|^2. \tag{8.43}$$

Once again, note that this cross section is a function of both the incident flux direction $\boldsymbol{k}_{\tilde{\alpha}\tilde{\beta}}$ and the scattered flux direction.

The radial functions (8.24) are normalized by unit density. It is more convenient to normalize them by the probability current. This requires dividing the right-hand side of Eq. (8.24) by the square root of the collision velocity $v^{1/2}_{\alpha\beta} = (\hbar k_{\alpha\beta}/\mu)^{1/2}$ of the particles in state $|\alpha\beta\rangle$. The boundary conditions for the channel states normalized by unit flux are

$$\begin{aligned} F_{\alpha\beta l m_l}(R \to \infty) = v^{-1/2}_{\alpha\beta} [\delta_{\alpha,\tilde{\alpha}} \delta_{\beta,\tilde{\beta}} \delta_{l,\tilde{l}} \delta_{m_l,\tilde{m}_l} e^{-i(k_{\alpha\beta}R - \pi l/2)} \\ - S_{\tilde{\alpha}\tilde{\beta}\tilde{l}\tilde{m}_l;\alpha\beta l m_l} e^{i(k_{\alpha\beta}R - \pi l/2)}]. \end{aligned} \tag{8.44}$$

The coefficients $P_{\tilde{\alpha}\tilde{\beta}\tilde{l}\tilde{m}_l;\alpha\beta l m_l}$ in Eq. (8.24) are the probability amplitudes for the transitions between states normalized by unit density. The coefficients $S_{\tilde{\alpha}\tilde{\beta}\tilde{l}\tilde{m}_l;\alpha\beta l m_l}$ in Eq. (8.44) are the probability amplitudes for the transitions between states normalized by unit flux. The coefficients $S_{\tilde{\alpha}\tilde{\beta}\tilde{l}\tilde{m}_l;\alpha\beta l m_l}$ are the elements of the S-matrix. It is more convenient to work with the S-matrix than with the probability amplitudes $P_{\tilde{\alpha}\tilde{\beta}\tilde{l}\tilde{m}_l;\alpha\beta l m_l}$, because the S-matrix is unitary and symmetric.

In order to obtain the scattering amplitude in terms of the S-matrix elements, we can use Eq. (8.44) and follow the arguments of Section 8.1.3. Note that comparing the coefficients in front of the incoming waves in Eqs. (8.28) and (8.44) introduces an additional factor $v^{1/2}_{\alpha\beta}$ into the coefficients (8.29), and consequently the factor $v^{1/2}_{\tilde{\alpha}\tilde{\beta}}/v^{1/2}_{\alpha\beta}$ into the scattering amplitude. The scattered part of the state can now be written as

$$|\psi^{\text{scattered}}_{\tilde{\alpha}\tilde{\beta}}\rangle = \sum_{\alpha} \sum_{\beta} \frac{e^{ik_{\alpha\beta}R}}{R} |\alpha\beta\rangle s_{\tilde{\alpha}\tilde{\beta};\alpha\beta}, \tag{8.45}$$

with $s_{\tilde{\alpha}\tilde{\beta};\alpha\beta}$ given by

$$s_{\tilde{\alpha}\tilde{\beta};\alpha\beta} = \frac{2\pi}{k_{\tilde{\alpha}\tilde{\beta}}} \left(\frac{k_{\tilde{\alpha}\tilde{\beta}}}{k_{\alpha\beta}}\right)^{1/2} q_{\tilde{\alpha}\tilde{\beta};\alpha\beta}$$

and

$$q_{\tilde{\alpha}\tilde{\beta};\alpha\beta} = \sum_{l'}\sum_{m_l'}\sum_{l}\sum_{m_l} i^{(l-l'+1)} Y^*_{lm_l}(\boldsymbol{k}_{\tilde{\alpha}\tilde{\beta}}/k_{\tilde{\alpha}\tilde{\beta}}) Y_{l'm_l'}(\boldsymbol{R}/R)$$
$$\times (\delta_{\tilde{\alpha},\alpha}\delta_{\tilde{\beta},\beta}\delta_{ll'}\delta_{m_l m_l'} - S_{\tilde{\alpha}\tilde{\beta}lm_l;\alpha\beta l'm_l'}), \tag{8.46}$$

where we have made the substitution $v^{1/2}_{\tilde{\alpha}\tilde{\beta}}/v^{1/2}_{\alpha\beta} = k^{1/2}_{\tilde{\alpha}\tilde{\beta}}/k^{1/2}_{\alpha\beta}$.

Note that in order for $p_{\tilde{\alpha}\tilde{\beta};\alpha\beta}$ and $s_{\tilde{\alpha}\tilde{\beta};\alpha\beta}$ to be the same, we must require that

$$P_{\tilde{\alpha}\tilde{\beta}lm_l;\alpha\beta l'm_l'} = \left(\frac{k_{\tilde{\alpha}\tilde{\beta}}}{k_{\alpha\beta}}\right)^{1/2} S_{\tilde{\alpha}\tilde{\beta}lm_l;\alpha\beta l'm_l'}. \tag{8.47}$$

This relation has a simple physical meaning. If the particles were moving with the same velocity before and after the collision, the probability amplitudes P and S would be the same. However, if the internal energy of the particles changes as a result of the collision, so does their velocity. In general, $k_{\tilde{\alpha}\tilde{\beta}} \neq k_{\alpha\beta}$. The coefficients S give the probability amplitudes for the transitions of particles between quantum states without regard for the change of velocity. Consider now a collision, in which the particles are promoted to quantum states of higher energy. Due to the conservation of energy, the relative velocity of the particles must decrease. If we simply count the number of particles passing through a spherical surface element per unit time, it may falsely appear that the scattering event has decreased the number of particles. We need to correct for this. This can be done by multiplying the scattering probability by the ratio $k_{\tilde{\alpha}\tilde{\beta}}/k_{\alpha\beta}$, which is the amount by which the scattered particles "slow down" or "accelerate" as a result of an inelastic (internal energy changing) collision.

The differential scattering cross section is

$$\frac{d\sigma}{d\Omega} = \frac{|j^{\alpha\beta}_{\text{scattered}}|R^2}{|j^{\tilde{\alpha}\tilde{\beta}}_{\text{incoming}}|} = \left(\frac{2\pi}{k_{\tilde{\alpha}\tilde{\beta}}}\right)^2 |q_{\tilde{\alpha}\tilde{\beta};\alpha\beta}|^2, \tag{8.48}$$

which shows that the differential scattering cross section is a fraction of the square of the de Broglie wavelength of the colliding particles.

The scattering amplitude is often written in terms of the T-matrix elements

$$q_{\tilde{\alpha}\tilde{\beta};\alpha\beta} = \frac{2\pi}{k_\alpha}\sum_{l'}\sum_{m_l'}\sum_{l}\sum_{m_l}\left(\frac{k_{\tilde{\alpha}\tilde{\beta}}}{k_{\alpha\beta}}\right)^{1/2} i^{(l-l'+1)} Y^*_{lm_l}(\boldsymbol{k}_{\tilde{\alpha}\tilde{\beta}}/k_{\tilde{\alpha}\tilde{\beta}}) Y_{l'm_l'}(\boldsymbol{R}/R)$$
$$\times T_{\tilde{\alpha}\tilde{\beta}lm_l;\alpha\beta l'm_l'}, \tag{8.49}$$

defined as

$$T_{\tilde{\alpha}\tilde{\beta}lm_l;\alpha\beta l'm_l'} = \delta_{\tilde{\alpha},\alpha}\delta_{\tilde{\beta},\beta}\delta_{ll'}\delta_{m_l m_l'} - S_{\tilde{\alpha}\tilde{\beta}lm_l;\alpha\beta l'm_l'}.$$

In order to obtain the *integral cross section* for the collision-induced transitions $|\tilde{\alpha}\tilde{\beta}\rangle \rightarrow |\alpha\beta\rangle$, we should integrate Eq. (8.48) over all orientations of the initial collision flux $\boldsymbol{k}_{\tilde{\alpha}\tilde{\beta}}$ and all orientations of $\boldsymbol{R}$. If any orientations of the initial flux are allowed, we should divide the result by 4π to account for the random initial orientation of $\boldsymbol{k}_{\tilde{\alpha}\tilde{\beta}}$ so the integral cross section in a gas of randomly moving particles is

$$\sigma_{\tilde{\alpha}\tilde{\beta}\to\alpha\beta} = \frac{1}{4\pi}\iint \mathrm{d}\Omega_{\mathrm{i}}\,\mathrm{d}\Omega \quad \frac{\mathrm{d}\sigma_{\tilde{\alpha}\tilde{\beta}\to\alpha\beta}(\theta_{\mathrm{i}},\phi_{\mathrm{i}},\theta,\phi)}{\mathrm{d}\Omega_{\mathrm{i}}\,\mathrm{d}\Omega}, \tag{8.50}$$

where $\mathrm{d}\Omega_{\mathrm{i}} = \sin\theta_{\mathrm{i}}\,\mathrm{d}\theta_{\mathrm{i}}\,\mathrm{d}\phi_{\mathrm{i}}$. Since the spherical harmonics depending on the same angles in Eq. (8.49) are orthonormal, the integration leads to

$$\sigma_{\tilde{\alpha}\tilde{\beta}\to\alpha\beta} = \frac{\pi}{k^2_{\tilde{\alpha}\tilde{\beta}}}\sum_{l}\sum_{m_l}\sum_{l'}\sum_{m_l'} |\delta_{ll'}\delta_{m_l m_l'}\delta_{\tilde{\alpha}\alpha'}\delta_{\tilde{\beta}\beta'} - S_{\tilde{\alpha}\tilde{\beta}lm_l;\alpha\beta'l'm_l'}|^2. \tag{8.51}$$

8.1.5 Scattering of Identical Molecules

Care must be taken when considering the collision events of identical particles. The total wavefunction of two identical particles must be symmetric (for bosons) or anitsymmetric (for fermions) under the permutation of the particles. For two linear molecules whose orientations are given by $\boldsymbol{r}_A$ and $\boldsymbol{r}_B$ and the center-of-mass separation by $\boldsymbol{R}$, the permutation operator $\hat{P}$ transforms: $\boldsymbol{r}_A \to \boldsymbol{r}_B$; $\boldsymbol{r}_B \to \boldsymbol{r}_A$; $\boldsymbol{R} \to -\boldsymbol{R}$. The properly symmetrized basis functions can be obtained by applying the operator $1 + \eta\hat{P}$ to the basis states $|\alpha\beta l m_l\rangle$ and normalizing the resulting states. The value of η is 1 for composite bosons and -1 for composite fermions.

The task of formulating the scattering theory for collision of identical molecules sounds simple enough: apply the symmetrization operator $1 + \eta\hat{P}$ to the product basis introduced in Section 8.1.4, use the symmetrized basis thus obtained to represent the total wave function, and repeat the derivations of the expressions for the cross sections. The problem is, however, not as simple as it seems. It is complicated enough that is has led to arguments in the literature and the final expressions for the cross sections that differ between authors. See, for example, the discussion of the expressions for the cross sections in terms of the T-matrix elements in the work by Huo and Green [309]. In what follows, I will present what I think is the right way of handling the problem. Some of my colleagues will disagree with the equations I derive here.

Let us start by defining the symmetrized basis. Note that the spherical harmonics change under the inversion of $\boldsymbol{R}$ as

$$Y_{lm_l}(-\boldsymbol{R}/R) = (-1)^l Y_{lm_l}(\boldsymbol{R}/R), \tag{8.52}$$

so

$$\hat{P}|lm_l\rangle = (-1)^l|lm_l\rangle \tag{8.53}$$

and the symmetrized basis states for the scattering calculations can be written as

$$|\phi^{\eta}_{\alpha\beta lm_l}\rangle = \frac{1}{\sqrt{2(1+\delta_{\alpha\beta})}}[|\alpha\beta\rangle + \eta(-1)^l|\beta\alpha\rangle]|lm_l\rangle. \tag{8.54}$$

Note that for $\alpha = \beta$ and $\eta(-1)^l = -1$, the states (8.54) do not exist. This means that the scattering states of bosons prepared in identical internal states can only have even-l partial waves, while fermions in identical internal states are allowed to have only odd-l partial waves. This restriction does not apply to collisions of particles in different internal states $\alpha \neq \beta$.

The states in Eq. (8.54) are normalized to unit density, whether the particles are in identical states or not. This is an important point because one might argue that the states of two identical particles in identical states must be normalized to 2. See, for example, the discussion in the famous book on Atomic Collisions by Mott and Massey [310]. We will take care of the normalization issue later on.

The eigenstates of the total Hamiltonian can now be represented by a basis set expansion in terms of the states (8.54)

$$|\Psi^{\eta}\rangle = R^{-1}\sum_{\alpha}\sum_{\beta\geq\alpha}\sum_{l}\sum_{m_l} F_{\eta,\alpha\beta lm_l}(R)|\phi^{\eta}_{\alpha\beta lm_l}\rangle. \tag{8.55}$$

Note the difference in the summation over α and β in Eq. (8.8) and in Eq. (8.55). In the former equation, the summation over α and β is unrestricted and should, in principle, include all quantum states of particle A and all quantum states of particle B. In Eq. (8.55), the quantum states α and β are no longer associated with specific particles A or B. The state $|\phi^{\eta}_{\alpha\beta lm_l}\rangle$ does not discriminate between the identity of the particles; all it describes is the scattering of two particles in states α and β. It can be a collision of A in state α and B in state β or A in state β and B in state α. Therefore, the sum in Eq. (8.55) is only over unique combinations of α and β.

The derivations of Sections 8.1.2 and 8.1.3 can now be repeated with Eq. (8.55) as a starting point. In particular, we must write

$$\begin{aligned} F_{\eta,\alpha\beta lm_l}(R\to\infty) = {}& \delta_{\alpha,\tilde{\alpha}}\delta_{\beta,\tilde{\beta}}\delta_{l,\tilde{l}}\delta_{m_l,\tilde{m}_l}\, \mathrm{e}^{-i(k_{\alpha\beta}R-\pi l/2)} \\ & - P^{\eta}_{\tilde{\alpha}\tilde{\beta}\tilde{l}\tilde{m}_l;\alpha\beta lm_l}\, \mathrm{e}^{i(k_{\alpha\beta}R-\pi l/2)}, \end{aligned} \tag{8.56}$$

where $P^{\eta}_{\tilde{\alpha}\tilde{\beta}\tilde{l}\tilde{m}_l;\alpha\beta lm_l}$ is the probability for the transition between collision channels $|\phi^{\eta}_{\alpha\beta lm_l}\rangle$ and $|\phi^{\eta}_{\tilde{\alpha}\tilde{\beta}\tilde{l}\tilde{m}_l}\rangle$ normalized to unit density. Noting that the symmetrized incoming plane wave can be written as

$$\begin{aligned} \left(\frac{1+\eta\hat{P}}{\sqrt{2(1+\delta_{\tilde{\alpha},\tilde{\beta}})}}\right)\{\exp[i\boldsymbol{k}_{\tilde{\alpha}\tilde{\beta}}\cdot\boldsymbol{R}]|\tilde{\alpha}\tilde{\beta}\rangle\} = {}& \frac{i2\pi}{k_{\tilde{\alpha}\tilde{\beta}}R}\sum_{l}\sum_{m_l} i^l Y^{*}_{lm_l}(\boldsymbol{k}_{\tilde{\alpha}\tilde{\beta}}/k_{\tilde{\alpha}\tilde{\beta}}) \\ & \times\{\exp[-i(k_{\tilde{\alpha}\tilde{\beta}}R-\pi l/2)] - \exp[i(k_{\tilde{\alpha}\tilde{\beta}}R-\pi l/2)]\}|\phi^{\eta}_{\tilde{\alpha}\tilde{\beta}lm_l}\rangle, \end{aligned} \tag{8.57}$$

we can write the symmetrized states

$$|\Psi^{\eta}_{\tilde{\alpha}\tilde{\beta}}\rangle = \sum_{\tilde{l}}\sum_{\tilde{m}_l} A_{\tilde{\alpha}\tilde{\beta}\tilde{l}\tilde{m}_l}|\Psi^{\eta}_{\tilde{\alpha}\tilde{\beta}\tilde{l}\tilde{m}_l}\rangle \tag{8.58}$$

with the same coefficients $A_{\tilde{\alpha}\tilde{\beta}\tilde{l}\tilde{m}_l}$ as given by Eq. (8.29).

This means that all of the subsequent arguments of Sections 8.1.2 and 8.1.3 can be applied to obtain the scattered part of the full state

$$|\psi^{\eta,\text{scattered}}_{\tilde{\alpha}\tilde{\beta}}\rangle = \frac{1}{R}\sum_{l}\sum_{m_l}\frac{i2\pi}{k_{\tilde{\alpha}\tilde{\beta}}}i^l Y^*_{lm_l}(\boldsymbol{k}_{\tilde{\alpha}\tilde{\beta}}/k_{\tilde{\alpha}\tilde{\beta}})$$
$$\times \sum_{\alpha}\sum_{\beta}\sum_{l'}\sum_{m'_l}(\delta_{\tilde{\alpha},\alpha}\delta_{\tilde{\beta},\beta}\delta_{ll'}\delta_{m_lm'_l} - P^{\eta}_{\tilde{\alpha}\tilde{\beta}lm_l;\alpha\beta l'm'_l})|\phi^{\eta}_{\tilde{\alpha}\tilde{\beta}lm_l}\rangle e^{i(k_{\alpha\beta}R-\pi l'/2)}. \tag{8.59}$$

Mimicking what we have done before, we want to be able to write the last equation as

$$|\psi^{\eta,\text{scattered}}_{\tilde{\alpha}\tilde{\beta}}\rangle \propto \sum_{\alpha}\sum_{\beta}\frac{e^{i(k_{\alpha\beta}R-\pi l'/2)}}{R}[|\alpha\beta\rangle + \eta|\beta\alpha\rangle]p^{\eta}_{\tilde{\alpha}\tilde{\beta},\alpha\beta}. \tag{8.60}$$

If we could do this, we could argue that the coefficients p^{η} are the probability amplitudes of producing spherical waves with pairs of particles in properly symmetrized states. Unfortunately, it doesn't appear to be possible to reduce Eq. (8.59) to the form of Eq. (8.60), at least, not in general. The complication arises because there is an l-dependent term that premultiplies the ket $|\beta\alpha\rangle$ in the symmetrized states. On the other hand, we can write the scattered part of the total state as follows:

$$|\psi^{\eta,\text{scattered}}_{\tilde{\alpha}\tilde{\beta}}\rangle = \frac{1}{\sqrt{2(1+\delta_{\alpha,\beta})}}\sum_{\alpha}\sum_{\beta}\frac{e^{i(k_{\alpha\beta}R-\pi l'/2)}}{R}$$
$$\times [|\alpha\beta\rangle p^{\eta}_{\tilde{\alpha}\tilde{\beta},\alpha\beta}(\boldsymbol{R}/R) + \eta|\beta\alpha\rangle p^{\eta}_{\tilde{\alpha}\tilde{\beta},\alpha\beta}(-\boldsymbol{R}/R)], \tag{8.61}$$

where the coefficients p^{η} are given by the following expression:

$$p^{\eta}_{\tilde{\alpha}\tilde{\beta};\alpha\beta} = \frac{2\pi}{k_{\tilde{\alpha}\tilde{\beta}}}\sum_{l'}\sum_{m'_l}\sum_{l}\sum_{m_l} i^{(l-l'+1)}Y^*_{lm_l}(\boldsymbol{k}_{\tilde{\alpha}\tilde{\beta}}/k_{\tilde{\alpha}\tilde{\beta}})Y_{l'm'_l}(\boldsymbol{R}/R) \tag{8.62}$$

$$\times (\delta_{\tilde{\alpha},\alpha}\delta_{\tilde{\beta},\beta}\delta_{ll'}\delta_{m_lm'_l} - P^{\eta}_{\tilde{\alpha}\tilde{\beta}lm_l;\alpha\beta l'm'_l}). \tag{8.63}$$

If $\alpha \neq \beta$, the factor

$$[|\alpha\beta\rangle p^{\eta}_{\tilde{\alpha}\tilde{\beta},\alpha\beta}(\boldsymbol{R}/R) + \eta|\beta\alpha\rangle p^{\eta}_{\tilde{\alpha}\tilde{\beta},\alpha\beta}(-\boldsymbol{R}/R)] \tag{8.64}$$

in Eq. (8.61) can be interpreted as an entangled state of molecules A and B in states α and β. The entanglement is produced by the scattering process. If $\alpha = \beta$, this state is reduced to

$$|\alpha\alpha\rangle[p^{\eta}_{\tilde{\alpha}\tilde{\beta},\alpha\alpha}(\boldsymbol{R}/R) + \eta p^{\eta}_{\tilde{\alpha}\tilde{\beta},\alpha\alpha}(-\boldsymbol{R}/R)]. \tag{8.65}$$

Equations (8.64) and (8.68) illustrate the fundamental difference between scattering into a channel with the two identical particles in different states and into the channel, where the state of the two particles is the same.

Now, the goal is to obtain an expression for the differential scattering cross sections written in terms of the probability amplitude coefficients $p^{\eta}_{\tilde{\alpha}\tilde{\beta},\alpha\beta}$. As before, we define the differential scattering cross section as the ratio of the number of particles scattered in a particular direction to the number of incoming particles per unit time.

Note that the number of incoming particles per unit time should be unaffected by whether the particles are in identical states or in distinguishable states. For example, if we have two colliding beams and simply count the number of particles entering the collision region per unit time, this count should be independent of whether the particles in the two beams are in the same internal state or in different internal states. However, – and here's the problem – if we calculate the average number of particles described by Eq. (8.57) for the cases $\tilde{\alpha} = \tilde{\beta}$ and $\tilde{\alpha} \neq \tilde{\beta}$, we will obtain different results. When $\tilde{\alpha} \neq \tilde{\beta}$, Eq. (8.57) reads

$$|\psi\rangle = \frac{1}{\sqrt{2}}(e^{ik_{\tilde{\alpha}\tilde{\beta}}\cdot R}|\tilde{\alpha}\tilde{\beta}\rangle + \eta e^{-ik_{\tilde{\alpha}\tilde{\beta}}\cdot R}|\tilde{\beta}\tilde{\alpha}\rangle), \tag{8.66}$$

so that $\langle\psi|\psi\rangle = 1$. When $\tilde{\alpha} = \tilde{\beta}$, we obtain

$$|\psi\rangle = \frac{1}{2}(e^{ik_{\tilde{\alpha}\tilde{\alpha}}\cdot R} + \eta e^{-ik_{\tilde{\alpha}\tilde{\alpha}}\cdot R})|\tilde{\alpha}\tilde{\alpha}\rangle \tag{8.67}$$

and the average value of $\langle\psi|\psi\rangle$ is 1/2. This is a consequence of the normalization factor we adopted to make the symmetrized basis states normalized to 1.

To correct for this, we will assume that the incoming collision flux of identical particles is the same as that of distinguishable particles given by Eq. (8.39) and multiply the differential cross sections for collisions of identical molecules in the *same internal state* by 2.

The situation must be treated differently for scattered particles because when one molecule is scattered in direction $\boldsymbol{R}$ and the other in direction $-\boldsymbol{R}$, one can't know which molecule travels where. Each molecule has to be considered as traveling in both directions simultaneously.

Let us use Eq. (8.37) to calculate the probability current of the scattered particles. To be explicit, let us consider the case of particles produced in the same state $\alpha = \beta$ and the case of particles produced in different states $\alpha \neq \beta$ separately.

To obtain the total probability current of particles in states α and β, note that $p^{\eta}_{\tilde{\alpha}\tilde{\beta},\alpha\alpha}(\boldsymbol{R}/R) = \eta p^{\eta}_{\tilde{\alpha}\tilde{\beta},\alpha\alpha}(-\boldsymbol{R}/R)$. This can be used to rewrite Eq. (8.64) as

$$\left[|\alpha\beta\rangle p^{\eta}_{\tilde{\alpha}\tilde{\beta},\alpha\beta}(\boldsymbol{R}/R) + |\beta\alpha\rangle p^{\eta}_{\tilde{\alpha}\tilde{\beta},\alpha\beta}(\boldsymbol{R}/R)\right], \tag{8.68}$$

which shows that, for the case $\alpha \neq \beta$, the flux scattered in the direction $\boldsymbol{R}$ can be thought of as a sum of two fluxes: that of particles $A(\alpha)$ and $B(\beta)$ and that of particles $A(\beta)$ and $B(\alpha)$. We thus have from Eqs. (8.37) and (8.59) the following expressions for the current

$$\begin{aligned} j^{\alpha\beta}_{\text{scattered}} &= \frac{\hbar k_{\alpha\beta}}{2\mu R^2}(|p^{\eta}_{\tilde{\alpha}\tilde{\beta},\alpha\beta}(\boldsymbol{R}/R)|^2 + |p^{\eta}_{\tilde{\alpha}\tilde{\beta},\alpha\beta}(\boldsymbol{R}/R)|^2) \\ &= \frac{\hbar k_{\alpha\beta}}{\mu R^2}|p^{\eta}_{\tilde{\alpha}\tilde{\beta},\alpha\beta}(\boldsymbol{R}/R)|^2. \end{aligned} \tag{8.69}$$

For particles produced in identical states $\alpha = \beta$, the probability current obtained from Eq. (8.59) is

$$\begin{aligned} j^{\alpha\alpha}_{\text{scattered}} &= \frac{\hbar k_{\alpha\alpha}}{4\mu R^2}|p^{\eta}_{\tilde{\alpha}\tilde{\beta},\alpha\alpha}(\boldsymbol{R}/R) + p^{\eta}_{\tilde{\alpha}\tilde{\beta},\alpha\alpha}(-\boldsymbol{R}/R)|^2 \\ &= \frac{\hbar k_{\alpha\alpha}}{\mu R^2}|p^{\eta}_{\tilde{\alpha}\tilde{\beta},\alpha\alpha}(\boldsymbol{R}/R)|^2. \end{aligned} \tag{8.70}$$

We thus conclude that the differential scattering cross section for identical molecules initially in states $(\tilde{\alpha}, \tilde{\beta})$ and, after scattering, in states (α, β) can be written as

$$\frac{\mathrm{d}\sigma}{\mathrm{d}\Omega} = (1 + \delta_{\tilde{\alpha}\tilde{\beta}})\frac{k_{\alpha\beta}}{k_{\tilde{\alpha}\tilde{\beta}}}\left(\frac{2\pi}{k_{\tilde{\alpha}\tilde{\beta}}}\right)^2 |f^{\eta}_{\tilde{\alpha}\tilde{\beta};\alpha\beta}|^2 \tag{8.71}$$

with the coefficients $f^{\eta}_{\tilde{\alpha}\tilde{\beta};\alpha\beta}$ defined as

$$p^{\eta}_{\tilde{\alpha}\tilde{\beta};\alpha\beta} = \frac{2\pi}{k_{\tilde{\alpha}\tilde{\beta}}} f^{\eta}_{\tilde{\alpha}\tilde{\beta};\alpha\beta}. \tag{8.72}$$

In order to obtain the integral cross section, we must integrate Eq. (8.71) over all orientations of the incident collision flux $\boldsymbol{k}_{\tilde{\alpha}\tilde{\beta}}$ and all scattered directions $\boldsymbol{R}$. As in the case of Eq. (8.51), the integration yields

$$\sigma_{\tilde{\alpha}\tilde{\beta}\to\alpha\beta} = \frac{(1 + \delta_{\tilde{\alpha}\tilde{\beta}})\pi}{k^2_{\tilde{\alpha}\tilde{\beta}}} \sum_{l}\sum_{m_l}\sum_{l'}\sum_{m'_l} |T^{\eta}_{\tilde{\alpha}\tilde{\beta}lm_l;\alpha\beta l'm'_l}|^2, \tag{8.73}$$

where

$$T^{\eta}_{\tilde{\alpha}\tilde{\beta}lm_l;\alpha\beta l'm'_l} = \delta_{\tilde{\alpha},\alpha}\delta_{\tilde{\beta},\beta}\delta_{ll'}\delta_{m_lm'_l} - S^{\eta}_{\tilde{\alpha}\tilde{\beta}lm_l;\alpha\beta l'm'_l}$$

is the T-matrix computed in the symmetrized basis.

8.1.6 Numerical Integration of Coupled-Channel Equations

The practical goal of the coupled-channel calculations is to compute the scattering S-matrix. Given the S-matrix elements, we can calculate any collision properties of molecular systems. There are multiple numerical methods developed for integrating the second-order differential equations of the form (8.11). Among the most popular ones are the Numerov method, the R-matrix method, and the log-derivative method. Like many practitioners in the field, I favor the log-derivative method. This method was originally developed by Johnson [311] and has gained much popularity after the improvements by Manolopoulos [312]. Here, I will describe the log-derivative method in its simplest form as proposed in the original paper by Johnson. This is certainly sufficient for many applications not pushing the limits of computational complexity. The reader interested in solving the scattering problems with more than 1000 coupled equations should consider the improved algorithm of Manolopoulos [312].

For simplicity, consider first the scattering problem that can be described entirely by only one term in the sum (8.8). If this is the case, Eq. (8.11) is reduced to

$$\left[\frac{\mathrm{d}^2}{\mathrm{d}R^2} - \frac{l(l+1)}{R^2} + k_{\alpha\beta}^2\right] F_{\alpha\beta l m_l}(R) = 2\mu\langle\alpha\beta l m_l|U|\alpha\beta l m_l\rangle F_{\alpha\beta l m_l}(R). \quad (8.74)$$

For convenience, we can rewrite this equation as

$$\left[\frac{\mathrm{d}^2}{\mathrm{d}R^2} + W(R)\right] F_{\alpha\beta l m_l}(R) = 0,$$

where

$$W(R) = k_{\alpha\beta}^2 - \frac{2\mu}{\hbar^2}\langle\alpha\beta l m_l|U|\alpha\beta l m_l\rangle - \frac{l(l+1)}{R^2}. \quad (8.75)$$

Instead of solving Eq. (8.74) directly for $F(R)$, we will introduce the log-derivative

$$y(R) = \left[\frac{\mathrm{d}F_{\alpha\beta l m_l}(R)}{\mathrm{d}R}\right] \frac{1}{F_{\alpha\beta l m_l}(R)}. \quad (8.76)$$

This change of variables converts the second-order differential equation (8.75) into the first-order differential equation

$$\frac{\mathrm{d}y(R)}{\mathrm{d}R} + W(R) + y^2(R) = 0, \quad (8.77)$$

which can be readily verified by replacing $y(R)$ as defined in Eq. (8.76) into Eq. (8.77).

We need to integrate Eq. (8.77) numerically in the interval of R from $R_{\mathrm{min}} \to 0$ to $R_{\mathrm{max}} \to \infty$. In practice, R_{min} is chosen to be small enough and R_{max} big enough to yield results that do not change upon further decreasing R_{min} and increasing R_{max}. The interval between R_{min} and R_{max} is granulized into $N+1$ descrete points, labeled by n. The starting value of n is $n = 0$ at $R = R_{\mathrm{min}}$ and the separation between the adjacent grid points is $R_{n+1} - R_n = h$. The value of y on this grid can be calculated as

$$y_n = (1 + hy_{n-1})^{-1} y_{n-1} - (h/3) w_n u_n, \tag{8.78}$$

where the coefficients w_n are

$$w_n = \begin{cases} W(R_n), & \text{if} \quad n = 0, 2, 4, \ldots, N \\ [1 + (h^2/6) W(R_n)]^{-1} W(R_n), & \text{if} \quad n = 1, 3, 5, \ldots, N-1 \end{cases}, \tag{8.79}$$

and the coefficients u_n come from the Simpson integration rule

$$u_n = \begin{cases} 1, & n = 0, N, \\ 4, & n = 1, 3, 5, \ldots, N-1, \\ 2, & n = 2, 4, 6, \ldots, N-2. \end{cases} \tag{8.80}$$

The integration begins at $R_{\min}$. This point should be chosen to be close to zero, deep in the classically forbidden region, where the wave function is expected to be very small. For typical molecular systems, it is sufficient to choose $R_{\min} < 1$ bohr. When $F(R) \to 0$, $y(R)$ diverges. The value of y at $n = 0$ must therefore be set to some very large number, for example, 10^{30}. The results of the calculations should be insensitive to the actual value of $y_{n=0}$, as long as it is deep in the classically forbidden region and is sufficiently large. Going outward from $n = 0$, the values of y_n can be calculated using Eq. (8.78), all the way until y_N at $R = R_{\max}$. The value of $R_{\max}$ should be sufficiently large, such that the end result does not depend on the actual value of $R_{\max}$.

Once $y_N(R_{\max})$ is computed, it can be matched to the log-derivative of the functions $F_{\alpha\beta l m_l}(R_{\max})$. In the limit of very large $R_{\max}$, the functions $F_{\alpha\beta l m_l}(R_{\max})$ are given by Eq. (8.44). In principle, we could use Eq. (8.44) to derive an analytical expression for y_N containing the S-matrix element and then match this expression to the numerically computed value of y_N in order to evaluate the S-matrix element. In practice, this is not the best idea since Eq. (8.44) is valid only at very large distances, where the centrifugal term $l(l+1)/R^2$ can be neglected. Instead of numerically propagating the solutions of Eq. (8.78) to such large distances, we can use the analytical solutions of Eq. (8.17), which are valid at shorter distances, namely at distances, where the interaction potential vanishes, but the centrifugal term still remains significant. At such distances, the coefficients $F_{\alpha\beta l m_l}(R_{\max})$ can be written as

$$F_{\alpha\beta l m_l}(R) = B[\hat{j}_l(kR) - K_l \hat{n}_l(kR)], \tag{8.81}$$

where $\hat{j}_l(kR)$ and $\hat{n}_l(kR)$ are the Riccati–Bessel functions, defined in terms of the spherical Bessel functions as

$$\hat{j}_l(x) = x j_l(x),$$
$$\hat{n}_l(x) = x n_l(x).$$

By taking the R-derivative of the functions (8.81) and equalizing F'/F with the numerically computed y_N at $R_{\max}$, we can obtain the value of K_l as follows:

$$K_l = [y_N \hat{n}_l - \hat{n}'_l]^{-1}[y_N \hat{j}_l - \hat{j}'_l]. \tag{8.82}$$

The value of the S-matrix element in Eq. (8.44) is related to K_l as following:

$$S_l = (1 + iK_l)(1 - iK_l)^{-1}. \tag{8.83}$$

The S-matrix element is what we are after, since it can be used to compute the scattering cross sections using Eq. (8.51). Of course, for the case of single-channel scattering, there is only one term in the sum (8.51), yielding the elastic scattering cross section.

The numerical procedure outlined above can be easily extended to the general case of multi-channel scattering. In general, Eq. (8.11) can be written in matrix–vector form

$$\mathbf{HF} = \mathbf{UF}, \tag{8.84}$$

where $\mathbf{H}$ is a diagonal matrix with the elements

$$\mathbf{H}_{\alpha\beta l m_l;\alpha\beta l m_l} = \left[\frac{\mathrm{d}^2}{\mathrm{d}R^2} - \frac{l(l+1)}{R^2} + k^2_{\alpha\beta}\right], \tag{8.85}$$

$\mathbf{F}$ is the column vector of the coefficients $F_{\alpha\beta l m_l}(R)$ and $\mathbf{U}$ is a square matrix with the elements $\langle \alpha\beta l m_l|U|\alpha'\beta' l' m'_l\rangle$. In order to apply the boundary conditions (8.44), we need to fix the incoming collision channel $\tilde{\alpha}\tilde{\beta}\tilde{l}\tilde{m}_l$. Since Eq. (8.11) is independent of the choice of the incoming collision channel, we can attach the label $\alpha\beta\tilde{l}\tilde{m}_l$ to each coefficient $F_{\alpha\beta l m_l}(R)$ and replace in Eq. (8.84) the column vector $\mathbf{F}$ with a square matrix containing $F^{\tilde{\alpha}\tilde{\beta}\tilde{l}\tilde{m}_l}_{\alpha\beta l m_l}$. Each row of this matrix corresponds to a different incoming collision channel $\tilde{\alpha}\tilde{\beta}\tilde{l}\tilde{m}_l$. In order to solve this matrix–matrix equation, we can construct the log-derivative matrix as

$$\mathbf{y} = \mathbf{F}'\mathbf{F}^{-1}, \tag{8.86}$$

which satisfies the following equation:

$$\frac{\mathrm{d}}{\mathrm{d}R}\mathbf{y}(R) + \mathbf{W}(R) + \mathbf{y}^2(R) = 0, \tag{8.87}$$

where

$$\mathbf{W}_{n,n'} = \delta_{n,n'}\left(k_n^2 - \frac{l(l+1)}{R^2}\right) - \frac{2\mu}{\hbar^2}\langle n|U|n'\rangle \tag{8.88}$$

with $n \equiv \alpha\beta l m_l$.

As in the case of the single-channel problem, the integration of Eq. (8.87) starts at $R = R_{\min}$, where

$$\mathbf{y} = a\mathbf{I}, \tag{8.89}$$

I is the identity matrix and a is a large number, for example, $a = 10^{30}$. The log-derivative matrix is propagated in the positive R-direction using Eq. (8.78), in which both y and u_n now become matrices. At the last grid point $R_N = R_{\max}$, the log-derivative matrix $\mathbf{y}_N$ is used to compute the K-matrix as

$$\mathbf{K} = [\mathbf{y}_N \mathbf{n} - \mathbf{n}']^{-1}[\mathbf{y}_N \mathbf{j} - \mathbf{j}'], \tag{8.90}$$

where **n**, **j**, **n′**, and **n′** are the diagonal matrices of the Riccati–Bessel functions and their derivatives

$$\mathbf{j}_{n,n'} = \delta_{n,n'} k_{\alpha\beta}^{-1/2} \hat{j}_l(k_{\alpha\beta} R_{\max}), \tag{8.91}$$

$$\mathbf{n}_{n,n'} = \delta_{n,n'} k_{\alpha\beta}^{-1/2} \hat{n}_l(k_{\alpha\beta} R_{\max}), \tag{8.92}$$

$$\mathbf{j}'_{n,n'} = \delta_{n,n'} k_{\alpha\beta}^{-1/2} \hat{j}'_l(k_{\alpha\beta} R_{\max}), \tag{8.93}$$

$$\mathbf{n}'_{n,n'} = \delta_{n,n'} k_{\alpha\beta}^{-1/2} \hat{n}'_l(k_{\alpha\beta} R_{\max}), \tag{8.94}$$

where, again, $n \equiv \alpha\beta l m_l$. The K-matrix is related to the S-matrix by the matrix version of Eq. (8.83)

$$\mathbf{S} = (\mathbf{I} + i\mathbf{K})(\mathbf{I} - i\mathbf{K})^{-1}. \tag{8.95}$$

In practice, it is easier to separate this equation into a sum of the real and imaginary parts

$$\mathbf{S} = (\mathbf{I} - \mathbf{K}^2)^{-1} + i2\mathbf{K}(\mathbf{I} + \mathbf{K}^2)^{-1}. \tag{8.96}$$

8.2 Interactions with External Fields

8.2.1 Coupled-Channel Equations in Arbitrary Basis

In order to include the interactions of molecules with external field into the coupled-channel equations, we first note that the basis states in expansion (8.8) need not be the eigenstates of molecule A and molecule B. Instead of using the eigenstates $|\alpha\rangle$ in Eq. (8.8), we can use any basis set related to $|\alpha\rangle$ by a unitary transformation. The same applies to $|\beta\rangle$.

Consider an arbitrary molecular basis $|a\rangle$ defined as

$$|a\rangle = \sum_{\alpha} C_{\alpha}^{a} |\alpha\rangle \tag{8.97}$$

and a basis $|b\rangle$ defined as

$$|b\rangle = \sum_{\beta} C_{\beta}^{b} |\beta\rangle. \tag{8.98}$$

The eigenstates of the total Hamiltonian (8.8) can be represented by a new basis set expansion

$$|\Psi\rangle = R^{-1} \sum_{a} \sum_{b} \sum_{l} \sum_{m_l} F_{ablm_l}(R) |ablm_l\rangle. \tag{8.99}$$

Using the arguments leading to Eq. (8.11), the coupled-channel equations can be written in the basis $|ablm_l\rangle$ as following:

$$\left[\frac{\mathrm{d}^2}{\mathrm{d}R^2} - \frac{l(l+1)}{R^2} + k^2\right] F_{ablm_l}(R)$$
$$= \frac{2\mu}{\hbar^2} \sum_{a'b'l'm_l'} \langle ablm_l | U + \hat{H}_A + \hat{H}_B | a'b'l'm_l' \rangle F_{a'b'l'm_l'}(R), \quad (8.100)$$

where

$$k^2 = \frac{2\mu}{\hbar^2} E. \quad (8.101)$$

Note that the matrices of $\hat{H}_A$ and $\hat{H}_B$ are not diagonal in the basis $|ab\rangle$, so we have included the matrix elements of these operators on the right-hand side of Eq. (8.100). These coupled equations can be integrated numerically, using, for example, the log-derivative procedure described in Section 8.1.6. As in any basis, we would start the integration at $R = R_{\min}$, where the log-derivative matrix is assumed to be diagonal with all nonzero elements set to a very large number. The log-derivative matrix can then be propagated to $R = R_{\max}$ as described in Section 8.1.6.

However, the boundary conditions cannot be applied directly in the basis (8.99), since the states $|a\rangle$ and $|b\rangle$ do not correspond to the eigenstates of the scattering molecules. We can correct for this deficiency by transforming the log-derivative matrix $\mathbf{y}_N$ at the last propagation point back to the representation $|\alpha\beta lm_l\rangle$, to obtain

$$\tilde{\mathbf{y}}_N = \mathbf{C}^\dagger \mathbf{y}_N \mathbf{C}, \quad (8.102)$$

where $\mathbf{C}$ is the matrix of the eigenvectors of the Hamiltonian $\hat{H}_A + \hat{H}_B$ in the basis $|ab\rangle$. Once this is done, the K-matrix and the S-matrix can be constructed from $\tilde{\mathbf{y}}_N$ as described in Section 8.1.6.

8.2.2 External Field Couplings

The interactions of molecules with external fields are included in the asymptotic Hamiltonians $\hat{H}_A$ and $\hat{H}_B$. In general, the interactions with external fields couple states of different total angular momenta. As a result, the basis states $|\alpha\rangle$ and $|\beta\rangle$ for molecules in external fields cannot be obtained analytically. This may complicate the evaluation of the matrix elements of the interaction potential in Eq. (8.11). Therefore, instead of using the eigenstates $|\alpha\rangle$ and $|\beta\rangle$, it may be easier to work with the basis that diagonalizes the molecular Hamiltonians at zero field. For example, the collision problem of a $^1\Sigma$ molecule in an electric field with a structureless atom can be conveniently described using the basis of rigid rotor states for the molecule. In this case, $|a\rangle \equiv |NM_N\rangle$ and $|b\rangle \equiv 1$, and the transformation (8.102) is a matrix of eigenvectors of the molecular Hamiltonian in the field.

Even more convenient may be to use the direct product basis of the eigenstates of different angular momentum operators. To make this point more general, consider the interaction of an open-shell atom with an open-shell molecule in a Σ electronic state. For these systems, the Hamiltonians of the separated particles A and B can be generally represented as

$$\hat{H}_{\rm A} = \hat{H}_{\rm A}^{S_{\rm A}} + \hat{H}_{\rm A}^{I_{\rm A}} + \hat{H}_{\rm A}^{L_{\rm A}} + \hat{V}_{\rm A}^{I_{\rm A},S_{\rm A}} + \hat{V}_{\rm A}^{S_{\rm A},L_{\rm A}} + \hat{V}_{\rm A}^{I_{\rm A},L_{\rm A}} + \hat{V}_{\rm A}^{I_{\rm A},L_{\rm A},S_{\rm A}},$$
$$H_{\rm B} = \hat{H}_{\rm B}^{S_{\rm B}} + \hat{H}_{\rm B}^{I_{\rm B}} + \hat{H}_{\rm B}^{L_{\rm B}} + \hat{V}_{\rm B}^{I_{\rm B},S_{\rm B}} + \hat{V}_{\rm B}^{S_{\rm B},L_{\rm B}} + \hat{V}_{\rm B}^{I_{\rm B},L_{\rm B}} + \hat{V}_{\rm B}^{I_{\rm B},L_{\rm B},S_{\rm B}}, \qquad (8.103)$$

where $S_{\rm A}$ and $S_{\rm B}$ denote the electron spin angular momenta, $I_{\rm A}$ and $I_{\rm B}$ are the nuclear spin angular momenta, and $L_{\rm A}$ and $L_{\rm B}$ denote the electron orbital angular momenta for open-shell atoms or the nuclear rotational angular momenta for Σ-state molecules. The first three terms in Eq. (8.103) depend only on S, I, or L, respectively, and the last four terms describe the interactions that give rise to fine and hyperfine structures of A and B. The terms $\hat{H}_{\rm A}^{S_{\rm A}}$, $\hat{H}_{\rm A}^{I_{\rm A}}$, $\hat{H}_{\rm B}^{S_{\rm B}}$, and $\hat{H}_{\rm B}^{I_{\rm B}}$ include the interaction with external magnetic fields and the terms $\hat{H}_{\rm A}^{L_{\rm A}}$ and $\hat{H}_{\rm B}^{L_{\rm B}}$ include the interaction of molecules with external electric fields.

Krems and Dalgarno [313] proposed to expand the total wave function in *direct* products of eigenfunctions of $\boldsymbol{l}^2$ and $\boldsymbol{l}_z$, and the eigenfunctions of each of $\hat{H}_{\rm A}^{S_{\rm A}}$, $\hat{H}_{\rm A}^{I_{\rm A}}$, $\hat{H}_{\rm A}^{L_{\rm A}}$, $\hat{H}_{\rm B}^{S_{\rm B}}$, $\hat{H}_{\rm B}^{I_{\rm B}}$, and $\hat{H}_{\rm B}^{L_{\rm B}}$ defined in the space-fixed quantization frame with the z-axis determined by the direction of an external field. For example, the close coupling expansion can be written as

$$\Psi = R^{-1} \sum_i F_i(R)\phi_i, \qquad (8.104)$$

where

$$\phi_i = |I_{\rm A}M_{I_{\rm A}}\rangle|I_{\rm B}M_{I_{\rm B}}\rangle|L_{\rm A}M_{L_{\rm A}}\rangle|L_{\rm B}M_{L_{\rm B}}\rangle|S_{\rm A}M_{S_{\rm A}}\rangle|S_{\rm B}M_{S_{\rm B}}\rangle|lm_l\rangle, \qquad (8.105)$$

and m_l, $M_{S_{\rm A}}$, $M_{S_{\rm B}}$, $M_{I_{\rm A}}$, $M_{I_{\rm B}}$, $M_{L_{\rm A}}$, $M_{L_{\rm B}}$ are the projections of $\boldsymbol{l}$, $\boldsymbol{S}_{\rm A}$, $\boldsymbol{S}_{\rm B}$, $\boldsymbol{I}_{\rm A}$, $\boldsymbol{I}_{\rm B}$, $\boldsymbol{L}_{\rm A}$, and $\boldsymbol{L}_{\rm B}$ on the space-fixed quantization axis. We assume for simplicity that the diatomic molecules contain only one atom with a nonzero nuclear spin. The generalization of the theory to describe molecules with two atoms carrying nonzero nuclear spin is straightforward.

The basis (8.105) is very convenient as it allows to evaluate the matrix elements of all operators entering Eq. (8.11) by a simple application of the Wigner–Eckart theorem (see Section 9.1). In order to see the utility of the fully uncoupled representation (8.105), consider the specific examples presented in Chapter 9.

8.3 The Arthurs–Dalgarno Representation

The numerical computation of the scattering S-matrices involves a series of matrix inversions. This is clear, for example, from Eq. (8.78). The dimension of the square matrices that need to be inverted at each integration step is $N \times N$, where N is the number of terms in the sum (8.8). The computation time required to invert a square Hermitian matrix increases as the third power of N. This limits the basis set expansion (8.8) to a finite number of states $\alpha\beta l m_l$, usually less than 10 000. The largest coupled-channel calculation I have ever done included about 25 000 coupled equations and it took about six months of computations to obtain the S-matrix in that work!

Working on rotationally inelastic scattering of rigid rotors [314], Arthurs and Dalgarno noted that one could decrease the computational complexity of the scattering problem by taking advantage of the conservation of total angular momentum. In the absence of external fields, the states α are the eigenstates of the square of the total angular momentum $\boldsymbol{f}_A$ and the z-component of the total angular momentum of molecule A. The angular momentum $\boldsymbol{f}_A$ is a vector sum of all nuclear and electronic spin angular momenta as well as the rotational angular momentum of molecule A. The energy states of molecule A can therefore be labeled by the quantum number f_A and the projection m_{f_A} of $\boldsymbol{f}_A$ on the quantization axis. Similarly, each state β can be associated with well-defined quantum numbers f_B and m_{f_B}. We can write explicitly $|\alpha\rangle \equiv |\bar{\alpha} f_A m_{f_A}\rangle$, where $\bar{\alpha}$ encapsulates all quantum numbers of particle A except f_A and m_{f_A}. Similarly, we can write $|\beta\rangle \equiv |\bar{\beta} f_B m_{f_B}\rangle$.

In the absence of external fields, the total angular momentum of the collision complex, defined as the vector sum of all angular momenta of particle A, all angular momenta of particle B, and the end-over-end rotational angular momentum $\boldsymbol{l}$ of the collision complex,

$$\boldsymbol{F} = \boldsymbol{f}_A + \boldsymbol{f}_B + \boldsymbol{l} \tag{8.106}$$

is conserved. What that means is that the Hamiltonian matrix of the collision system cannot contain elements coupling states of different total angular momenta.

Instead of using the states $|\alpha\beta l m_l\rangle$ as a basis for the expansion (8.8), Arthurs and Dalgarno proposed to use the states

$$|\bar{\alpha}\bar{\beta}F(f_A f_B l)M_F\rangle = \sum_{f_{AB}} \sum_{m_{f_A}} \sum_{m_{f_B}} \sum_{m_l} C^{F,M_F}_{f_A, f_B, l, m_{f_A}, m_{f_B}, m_l} |\bar{\alpha} f_A m_{f_A} \bar{\beta} f_B m_{f_B} l m_l\rangle, \tag{8.107}$$

where $|f_A - f_B| \leq f_{AB} \leq f_A + f_B$ and the coefficients of the unitary transformation are products of the Clebsch–Gordan coefficients

$$C^{F,M_F}_{f_A,f_B,l,m_{f_A},m_{f_B},m_l} = C^{f_{AM},m_{f_{AB}}}_{f_A m_{f_A} f_B m_{f_B}} C^{FM_F}_{f_{AB} m_{f_{AB}} l m_l}. \quad (8.108)$$

Note that $m_{f_{AB}} = m_{f_A} + m_{f_B}$ and that $M_F = m_{f_A} + m_{f_B} + m_l$.

The expansion (8.8) can now be rewritten as

$$|\Psi\rangle = R^{-1} \sum_F \sum_{M_F} \sum_{\bar{\alpha}} \sum_{\bar{\beta}} \sum_l \sum_{f_A} \sum_{f_B} F_{\bar{\alpha}\bar{\beta}f_Af_Bl;FM_F}(R)|\bar{\alpha}\bar{\beta}F(f_Af_Bl)M_F\rangle. \quad (8.109)$$

Because the transformation in Eq. (8.107) is unitary, the number of states $|\bar{\alpha}\bar{\beta}F(f_Af_Bl)M_F\rangle$ is exactly the same as the number of states $|\alpha\beta l m_l\rangle$. However, the total Hamiltonian is diagonal in F and M_F quantum numbers. Thus, the transformation (8.107) block-diagonalizes the scattering problem. In other words, we can define states

$$|\Psi_{FM_F}\rangle = R^{-1} \sum_{\bar{\alpha}} \sum_{\bar{\beta}} \sum_l \sum_{f_A} \sum_{f_B} F_{\bar{\alpha}\bar{\beta}f_Af_Bl;FM_F}(R)|\bar{\alpha}\bar{\beta}F(f_Af_Bl)M_F\rangle \quad (8.110)$$

that correspond to fixed values of F and M_F and solve the scattering problem for each of these states independently. The total angular momentum representation of the scattering states (8.110) is often called the Arthurs–Dalgarno representation. There are, by far, fewer states in the sum (8.110) than in (8.8). Thus, the Arthurs–Dalgarno representation replaces a large number of coupled differential equations by a set of smaller coupled-channel problems.

The savings offered by the total angular momentum representation are huge! Consider a hypothetical problem with 10 coupled equations that takes one unit of time to integrate. The same problem with 100 coupled equations would take 10^3 units of time to integrate. If the 100-channel problem can be split into 10 independent problems each with 10 coupled equations, the benefits are obvious. For this reason, the Arthurs–Dalgarno representation should always be used for scattering problems in the absence of external fields.

In the presence of an external field, the total angular momentum is not conserved. It may therefore appear that the Arthurs–Dalgarno representation is not useful for the scattering problems in fields. However, if expressed in the total angular momentum representation, the matrix of the full Hamiltonian of the collision system in an external field has a simple tridiagonal structure shown in Figure 8.1. Each diagonal block of the matrix corresponds to a well-defined quantum number F. The adjacent blocks contain the matrix elements $\langle F|\hat{H}|F \pm 1\rangle$ induced by the fields. The structure of the matrix suggests that the presence of block $n + k$ in the matrix will be of little consequence for the dynamics of states within block n, if k is sufficiently large. The convergence of the calculations with respect to the basis set may be sought by increasing the number of

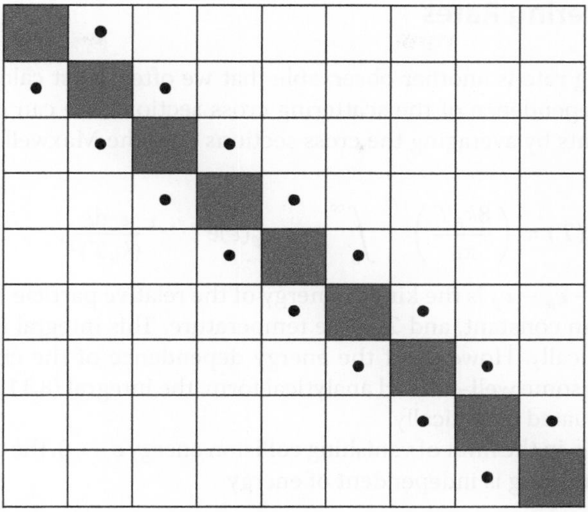

Figure 8.1 Schematic structure of the Hamiltonian matrix in the total angular momentum representation (8.109) for two interacting particles in an external field. Each square of the table represents the block of the matrix elements corresponding to a set of two quantum numbers: F and F'. The empty squares show the blocks of the matrices, in which all matrix elements are zero. The shaded squares show the nonvanishing blocks of the matrix. The bullets show the blocks of the matrix populated by the field-induced couplings. In the absence of an external field, the Hamiltonian matrix is diagonal in F so only the blocks with $F = F'$ are nonzero.

F-blocks. It was shown in a number of recent works [315, 316] that this way of truncating the basis set is indeed more efficient, leading to fully converged calculations with fewer coupled equations than when expansion (8.8) is used.

We should also mention that, if the field is directed along the quantization axis, the full Hamiltonian is diagonal in the M_F quantum number. That is why, one should always group states with the same value of M_F together, whether Eq. (8.8) or Eq. (8.109) is used. This point will be made clear in Chapter 9, where we consider specific examples.

Finally, one more note on the choice of the basis (8.8) vs (8.109). While the expansion (8.109) is likely to always lead to fewer coupled equations, the use of the basis (8.8) simplifies the expressions for the matrix elements of the full Hamiltonian, and consequently the preparation of the computer codes. This will be illustrated in Chapter 9. I often write the scattering codes in both representations in order to verify the validity of the codes. At zero field, the results with both representations must be identical. At nonzero fields, the results must be the same if the calculations are fully converged with respect to the basis set size.

8.4 Scattering Rates

The scattering rate is another observable that we often must calculate. Given the energy dependence of the scattering cross sections, we can calculate the rate coefficients by averaging the cross sections over the Maxwell–Boltzmann distribution

$$R_{\tilde{\alpha}\tilde{\beta}\to\alpha\beta}(T) = \left(\frac{8k_{\rm B}T}{\pi\mu}\right)^{1/2} \int_0^\infty \sigma_{\tilde{\alpha}\tilde{\beta}\to\alpha\beta}(\varepsilon)\mathrm{e}^{-\varepsilon/k_{\rm B}T}\frac{\varepsilon\, \mathrm{d}\varepsilon}{(k_{\rm B}T)^2}, \tag{8.111}$$

where $\varepsilon = E - \epsilon_{\tilde{\alpha}} - \epsilon_{\tilde{\beta}}$ is the kinetic energy of the relative particle motion, $k_{\rm B}$ is the Boltzmann constant, and T is the temperature. This integral is often evaluated numerically. However, if the energy dependence of the cross sections $\sigma_{\tilde{\alpha}\tilde{\beta}\to\alpha\beta}(\varepsilon)$ has some well-defined analytical form, the integral (8.111) can sometimes be evaluated analytically.

For example, in the limit of vanishing collision energy $\varepsilon \to 0$, the cross section for elastic scattering is independent of energy

$$\sigma_{\tilde{\alpha}\tilde{\beta}\to\tilde{\alpha}\tilde{\beta}} \sim \text{const}, \tag{8.112}$$

and the cross sections for inelastic scattering are inversely proportional to the collision velocity

$$\sigma_{\tilde{\alpha}\tilde{\beta}\to\alpha\beta} \sim \frac{1}{\varepsilon^{1/2}}. \tag{8.113}$$

These expressions are known as the Wigner threshold laws [317] (see also an earlier paper by Bethe and Placzek [318]). The inelastic process (8.113) is, of course, only possible if it is exoergic, i.e. if the energy of the final state $(\alpha\beta)$ is lower than the energy of the initial state $(\tilde{\alpha}\tilde{\beta})$.

If the scattering cross section is expressed by the power law

$$\sigma_{\tilde{\alpha}\tilde{\beta}\to\alpha\beta} = A\varepsilon^a \tag{8.114}$$

with A and a being some constants, the integral (8.111) can be evaluated analytically to yield

$$R_{\tilde{\alpha}\tilde{\beta}\to\alpha\beta} = A\left(\frac{8}{\pi\mu}\right)(k_{\rm B}T)^{a+1/2}(a+1)! \tag{8.115}$$

This shows that in the limit of zero temperature, where the elastic scattering cross section is given by Eq. (8.112), the rate for elastic scattering is vanishing as

$$R_{\tilde{\alpha}\tilde{\beta}\to\tilde{\alpha}\tilde{\beta}} \sim (T)^{1/2}. \tag{8.116}$$

At the same time, for inelastic scattering ($\tilde{\alpha} \neq \alpha$ and/or $\tilde{\beta} \neq \beta$) described by the cross section of the form (8.113), we have

$$R_{\tilde{\alpha}\tilde{\beta}\to\alpha\beta} \sim \text{const}. \tag{8.117}$$

This is a remarkable result. It implies that the rates of exoergic inelastic scattering and chemical reactions are finite at zero temperature. As recent calculations show [319], these zero-temperature rates for inelastic scattering and chemical reactions of diatomic molecules can be very large, in some instances, larger than at room temperature! This suggests an interesting regime of chemical dynamics, where the collision energy is much smaller than perturbations induced by external electromagnetic fields. We will discuss the effects of external fields on collision dynamics of molecules at such low temperatures in Chapter 11.

9

Matrix Elements of Collision Hamiltonians

This chapter is written for those interested in the practical implementation of the scattering theory described in Chapter 8. It also serves as a good example of the power of angular momentum algebra in application to quantum scattering theory.

Chapter 8 discussed how to solve the coupled-channel equations (8.11) in order to compute the probabilities for elastic and inelastic scattering of molecules. However, in order to solve these equations, we must first define them completely. Equation (8.11) is parametrized by the matrix elements of the interaction potential U (and other operators in the Hamiltonian). In the present chapter, we will discuss how to evaluate these matrix elements.

I will present several specific examples pertinent to the scattering problems for molecules in external fields. I hope that these examples will serve as a tutorial for those who need to evaluate the matrix elements of various Hamiltonians in various basis sets.

I will begin with the formulation based on the fully uncoupled representation of the form (8.105). This representation simplifies the evaluation of the matrix elements of all operators in the Hamiltonian to a great extent. Later in this chapter, we will also consider the matrix elements of the field-dependent interactions in the total angular momentum representation (8.109). As discussed in Section 8.3, the advantage of the total angular momentum representation is the reduced number of coupled equations required for fully converged calculations.

As discussed in Chapter 8, the Hamiltonian of the collision complex of two particles A and B can be generally written in the following form:

$$\hat{H} = -\frac{\hbar^2}{2\mu}\Delta_R + U + \hat{H}_A + \hat{H}_B, \tag{9.1}$$

where U is an operator describing the interaction between A and B. The Hamiltonian (9.1) is the starting point for all scattering calculations. The particular form of the operators U, $\hat{H}_A$, and $\hat{H}_B$ depends on the specific

Molecules in Electromagnetic Fields: From Ultracold Physics to Controlled Chemistry, First Edition. Roman V. Krems.

problem under consideration. Within the assumptions adopted in the present manuscript, the effects of external fields are included entirely in $\hat{H}_A$ and $\hat{H}_B$, and the R-dependence of the external-field-induced interactions is neglected.

Given the formulation of the scattering problem in Chapter 8, the practical calculations involve the following steps:

i) Choosing a specific basis;
ii) Evaluation of the matrix elements of the interaction potential U and the Hamiltonians $\hat{H}_A$ and $\hat{H}_B$ parametrizing Eq. (8.100);
iii) Numerical integration of the coupled equations, for example, using the log-derivative method of Section 8.1.6;
iv) Transformation of the log-derivative matrix to the representation, in which $\hat{H}_A$ and $\hat{H}_B$ are diagonal; this transformation is numerically obtained by diagonalizing the matrix of $\hat{H}_A + \hat{H}_B$ in the chosen basis;
v) Computation of the scattering S-matrix.

Steps (iii) – (v) are described in Chapter 8. They are general for any scattering problem. Here, we present examples of the evaluation of the matrix elements of U, $\hat{H}_A$, and $\hat{H}_B$ for specific systems.

9.1 Wigner–Eckart Theorem

Throughout this chapter, we will use one indispensable tool: the Wigner–Eckart theorem. It is impossible to overstate the usefulness of this theorem, not only for scattering theory, but in general when working with angular momentum algebra. I use it every day and I can't stop admiring it. The Wigner–Eckart theorem is the following expression [9, 10]:

$$\langle jm|\hat{T}_q^k|j'm'\rangle = (-1)^{j-m}\begin{pmatrix} j & k & j' \\ -m & q & m' \end{pmatrix}\langle j||\hat{T}^k||j'\rangle. \tag{9.2}$$

Here, $|jm\rangle$ is an angular momentum state, i.e. an eigenstate of $\boldsymbol{j}^2$ and j_Z, where $\boldsymbol{j}$ is some angular momentum. The thing in the brackets is a $3j$-symbol, which we know how to evaluate. There are many subroutines online, or in Mathematica, or in Matlab, which, given the six numbers $j, -m, k, q, j', m'$ will produce the value of the $3j$-symbol in Eq. (9.2). The last part

$$\langle j||\hat{T}^k||j'\rangle \tag{9.3}$$

is the reduced matrix element of the corresponding spherical tensor. It is just a number, which we need to learn how to evaluate. Note that the reduced matrix element is independent of the projection quantum numbers m and m'. So if we know the reduced matrix element (one number), we know all the matrix elements in Eq. (9.2).

The spherical tensor is an operator that has a rank k and $2k + 1$ components. The components of a spherical tensor will be denoted by $\hat{T}^k_q$, with $q = -k, -k + 1, \ldots, k - 1, k$. As discussed in more detail in Appendix E, spherical tensors are defined as operators that transform under rotation $\boldsymbol{R}$ of the coordinate system as [9]

$$\boldsymbol{R}\hat{T}^k_q\boldsymbol{R}^{-1} = \sum_{q'} D^k_{q'q}\hat{T}^k_{q'}, \tag{9.4}$$

where $D^k_{q'q}$ are the Wigner's D-functions defined in Appendix D. In other words, a spherical tensor is an operator, whose rank k does not change upon rotation of the coordinate system, and whose components q in the rotated coordinate system can be written as linear combinations of the components q' of the same tensor in the original, unrotated coordinate system, with the coefficients in these linear combinations represented by the D-functions.

It is easy to see that spherical harmonics Y_{lm} are spherical tensors of rank l, with m labeling the $2l + 1$ components. This is great, because if we can represent the operators appearing in the Hamiltonian as sums over products of spherical harmonics, we can use the Wigner–Eckart theorem to evaluate the matrix elements of these operators. Let me revisit the simple example, which we considered in Section 2.4.1, namely, the interaction of a rigid rotor with a DC electric field.

The field-free states of the rigid rotor are the spherical harmonics Y_{NM_N} and the interaction of the rigid rotor with the electric field is given by (cf. Eqs. (2.34) and (2.35))

$$\hat{V} = -Ed_0\sqrt{\frac{4\pi}{3}}Y_{1,0}. \tag{9.5}$$

in the basis of the rigid-rotor states $Y_{NM_N} \equiv |NM_N\rangle$ can then be written using the Wigner–Eckart theorem as follows:

$$\begin{aligned}\langle NM_N|\hat{V}|N'M'_N\rangle &= -Ed_0\langle NM_N|Y_{1,0}|N'M'_N\rangle \\ &= (-1)^{N-M_N}\begin{pmatrix} N & 1 & N' \\ -M_N & 0 & M'_N \end{pmatrix}\langle N||\hat{Y}^1||N'\rangle. \end{aligned}\tag{9.6}$$

The Wigner–Eckart theorem immediately shows that the matrix elements will be zero unless $M_N = M'_N$. This follows from the properties of the $3j$-symbols (cf. page 44 and Appendix B).

Equation (9.6) contains the reduced matrix element $\langle N||\hat{Y}^1||N'\rangle$. What is the value of this matrix element? To answer this question, all we need to do is evaluate one integral over a product of three spherical harmonics $Y_{l_1m_1}$, $Y_{l_2m_2}$, and $Y_{l_3m_3}$ with specific values of m_1, m_2, and m_3 and compare the result with the expression given by the Wigner–Eckart theorem. For example, consider the case of $m_1 = m_2 = m_3 = 0$. In this case, the integral can be evaluated analytically, as shown for example in Ref. [9] or discussed in Appendix D, to yield

$$\langle l_1 0 | Y_{l_2,0} | l_3 0 \rangle = \left[\frac{(2l_1+1)(2l_2+1)(2l_3+1)}{4\pi} \right]^{1/2} \times \begin{pmatrix} l_1 & l_2 & l_3 \\ 0 & 0 & 0 \end{pmatrix} \begin{pmatrix} l_1 & l_2 & l_3 \\ 0 & 0 & 0 \end{pmatrix}. \tag{9.7}$$

Comparing this equation with what we would obtain using the Wigner–Eckart theorem for the matrix element $\langle l_1 0 | Y_{l_2,0} | l_3 0 \rangle$, we must conclude that

$$\langle l_1 || Y_{l_2} || l_3 \rangle = (-1)^{l_1} \left[\frac{(2l_1+1)(2l_2+1)(2l_3+1)}{4\pi} \right]^{1/2} \begin{pmatrix} l_1 & l_2 & l_3 \\ 0 & 0 & 0 \end{pmatrix}. \tag{9.8}$$

This illustrates a general strategy for evaluating the reduced matrix elements of a spherical tensor:

- Consider a specific matrix elements $\langle jm | \hat{T}^k_q | j'm' \rangle$ with any fixed values of m, q, and m';
- Evaluate this matrix element, analytically whenever possible, or numerically;
- Divide the result by the factor

$$(-1)^{j-m} \begin{pmatrix} j & k & j' \\ -m & q & m' \end{pmatrix}. \tag{9.9}$$

In the subsequent few sections, I will illustrate the application of the Wigner–Eckart theorem to evaluating fairly complex matrix elements needed for solving the scattering problem of open-shell molecules in electric and magnetic fields.

9.2 Spherical Tensor Contraction

Section 9.1 suggests a general strategy that we will use in this chapter and that I would recommend for the evaluation of matrix elements of general Hamiltonians. The strategy is

- Write the operators in the Hamiltonian as sums over products of spherical tensors, each depending on a different degree of freedom;
- Use the basis set of direct products of states depending on different degrees of freedom;
- Use the Wigner–Eckart theorem to evaluate the matrix elements of the corresponding spherical tensors.

Sometimes, it will be desired to evaluate the elements of the Hamiltonian matrix in some other than the direct product basis. For example, later in this chapter, we will discuss the matrix elements of the Hamiltonian in the total angular momentum representation. However, the strategy formulated earlier can still be used for the evaluation of these matrix elements. It just needs to be supplemented by a transformation to the desired basis, once the matrix of the Hamiltonian in the basis of direct products is evaluated.

9.3 Collisions in a Magnetic Field

9.3.1 Collisions of ^1S-Atoms with $^2\Sigma$-Molecules

Let us jump straight to a specific example of how one would evaluate the matrix elements of the Hamiltonian for the system of a closed-shell structureless atom (such as He) colliding with a $^2\Sigma$ molecule – the simplest open-shell molecule – in the presence of a magnetic field. In this case, the atom–molecule interaction potential is spin-independent, and the Hamiltonian of the ^1S-atom–$^2\Sigma$-molecule system in a magnetic field can be written as

$$\hat{H} = -\frac{\hbar^2}{2\mu} R^{-2} \frac{\partial}{\partial R} R^2 \frac{\partial}{\partial R} + \frac{\hbar^2 \boldsymbol{l}^2}{2\mu R^2} + \hat{H}_A + U(\boldsymbol{R}, \boldsymbol{r}) - V_{\mathrm{m}}(r), \tag{9.10}$$

where $U(\boldsymbol{R}, \boldsymbol{r})$ is the atom–molecule interaction potential, and the Hamiltonian $\hat{H}_A$ represents the vibrating and rotating molecule in a magnetic field [2]

$$\hat{H}_A = -\frac{\hbar^2}{2\mu_{\mathrm{m}}} r^{-2} \frac{\partial}{\partial r} r^2 \frac{\partial}{\partial r} + \frac{\hbar^2 \boldsymbol{N}^2}{2\mu_{\mathrm{m}} r^2} + \gamma \boldsymbol{N} \cdot \boldsymbol{S} + V_{\mathrm{m}}(r) + 2\mu_0 \boldsymbol{B} \cdot \boldsymbol{S}, \tag{9.11}$$

where

- r is the distance between the two atoms in the diatomic molecule
- μ_{m} is the reduced mass of the molecule
- $\boldsymbol{N}$ is the rotational angular momentum of the molecule
- $\boldsymbol{S}$ is the spin angular momentum of the molecule
- γ is the spin–rotation interaction constant
- $V_{\mathrm{m}}(r)$ is the intramolecular potential determining the vibrational motion of the diatomic molecule
- $\boldsymbol{B}$ is the vector of the magnetic field
- μ_0 is the Bohr magneton.

Note that there is no Hamiltonian $\hat{H}_B$ in Eq. (9.10) because we are dealing with a structureless atom.

Within the rigid-rotor approximation, the intramolecular distance r is fixed to r_0 so that $V_{\mathrm{m}}(r) \Rightarrow V(r_0)$ and $U(\boldsymbol{R}, \boldsymbol{r}) \Rightarrow U(\boldsymbol{R}, \boldsymbol{r}_0)$, and the Hamiltonian (9.11) is reduced to

$$\hat{H}_A = \frac{\hbar^2 \boldsymbol{N}^2}{2\mu_{\mathrm{m}} {r_0}^2} + \gamma \boldsymbol{N} \cdot \boldsymbol{S} + 2\mu_0 \boldsymbol{B} \cdot \boldsymbol{S}, \tag{9.12}$$

where the constant term $V(r_0)$ has been eliminated by redefining the zero of energy. The effects of external fields are often the most pronounced for molecules in low-energy vibrational and rotational states. For these states, the rigid-rotor approximation is almost always valid.

The quantization Z-axis is chosen in the direction of $\boldsymbol{B}$ so that the last term in Eq. (9.11) can be rewritten as

$$2\mu_0 \boldsymbol{B} \cdot \boldsymbol{S} = 2\mu_0 B S_Z,$$

where S_Z is the operator giving the Z-component of the electron spin. The Hamiltonian (9.12) and its eigenstates were discussed in Section 3.1.

As proposed in the Section 9.2, the collision problem can be formulated with the basis of direct products

$$|\phi\rangle = |\chi_v^N\rangle |NM_N\rangle |SM_S\rangle |lm_l\rangle, \tag{9.13}$$

where M_N and M_S denote the projections of $\boldsymbol{N}$ and $\boldsymbol{S}$ on the magnetic field direction and $|\chi_v^N\rangle$ are the vibrational states of the diatomic molecule labeled by the vibrational quantum number v and the quantum number N of the rotational angular momentum. Here, I am using ϕ as a quantity that labels collectively the basis states on the right-hand side of Eq. (9.13). Within the rigid-rotor approximation $|\chi_v^N\rangle = 1$ and the vibrational quantum number v goes out of the picture. The eigenstates of the full Hamiltonian can be represented as

$$|\Psi\rangle = \frac{1}{R} \sum_{\phi} F_\phi(R) |\phi_i\rangle, \tag{9.14}$$

where the functions F_ϕ are the same as F_{ablm_l} in Eq. (8.100), with

$$|a\rangle \equiv |\chi_v^N\rangle |NM_N\rangle |SM_S\rangle. \tag{9.15}$$

In order to preserve generality, we will not use the rigid-rotor approximation at this time and will continue to include the vibrational states explicitly.

The interaction potential $U(\boldsymbol{R}, \boldsymbol{r})$ is a function of three variables: the distance R between the center of mass of the molecule and the center of mass of the atom, the intramolecular distance r, and the angle θ between the vectors $\boldsymbol{R}$ and $\boldsymbol{r}$. We should have written the variables of the interaction potential more correctly as $U(R, r, \cos\theta)$ or as $U(R, r, \boldsymbol{R} \cdot \boldsymbol{r}/Rr)$. I prefer to write the interaction potentials as $U(\boldsymbol{R}, \boldsymbol{r})$ in anticipation of the next step, where we will decompose this function into a sum over operators that depend on the two vectors separately.

Any function of an angle between two vectors can be expanded in Legendre polynomials

$$U(\boldsymbol{R}, \boldsymbol{r}) = \sum_{\lambda} U_\lambda(R, r) P_\lambda(\cos\theta), \tag{9.16}$$

where the expansion coefficients $U_\lambda(R, r)$ can be found by evaluating the integrals

$$U_\lambda(R, r) = \frac{2\lambda + 1}{2} \int_{\theta=0}^{\theta=\pi} U(\boldsymbol{R}, \boldsymbol{r}) P_\lambda(\cos\theta) \mathrm{d}\cos\theta. \tag{9.17}$$

These integrals can be evaluated numerically using, for example, Gauss–Legendre quadratures.

The expansion (9.16) can be made more useful by means of the spherical harmonics addition theorem [9]

$$P_\lambda(\cos\theta) = \frac{4\pi}{2\lambda+1}\sum_{m_\lambda}(-1)^{m_\lambda}Y_{\lambda,-m_\lambda}(\hat{R})Y_{\lambda,m_\lambda}(\hat{r}), \tag{9.18}$$

where we introduced the unit vectors $\hat{R} = \boldsymbol{R}/R$ and $\hat{r} = \boldsymbol{r}/r$. The last equation allows us to represent the atom–molecule interaction potential

$$U(\boldsymbol{R},\boldsymbol{r}) = \sum_\lambda\left(\frac{4\pi}{2\lambda+1}\right)U_\lambda(R,r)\sum_{m_\lambda}(-1)^{m_\lambda}Y_{\lambda,-m_\lambda}(\hat{R})Y_{\lambda,m_\lambda}(\hat{r}), \tag{9.19}$$

as a sum over products of operators depending on different variables. This is an example of what I meant when I said in Section 9.2 that any operator can be represented as a sum over products of spherical tensors, each acting on different parts of the composite Hilbert space. This is particularly useful if we are to work with the direct product basis (9.13).

This is great because with Eq. (9.19) the matrix elements of the interaction potential in the basis (9.13) can be evaluated by the direct application of the Wigner–Eckart theorem as follows:

$$\begin{aligned}&\langle\chi_v^N(r)|\langle NM_N|\langle SM_S|\langle lm_l|U(\boldsymbol{R},\boldsymbol{r})|\chi_{v'}^{N'}(r)\rangle|N'M_N'\rangle|SM_S'\rangle|l'm_l'\rangle\\&\quad=\delta_{M_SM_S'}\langle\chi_v^N(r)|\langle NM_N|\langle lm_l|U(\boldsymbol{R},\boldsymbol{r})|\chi_{v'}^{N'}(r)\rangle|N'M_N'\rangle|l'm_l'\rangle\end{aligned} \tag{9.20}$$

$$\begin{aligned}&\langle\chi_v^N(r)|\langle NM_N|\langle lm_l|U(\boldsymbol{R},\boldsymbol{r})|\chi_{v'}^{N'}(r)\rangle|N'M_N'\rangle|l'm_l'\rangle\\&\quad=\sum_\lambda\left(\frac{4\pi}{2\lambda+1}\right)\langle\chi_v^N|U_\lambda(R,r)|\chi_{v'}^{N'}\rangle\\&\qquad\times\sum_{m_\lambda}(-1)^{m_\lambda}\langle NM_N|Y_{\lambda,m_\lambda}(\hat{r})|N'M_N'\rangle\langle lm_l|Y_{\lambda,-m_\lambda}(\hat{R})|l'm_l'\rangle.\end{aligned} \tag{9.21}$$

To write Eq. (9.20), we used the fact that the interaction potential between a closed-shell atom and an open-shell molecule is independent of the electron spin degrees of freedom, so it must be diagonal in the spin quantum number M_S. The integrals $\langle\chi_v^N|U_\lambda(R,r)|\chi_{v'}^{N'}\rangle$ over the vibrational wave functions of the molecule must be evaluated numerically at each fixed value of the atom–molecule separation R. However, if the rigid-rotor approximation is used, we must replace these integrals with values of the expansion coefficients of the interaction potential

$$\langle\chi_v^N|U_\lambda(R,r)|\chi_v^N\rangle \Rightarrow U_\lambda(R,r_0) \tag{9.22}$$

evaluated at $r = r_0$. The matrix elements $\langle NM_N|Y_{\lambda,m_\lambda}(\hat{r})|N'M_N'\rangle$, and $\langle lm_l|Y_{\lambda,-m_\lambda}(\hat{R})|l'm_l'\rangle$ can be evaluated using the Wigner–Eckart theorem (9.2) applied to the integrals over products of three spherical harmonics, as discussed in Section 9.2.

For the record, the final result for the matrix elements of the interaction potential parametrizing the coupled-channel equations (8.11) in the uncoupled basis (9.13) is

$$\langle \chi_v^N(r)|\langle NM_N|\langle SM_S|\langle lm_l|U(\boldsymbol{R},\boldsymbol{r})|\chi_{v'}^{N'}(r)\rangle|N'M'_N\rangle|SM'_S\rangle|l'm'_l\rangle$$
$$= \delta_{M_SM'_S}(-1)^{-m_l-M_N}[(2l+1)(2l'+1)(2N+1)(2N'+1)]^{1/2}$$
$$\times \sum_\lambda \langle \chi_v^N|U_\lambda(R,r)|\chi_{v'}^{N'}\rangle \begin{pmatrix} l & \lambda & l' \\ 0 & 0 & 0 \end{pmatrix} \begin{pmatrix} N & \lambda & N' \\ 0 & 0 & 0 \end{pmatrix}$$
$$\times \sum_{m_\lambda} (-1)^{m_\lambda} \begin{pmatrix} l & \lambda & l' \\ -m_l & -m_\lambda & m'_l \end{pmatrix} \begin{pmatrix} N & \lambda & N' \\ -M_N & m_\lambda & M_N \end{pmatrix}, \tag{9.23}$$

where the symbols in parentheses are the 3j-symbols.

The matrices of $\boldsymbol{N}^2$, $\boldsymbol{l}^2$, and $2\mu_0 BS_z$ are diagonal in the basis (9.13):

$$\langle \chi_v^N(r)|\langle NM_N|\langle SM_S|\langle lm_l|\mathbf{N}^2|\chi_{v'}^{N'}(r)\rangle|N'M'_N\rangle|SM'_S\rangle|l'm'_l\rangle$$
$$= \delta_{vv'}\delta_{NN'}\delta_{M_NM'_N}\delta_{M_SM'_S}\delta_{l'l'}\delta_{m_lm'_l}N(N+1), \tag{9.24}$$

$$\langle \chi_v^N(r)|\langle NM_N|\langle SM_S|\langle lm_l|\boldsymbol{l}^2|\chi_{v'}^{N'}(r)\rangle|N'M'_N\rangle|SM'_S\rangle|l'm'_l\rangle$$
$$= \delta_{vv'}\delta_{NN'}\delta_{M_NM'_N}\delta_{M_SM'_S}\delta_{l'l'}\delta_{m_lm'_l}l(l+1), \tag{9.25}$$

$$\langle \chi_v^N(r)|\langle NM_N|\langle SM_S|\langle lm_l|S_z|\chi_{v'}^{N'}(r)\rangle|N'M'_N\rangle|SM'_S\rangle|l'm'_l\rangle$$
$$= \delta_{vv'}\delta_{NN'}\delta_{M_NM'_N}\delta_{M_SM'_S}\delta_{l'l'}\delta_{m_lm'_l}M_S. \tag{9.26}$$

This is one of the advantages of the fully uncoupled basis! If we used a coupled basis, the matrix elements of some of these operators would have been fairly complex. More on this in the subsequent sections.

To complete the derivation of the coupled equations (8.100), we need to compute the matrix elements of the $\boldsymbol{N}\cdot\boldsymbol{S}$ operator in $\hat{H}_A$. We could again write this operator as a sum over products of spherical tensors, one acting on the states $|NM_N\rangle$ and one – on the states $|SM_S\rangle$. However, to illustrate a slightly different approach, we can use the following identity for the scalar product of two angular momentum vectors:

$$\gamma\boldsymbol{N}\cdot\boldsymbol{S} = \gamma\left[N_zS_z + \frac{1}{2}(N_+S_- + N_-S_+)\right]. \tag{9.27}$$

This identity leads to the following expression for the matrix elements:

$$\langle \chi_v^N(r)|\langle NM_N|\langle SM_S|\langle lm_l|\gamma\mathbf{N}\cdot\mathbf{S}|\chi_{v'}^{N'}(r)\rangle|N'M'_N\rangle|SM'_S\rangle|l'm'_l\rangle$$
$$= \delta_{vv'}\delta_{ll'}\delta_{m_lm'_l}\delta_{NN'}\delta_{M_NM'_N}\delta_{M_SM'_S}\gamma M_N M_S$$
$$+ \delta_{vv'}\delta_{ll'}\delta_{m_lm'_l}\delta_{NN'}\delta_{M_NM'_N\pm1}\delta_{M_SM'_S\mp1}$$
$$\times \frac{\gamma}{2}[N(N+1) - M'_N(M'_N \pm 1)]^{1/2}[S(S+1) - M'_S(M'_S \mp 1)]^{1/2} \tag{9.28}$$

We now have all the ingredients to set up the coupled-channel equations (8.100). Because the matrix of the spin–rotation interaction is nondiagonal, we need to introduce a transformation **C** that diagonalizes the matrix of the molecular Hamiltonian $\hat{H}_A$. This transformation does not mix states with different values of N. Thus, N is a good quantum number at $R = \infty$. Since there are no elements in the matrix of the $\boldsymbol{N} \cdot \boldsymbol{S}$ operator that couple states with different values of $M_S + M_N$, the transformation **C** conserves the sum. It follows from Eq. (9.23) that the matrix elements of the interaction potential that couple the states with M_N and $M_N \pm \Delta M_N$ also couple the states with m_l and $m_l \mp \Delta M_N$ and there are no couplings between the states corresponding to different values of $M_N + m_l$. Because the matrix of the electrostatic interaction is diagonal in M_S and the matrix of the spin–rotation interaction is diagonal in m_l, the sum $M_J = M_N + M_S + m_l$ is conserved in a collision. This reflects the conservation of the total angular momentum projection on the Z-axis.

In order to take advantage of the conservation of the total angular momentum projection, the basis states (9.13) should be grouped together in sets corresponding to different values of M_J. We can rewrite the expansion (9.14) as

$$|\Psi\rangle = \sum_{M_J} |\Psi_{M_J}\rangle \tag{9.29}$$

with $|\Psi_{M_J}\rangle$ defined as

$$|\Psi_{M_J}\rangle = \frac{1}{R} \sum_{v} \sum_{N} \sum_{l} \sum_{M_N} \sum_{M_S} \sum_{m_l} |\chi_v^N\rangle |NM_N\rangle |SM_S\rangle |lm_l\rangle F_{vNM_NM_Slm_l} \tag{9.30}$$

with the sums over the projections restricted to terms that satisfy $m_l + M_N + M_S = M_J$. Because the matrix elements of the Hamiltonian between states of different M_J are zero

$$\langle \Psi_{M_J} | \hat{H} | \Psi_{M_J'} \rangle \propto \delta_{M_J, M_J'} \tag{9.31}$$

the calculations can then be carried out for different values of M_J independently.

9.3.2 Collisions of ^1S-Atoms with $^3\Sigma$-Molecules

Let us complicate the system considered in the previous subsection by adding a second unpaired electron to the molecule. This results in an additional term in the Hamiltonian of the molecule, describing the spin–spin interaction. The spin–spin interaction can be written in the following form:

$$V_{SS} = \frac{2}{3} \lambda_{SS} \left[\frac{4\pi}{5}\right]^{1/2} \sqrt{6} \sum_q (-1)^q Y_{2-q}(\hat{r}) [\boldsymbol{S} \otimes \boldsymbol{S}]_q^{(2)}, \tag{9.32}$$

where λ_{SS} is the spin–spin interaction constant and $[\boldsymbol{S} \otimes \boldsymbol{S}]^{(2)}$ is a tensorial product of $\boldsymbol{S}$ with itself [2]. The basis functions are the same as in the previous case defined by Eq. (9.13). To define the collision problem completely, we need to determine the matrix elements of the V_{SS} operator in the fully uncoupled basis. This is a fairly complex operator so we will consider each step in the derivation of the matrix elements in detail.

By sandwiching the V_{SS} operator between the basis states (9.13), we obtain

$$\langle \chi_v^N(r)|\langle NM_N|\langle SM_S|\langle lm_l|V_{SS}|\chi_{v'}^{N'}(r)\rangle|N'M_N'\rangle|SM_S'\rangle|l'm_l'\rangle$$
$$= \delta_{vv'}\delta_{ll'}\delta_{m_lm_l'}\frac{2}{3}\lambda_{SS}\left[\frac{4\pi}{5}\right]^{1/2}\sqrt{6}\times\sum_q(-1)^q\langle NM_N|Y_{2-q}|N'M_N'\rangle$$
$$\times\langle SM_S|[\boldsymbol{S}\otimes\boldsymbol{S}]_q^{(2)}|SM_S'\rangle, \tag{9.33}$$

where, with the help of the Wigner–Eckart theorem, we can write

$$\langle NM_N|Y_{2-q}|N'M_N'\rangle = (-1)^{-M_N}[(2N+1)(2N'+1)]^{1/2}$$
$$\times\left[\frac{5}{4\pi}\right]^{1/2}\begin{pmatrix} N & 2 & N' \\ -M_N & -q & M_N' \end{pmatrix}\begin{pmatrix} N & 2 & N' \\ 0 & 0 & 0 \end{pmatrix} \tag{9.34}$$

and

$$\langle SM_S|[\boldsymbol{S}\otimes\boldsymbol{S}]_q^{(2)}|SM_S'\rangle = (-1)^{S-M_S}\begin{pmatrix} S & 2 & S \\ -M_S & q & M_S' \end{pmatrix}\langle S||[\mathbf{S}\otimes\mathbf{S}]^{(2)}||S\rangle. \tag{9.35}$$

It immediately follows from Eqs. (9.34) and (9.35) that the spin–spin interaction mixes states with different values of M_S and M_N but conserves the sum $M_S + M_N$. In other words, the projection of the total angular momentum of the molecule is conserved in the absence of a third body. We can also see that the V_{SS} operator couples states with different values of N so N is not a good quantum number, even when the molecule is separated from the atom.

The question now is, how do we evaluate the reduced matrix element in the last equation? I will use this example to illustrate a general approach for evaluating the reduced matrix elements of composite tensors formed by tensor products of spherical tensors.

We can start by writing out the tensor product

$$[\mathbf{S}\otimes\mathbf{S}]_q^{(2)} = \sum_{q_1}\sum_{q_2}\langle 1q_1 1q_2|2q\rangle\hat{S}_{q_1}^{(1)}\hat{S}_{q_2}^{(1)}, \tag{9.36}$$

where $\langle 1q_1 1q_2|2q\rangle$ is a Clebsch–Gordan coefficient. This is a special case of a general expression for the tensor product of spherical tensors $T_{q_1}^{k_1}$ and $T_{q_2}^{k_2}$ yielding a spherical tensor $\hat{T}_q^k$ or rank k [9]

$$\hat{T}_q^k = \sum_{q_1}\sum_{q_2}\langle k_1q_1k_1q_2|kq\rangle\hat{T}_{q_1}^{k_1}\hat{T}_{q_2}^{k_2}. \tag{9.37}$$

The operators $S_q^{(1)}$ are the components of the rank-1 spherical tensor representing spin angular momentum. Any angular momentum is a vector operator so it is a rank-1 spherical tensor, whose components are given by [9]

$$\hat{S}_1 = -\frac{1}{\sqrt{2}}\hat{S}_+ \tag{9.38}$$

$$\hat{S}_{-1} = \frac{1}{\sqrt{2}}\hat{S}_- \tag{9.39}$$

$$\hat{S}_0 = \hat{S}_Z, \tag{9.40}$$

where $\hat{S}_\pm$ are the raising/lowering operators and $\hat{S}_Z$ is the Z-component of angular momentum $\boldsymbol{S}$.

With this, we can try to evaluate the matrix elements of the operator $[\mathbf{S}\otimes\mathbf{S}]_q^{(2)}$ in the basis of states $|SM_S\rangle$. We have

$$\langle SM_S|[\mathbf{S}\otimes\mathbf{S}]_q^{(2)}|SM'_S\rangle = \sum_{q_1}\sum_{q_2}\langle 1q_1 1q_2|2q\rangle\langle SM_S|\hat{S}_{q_1}^{(1)}\hat{S}_{q_2}^{(1)}|SM'_S\rangle. \tag{9.41}$$

We next use the resolution of identity

$$\hat{I} = \sum_{M''_S}|SM''_S\rangle\langle SM''_S| \tag{9.42}$$

to write

$$\langle SM_S|[\mathbf{S}\otimes\mathbf{S}]_q^{(2)}|SM'_S\rangle = \sum_{q_1}\sum_{q_2}\langle 1q_1 1q_2|2q\rangle \times\sum_{M''_S}\langle SM_S|\hat{S}_{q_1}^{(1)}|SM''_S\rangle\langle SM''_S|\hat{S}_{q_2}^{(1)}|SM'_S\rangle. \tag{9.43}$$

The integrals of the type $\langle SM_S|\hat{S}_{q_1}^{(1)}|SM'_S\rangle$ can be handled with the Wigner–Eckart theorem

$$\langle SM_S|\hat{S}_{q_1}^{(1)}|SM'_S\rangle = (-1)^{S-M_S}\begin{pmatrix} S & 1 & S \\ -M_S & q_1 & M'_S\end{pmatrix}\langle S||\hat{S}^{(1)}||S\rangle. \tag{9.44}$$

In this case, it is easy to find the reduced matrix element. Consider a matrix element of the operator $S_0^{(1)} = \hat{S}_Z$. We have

$$\langle SM_S|\hat{S}_0^{(1)}|SM_S\rangle = M_S = (-1)^{S-M_S}\begin{pmatrix} S & 1 & S \\ -M_S & 0 & M_S\end{pmatrix}\langle S||\hat{S}^{(1)}||S\rangle. \tag{9.45}$$

The value of this specific 3j-symbol is [9]

$$\begin{pmatrix} S & 1 & S \\ -M_S & 0 & M_S\end{pmatrix} = (-1)^{S-M_S}\frac{M_S}{[(2S+1)S(S+1)]^{1/2}}, \tag{9.46}$$

so

$$\langle S||\hat{S}^{(1)}||S\rangle = [(2S+1)S(S+1)]^{1/2}. \tag{9.47}$$

We can now return to Eq. (9.43) to write it as

$$\langle SM_S|[\mathbf{S}\otimes\mathbf{S}]_q^{(2)}|SM_S'\rangle = \sqrt{5}[(2S+1)S(S+1)](-1)^{2S-M_S+q}$$
$$\sum_{q_1}\sum_{q_2}\sum_{M_S''}(-1)^{-M_S''}\begin{pmatrix}1 & 1 & 2\\ q_1 & q_2 & -q\end{pmatrix}\begin{pmatrix}S & 1 & S\\ -M_S & q_1 & M_S''\end{pmatrix}\begin{pmatrix}S & 1 & S\\ -M_S'' & q_2 & M_S'\end{pmatrix}, \tag{9.48}$$

where we have converted the Clebsch–Gordan coefficient into the 3*j*-symbol using the standard relation [9] and applied the Wigner–Eckart theorem to the integrals on the right-hand side. If we compare the latest equation with (9.35), we can obtain the reduced matrix element in Eq. (9.35).

To write a nicer expression for the reduced matrix element, we will use the following expression deriving from the orthogonality of the 3*j*-symbols:

$$\sum_{M_S}\sum_{M_S'}\begin{pmatrix}S & 2 & S\\ -M_S & q & M_S'\end{pmatrix}\begin{pmatrix}S & 2 & S\\ -M_S & q & M_S'\end{pmatrix} = 1/(2q+1). \tag{9.49}$$

This gives

$$\langle S||[\mathbf{S}\otimes\mathbf{S}]^{(2)}||S\rangle = \sum_{q}\sum_{M_S}\sum_{M_S'}(-1)^{S-M_S}\begin{pmatrix}S & 2 & S\\ -M_S & q & M_S'\end{pmatrix}$$
$$\times\langle SM_S|[\mathbf{S}\otimes\mathbf{S}]_q^{(2)}|SM_S\rangle. \tag{9.50}$$

Using Eq. (9.48) for the matrix elements on the right-hand side, we have

$$\langle S||[\mathbf{S}\otimes\mathbf{S}]^{(2)}||S\rangle = \sqrt{5}[(2S+1)S(S+1)]$$
$$\sum_{q_1}\sum_{q_2}\sum_{q}\sum_{M_S}\sum_{M_S'}\sum_{M_S''}(-1)^{3S-2M_S-M_S''+q}\begin{pmatrix}S & 2 & S\\ -M_S & q & M_S'\end{pmatrix}\begin{pmatrix}1 & 1 & 2\\ q_1 & q_2 & -q\end{pmatrix}$$
$$\times\begin{pmatrix}S & 1 & S\\ -M_S & q_1 & M_S''\end{pmatrix}\begin{pmatrix}S & 1 & S\\ -M_S'' & q_2 & M_S'\end{pmatrix}. \tag{9.51}$$

Noting that $q = M_S - M_S'$, we can rewrite this lengthy expression in a more compact form

$$\langle S||[\mathbf{S}\otimes\mathbf{S}]^{(2)}||S\rangle = \sqrt{5}[(2S+1)S(S+1)]\begin{Bmatrix}1 & 1 & 2\\ S & S & S\end{Bmatrix}, \tag{9.52}$$

where the symbol in the curly braces is a 6*j*-symbol. When $S = 1$, the 6*j*-symbol is equal to $1/6$ and the double-bar matrix element is equal to $\sqrt{5}$.

To obtain Eq. (9.52), I used the definition of the 6j-symbol, which can be written in terms of the 3j-symbols as follows [9, 10]:

$$\begin{Bmatrix} j_1 & j_2 & j_3 \\ j_4 & j_5 & j_6 \end{Bmatrix} = \sum_{m_1,m_2,m_3,m_4,m_5,m_6} (-1)^{j_4+j_5+j_6-m_4-m_5-m_6}$$
$$\times \begin{pmatrix} j_1 & j_2 & j_3 \\ m_1 & m_2 & m_3 \end{pmatrix} \begin{pmatrix} j_1 & j_5 & j_6 \\ m_1 & -m_5 & m_6 \end{pmatrix}$$
$$\times \begin{pmatrix} j_4 & j_2 & j_6 \\ m_4 & m_2 & -m_6 \end{pmatrix} \begin{pmatrix} j_4 & j_5 & j_3 \\ -m_4 & m_5 & m_3 \end{pmatrix}. \tag{9.53}$$

9.4 Collisions in an Electric Field

As an example of molecular collisions in a DC electric field, we will consider the scattering problem of molecules in a $^2\Pi$ electronic state with a structureless atom. As discussed in Section 2.4.4, molecules in a Π electronic state possess states of the opposite parity lying close in energy so the collision dynamics of $^2\Pi$ molecules should be expected to be sensitive to external electric fields.

9.4.1 Collisions of $^2\Pi$ Molecules with ^1S Atoms

This is an interesting case because the proper description of the interaction of a structureless atom with a diatomic molecule in a Π state requires two three-dimensional potential energy surfaces. Using the conventional notation, we will refer to these surfaces as A' and A'' [320].

Once again, we will write the full Hamiltonian of the collision complex of a molecule in the $^2\Pi$ electronic state with a structureless atom as

$$\hat{H} = -\frac{\hbar^2}{2\mu} R^{-2} \frac{\partial}{\partial R} R^2 \frac{\partial}{\partial R} + \frac{\hbar^2 l^2(\theta,\phi)}{2\mu R^2} + U(R,r,\theta) + \hat{H}_A, \tag{9.54}$$

where $U(R, r, \theta)$ is the operator of the interaction potential between the molecule and the atom. We now prefer to write the interaction potential as a function of two distances and an angle between the vectors $\boldsymbol{R}$ and $\boldsymbol{r}$.

The Hamiltonian $\hat{H}_A$ describes an isolated molecule in a $^2\Pi$ electronic state placed in a DC electric field. The eigenenergies and eigenstates of this Hamiltonian were discussed in Section 2.5, and we will not repeat the discussion here.

The fully uncoupled angular basis set for this problem can be defined as a direct product of the parity-adapted Hund's case (a) states and the eigenstates of

l^2 and l_z. As discussed in Section 2.4.4, the parity-adapted states of the molecule can be constructed for the Hund's case (a) states as follows:

$$|JM\bar{\Omega}\epsilon\rangle = \frac{1}{2}\{|JM\bar{\Omega}\rangle|\Lambda = 1, \Sigma = \bar{\Omega} - 1\rangle + \epsilon(-)^{J-1/2}|JM - \bar{\Omega}\rangle|\Lambda = -1, \Sigma = -\bar{\Omega} + 1\rangle\}, \tag{9.55}$$

so the fully uncoupled basis for the scattering calculations can be defined as

$$|JM\bar{\Omega}\epsilon\rangle|lm_l\rangle. \tag{9.56}$$

In practice, it may actually be easier to evaluate the matrix elements in the basis of products of parity-unadapted states $|JM\Omega\rangle$ and the partial wave states $|lm_l\rangle$. The matrix elements of the Hamiltonian in the parity-adapted basis (9.56) can then be obtained by combining the appropriate matrix elements in the basis $|JM\Omega\rangle|lm_l\rangle$.

As in Section 9.4, the matrix elements of the Hamiltonian $\hat{H}_A$ in basis (9.56) are diagonal in l and m_l

$$\langle JM\bar{\Omega}\epsilon|\langle lm_l|\hat{H}_A|J'M'\bar{\Omega}'\epsilon'\rangle|l'm_l'\rangle = \delta_{ll'}\delta_{m_lm_l'}\langle JM\bar{\Omega}\epsilon|\hat{H}_A|J'M'\bar{\Omega}'\epsilon'\rangle \tag{9.57}$$

with the matrix elements appearing on the right-hand side of this equation given in Section 2.5.

Since the interaction potential now involves two surfaces, it cannot be simply expanded in Legendre polynomials as was the case for molecules in the Σ state. To account for the proper angular behavior of the potential, the *matrix elements* of the interaction potential between the states with well-defined values of Λ must be expanded in reduced Wigner's D-functions [320]

$$V_{\Lambda\Lambda'}(R,\theta) = \sum_{\lambda} D^{\lambda *}_{0,\Lambda'-\Lambda}(0,\theta,0)V_{\lambda,\Lambda'-\Lambda}(R). \tag{9.58}$$

Since in the case of Π-state molecules $|\Lambda| = |\Lambda'| = 1$, only the matrix elements in Eq. (9.58) with $\Lambda' - \Lambda = 0$ or $\Lambda' - \Lambda = \pm 2$ are different from zero. They can be obtained by expanding the half-sum and half-difference of the two ground-state potential energy surfaces of A' and A'' symmetry [320]

$$\frac{1}{2}(V_{A'} + V_{A''}) = \sum_{\lambda=0}^{\lambda_{\max}} P_\lambda(\cos\theta)V_{\lambda 0}(R) \tag{9.59}$$

$$\frac{1}{2}(V_{A''} - V_{A'}) = \sum_{\lambda=2}^{\lambda_{\max}} d^\lambda_{02}(\cos\theta)V_{\lambda 2}(R), \tag{9.60}$$

where $d^\lambda_{0\mu}(\cos\theta)$ are the reduced Wigner's D-functions and $P_{\lambda\mu}(\cos\theta)$ are the associated Legendre polynomials, which are related through [9]

$$d^\lambda_{0\mu}(\cos\theta) = \left[\frac{(\lambda-\mu)!}{(\lambda+\mu)!}\right]^{1/2} P_{\lambda\mu}(\cos\theta). \tag{9.61}$$

The matrix elements of the interaction potential (9.58) can then be evaluated using the generalized spherical harmonics addition theorem [9, 320]

$$d^{\lambda}_{0\mu}(\cos\theta) = (-)^{-\mu} d^{\lambda}_{0,-\mu}(\cos\theta)$$
$$= (-)^{\mu} \sum_{m_\lambda} \left[\frac{4\pi}{2\lambda+1}\right]^{1/2} D^{\lambda}_{m_\lambda \mu}(\bar{\alpha}, \bar{\beta}, 0) Y_{\lambda m_\lambda}(\hat{R}). \qquad (9.62)$$

First, it is useful to evaluate the matrix elements of the interaction potential between states $|JM\Omega\rangle$, which are not parity adapted. Recalling that $|JM\Omega\rangle$ are related to Wigner's D-functions (see Appendix D), we can exploit the expressions for the integrals over the products of three D-functions, to obtain

$$\langle JM\Omega|\langle\Lambda\Sigma|\langle lm_l|\hat{V}(R,\theta)|J'M'\Omega'\rangle|\Lambda'\Sigma'\rangle|l'm_l'\rangle$$
$$= \delta_{\Sigma\Sigma'}[(2J+1)(2J'+1)(2l+1)(2l'+1)]^{1/2}$$
$$\times (-)^{m_l+M'-\Omega'} \sum_{\lambda,m_\lambda} \begin{pmatrix} J & \lambda & J' \\ M & m_\lambda & -M' \end{pmatrix} \begin{pmatrix} J & \lambda & J' \\ \Omega & \Lambda'-\Lambda & -\Omega' \end{pmatrix}$$
$$\times \begin{pmatrix} l & \lambda & l' \\ -m_l & m_\lambda & m_l' \end{pmatrix} \begin{pmatrix} l & \lambda & l' \\ 0 & 0 & 0 \end{pmatrix}$$
$$\times V_{\lambda,\Lambda'-\Lambda}(R). \qquad (9.63)$$

The expansion coefficients have the property $V_{\lambda,\Lambda'-\Lambda}(R) = V_{\lambda,\Lambda-\Lambda'}(R)$ [320].

Note that the 3-j symbols in Eq. (9.63) vanish unless $m_\lambda = M' - M = m_l - m_l'$. Thus, the electrostatic interaction potential only couples the states with the same total angular momentum projection $M_{\text{tot}} = M + m_l = M' + m_l'$. Now we need to combine the above matrix elements to write the expression for the interaction potential in the parity-adapted basis:

$$\langle JM\bar{\Omega}\epsilon|\langle lm_l|\hat{V}(R,\theta)|J'M'\bar{\Omega}'\epsilon'\rangle|l'm_l'\rangle$$
$$= [(2J+1)(2J'+1)(2l+1)(2l'+1)]^{1/2}(-)^{m_l+M'-\bar{\Omega}'} \sum_{\lambda,m_\lambda} \frac{1}{2}[1+\epsilon\epsilon'(-1)^{\lambda}]$$
$$\times \begin{pmatrix} J & \lambda & J' \\ M & m_\lambda & -M' \end{pmatrix} \begin{pmatrix} l & \lambda & l' \\ -m_l & m_\lambda & m_l' \end{pmatrix} \begin{pmatrix} l & \lambda & l' \\ 0 & 0 & 0 \end{pmatrix}$$
$$\times \left[\begin{pmatrix} J & \lambda & J' \\ \bar{\Omega} & 0 & -\bar{\Omega}' \end{pmatrix} V_{\lambda 0}(R) - \epsilon'(-)^{J'-1/2} \begin{pmatrix} J & \lambda & J' \\ \bar{\Omega} & -2 & \bar{\Omega}' \end{pmatrix} V_{\lambda 2}(R)\right]. \qquad (9.64)$$

This equation contains a lot of interesting information. We recall that the 3j-symbols

$$\begin{pmatrix} j_1 & j_2 & j_3 \\ m_1 & m_2 & m_3 \end{pmatrix} = 0 \text{ unless } m_1 + m_2 + m_3 = 0 \qquad (9.65)$$

and

$$\begin{pmatrix} j_1 & j_2 & j_3 \\ 0 & 0 & 0 \end{pmatrix} = 0 \text{ unless } j_1 + i_2 + j_3 \text{ is an even integer.} \qquad (9.66)$$

This allows us to conclude the following:

- As was discussed in Section 2.5, the rotational energy levels of a $^2\Pi$ molecule are represented by two manifolds, corresponding to two values of $\bar{\Omega} = 1/2$ and $\bar{\Omega} = 3/2$. The 3*j*-symbols in Eq. 9.64 show that the rotational levels in the same manifold ($\bar{\Omega}' = \bar{\Omega}$) are coupled by the $V_{\lambda 0}$ term, i.e. by the half-sum of the two potential energy surfaces A' and A''.
- By the same arguments, the rotational states belonging to different manifolds $\bar{\Omega}$ are coupled by the $V_{\lambda 2}$ term, i.e. by the half-difference of the two potential energy surfaces A' and A''.
- Due to the factor $\frac{1}{2}[1 + \epsilon\epsilon'(-1)^\lambda]$, the couplings between the *molecular* states of the same parity are induced by the anisotropic terms with even λ, while the couplings between the molecular states of opposite parity are induced by the anisotropic terms with odd λ. Note that the parity of the full collision system is determined by the parity of the molecular states of the diatomic molecule and the parity of the states $|lm_l\rangle$. Since $|lm_l\rangle$ are spherical harmonics, the parity of the states $|lm_l\rangle$ is determined by the value of l (even vs odd). The presence of the 3*j*-symbol

$$\begin{pmatrix} l & \lambda & l' \\ 0 & 0 & 0 \end{pmatrix}$$

in combination with Eq. (9.66) and the factor $\frac{1}{2}[1 + \epsilon\epsilon'(-1)^\lambda]$ ensure that the interaction potential operator does not couple states of the collision complex that have different overall parity.

9.5 Atom–Molecule Collisions in a Microwave Field

The scattering problem of molecules in the presence of off-resonant microwave fields can be formulated using the Floquet theory described in Section 4.2 or, equivalently, the quantum description of the molecule–field interaction discussed in Section 4.8. The two approaches give the same results. For illustration purposes, I will use the fully quantum description of the molecule–field interaction in the present section.

Our starting point is, once again, the Hamiltonian given by Eq. (9.54). The operator $\hat{H}_A$ now includes the interaction of the molecule with a microwave field. In order to simplify the equations, we will assume that the microwave field has only one frequency ω and that it is linearly polarized. We will also assume that the frequency of the microwave field is far detuned from the vibrational transitions in the molecule and that the molecule can be treated as a rigid rotor of length r_0. We will relax this assumption at the end of this section.

As described in Section 4.8, the Hamiltonian of a rigid rotor in a microwave field can be represented as

$$\hat{H}_A = B_e\hat{N}^2 + \hat{V}_{\text{mol,f}} + \hbar\omega(\hat{a}\hat{a}^\dagger - \bar{N}), \tag{9.67}$$

where B_e is the rotational constant of the molecule, $\hat{N}$ is the rotational angular momentum of the molecule, and $\bar{N}$ is the average number of photons in the field. The interaction of the molecule with a linearly polarized microwave field is

$$\hat{V}_{\text{mol,f}} = -\frac{\Omega}{2\sqrt{\bar{N}}}(\hat{a} + \hat{a}^\dagger)\cos\chi, \tag{9.68}$$

where χ is the angle between the molecular axis and the polarization vector of the AC field, the operators $\hat{a}^\dagger$ and $\hat{a}$ create and annihilate microwave photons, and Ω is the strength of the field-induced coupling between the rotational states given by the product of the field strength and the permanent dipole moment of the molecule.

As discussed in Chapter 4, the energy states of the molecule in a microwave field can be obtained by diagonalizing the Hamiltonian (9.67) in the field-dressed basis $|NM_N\rangle|\bar{N}+n\rangle$, where $|\bar{N}+n\rangle$ are the photon number states. Typically, the number of photons (n) participating in the atom–molecule collision dynamics is much smaller than the average number of photons in the field, $n \ll \bar{N}$. In this case, the matrix elements the molecule–field interaction (9.68) are independent of $\bar{N}$.

The eigenstates of the molecule–field Hamiltonian (9.67) can be generally written as coherent superpositions of the products of molecular rotational states and the photon number states

$$|\nu K\rangle = \sum_{N,M_N}\sum_{n=-n_{\max}}^{n_{\max}} C_{NM_N\ n,\nu K}|NM_N\rangle|\bar{N}+n\rangle, \tag{9.69}$$

where the indices ν and K label the field-dressed states and the coefficients $C_{NM_N\ n,\nu K}$ depend on both the strength of the interaction with the field Ω and the field frequency ω. As was shown in Chapter 4, the field-dressed states are arranged in manifolds separated by multiples of the photon energy $\hbar\omega$. The quantum number ν is used to label the state of the molecule within a specific photon manifold, and the quantum number K is used to label the photon manifolds. If the field is switched off adiabatically, each coherent superposition (9.72) should produce molecules populating a particular rotational state, with which the field-dressed state correlates in the $\Omega \to 0$ limit. If the field is switched off rapidly, molecules will remain in coherent superpositions of rotational states.

The most convenient basis for the scattering calculations is the direct product of the rotational states, the photon number states, and the partial wave states

$$|NM_N\rangle|\bar{N}+n\rangle|lm_l\rangle. \tag{9.70}$$

The matrix of the interaction potential $U(R, r, \theta)$ in this basis is diagonal in the photon number states,

$$\begin{aligned}&\langle NM_N|\langle\bar{N}+n|\langle lm_l|U|N'M'_N\rangle|\bar{N}+n'\rangle|l'm'_l\rangle\\&=\delta_{n,n'}\langle NM_N|\langle lm_l|U|N'M'_N\rangle|l'm'_l\rangle,\end{aligned} \tag{9.71}$$

and the matrix elements $\langle NM_N|\langle lm_l|U|N'M'_N\rangle|l'm'_l\rangle$ are given by an equation analogous to Eq. (9.23).

If the frequency of the microwave field is comparable with the frequency of the vibrational transitions in the molecule or the microwave field is very strong (with the strength of the coupling approaching the vibrational energy level separation), it is necessary to include the vibrational states in the basis. This can be achieved by dressing the basis (9.70) with the states $|\chi_v^N\rangle$. The eigenstates of the molecule will then be the coherent superpositions of the ro-vibrational states

$$|\nu K\rangle = \sum_v \sum_{N,M_N} \sum_{n=-n_{\max}}^{n_{\max}} C_{v,NM_N\ n,\nu K}|\chi_v^N\rangle|NM_N\rangle|\bar{N}+n\rangle, \tag{9.72}$$

the interaction potential $U(R, r, \theta)$ will depend explicitly on the intramolecular distance r, and the matrix elements of the interaction potential will be given by the same expressions as in Eq. (9.23).

9.6 Total Angular Momentum Representation for Collisions in Fields

As discussed in Section 8.3, the total angular momentum basis reduces a coupled-channel scattering problem in the absence of external fields to a set of uncoupled problems involving a smaller number of coupled equations. This cannot be done for collision problems with external fields because the field-induced interactions couple states of different total angular momenta. However, the total angular momentum basis may still be advantageous over the fully uncoupled basis such as the basis of Eq. (9.13). The advantage comes from the tri-diagonal structure of the full Hamiltonian of molecular systems in external fields (see Figure 8.1). The tri-diagonal structure of the matrix suggests that the convergence of the calculations with respect to the basis set may be sought by increasing the number of total angular momentum blocks. As was shown in multiple recent calculations [315, 316], this way of truncating the basis set is more efficient than simply limiting the number of states in the uncoupled (direct product) basis.

To take advantage of the tri-diagonal structure of the Hamiltonian depicted in Figure 8.1, we should formulate the collision problem by representing the total states of the colliding species as the following expansion

$$|\Psi^M\rangle = \frac{1}{R}\sum_{\alpha_A,\alpha_B}\sum_J F^M_{\alpha_A\alpha_B J}(R)|JMjl(j_A j_B)\rangle, \tag{9.73}$$

where $|JMjl(j_A j_B)\rangle$ are the eigenstates of J^2 and $\hat{J}_Z$, J is the *total* angular momentum of the collision complex A–B, and $\hat{J}_Z$ is its Z-component in the *space-fixed* coordinate frame. The Z-axis of this coordinate frame is directed along the field vector. The quantum numbers α_A and α_B describe collectively all the quantum numbers of particles A and B, respectively. Because the projection of the total angular momentum on the field direction is conserved, the states are parametrized by a fixed value of M, as discussed in Chapter 8.

The total angular momentum is defined as the vector sum

$$\boldsymbol{J} = \boldsymbol{j} + \boldsymbol{l}, \tag{9.74}$$

with $\boldsymbol{j}$ defined as the vector sum

$$\boldsymbol{j} = \boldsymbol{j}_A + \boldsymbol{j}_B, \tag{9.75}$$

where $\boldsymbol{j}_A$ and $\boldsymbol{j}_B$ are the total angular momenta of the colliding particles A and B. With these definitions, the states $|JMjl(j_A j_B)\rangle$ can be written in terms of direct product states as

$$|JMjl(j_A j_B)\rangle = \sum_{m_A,m_B,m_l} C^{JM}_{jm_j lm_l} C^{jm}_{j_A m_A j_B m_B} |j_A m_A\rangle |j_B m_B\rangle |l m_l\rangle. \tag{9.76}$$

Here, I use a compact notation for the Clebsch–Gordan coefficients

$$C^{g\gamma}_{a\alpha b\beta} \equiv \langle g\gamma | a\alpha b\beta\rangle. \tag{9.77}$$

If the species A and/or B possess complex internal structure described by more than one angular momentum, the states $|j_A m_A\rangle$ and/or $|j_B m_B\rangle$ can be written as sums over the fully uncoupled direct product states. For example, if both A and B are molecules in a $^2\Sigma$ electronic state, the molecular states can be written as

$$|j_A m_A\rangle = \sum_{M_{N_A}}\sum_{M_{S_A}} C^{j_A m_A}_{N_A M_{N_A} S_A M_{S_A}} |N_A M_{N_A}\rangle |S_A M_{S_A}\rangle \tag{9.78}$$

with a similar equation for $|j_B m_B\rangle$. Note that all of the projections in all of these angular momentum states are with respect to the field direction.

If we substitute Eq. (9.78) and the equivalent expansion for $|j_B m_B\rangle$ into Eq. (9.76), we will have an expression for the total angular momentum states written in terms of the fully uncoupled basis states. The resulting equation can be used to express the matrix elements in the coupled basis $|JMjl(j_A j_B)\rangle$ in terms of the matrix elements in the fully uncoupled basis, which can be calculated by the repeated application of the Wigner–Eckart theorem as described earlier in this chapter.

For example, for the case of two molecules in a ${}^2\Sigma$ electronic state, we have

$$|JMjl(j_Aj_B)\rangle = \sum_{\text{all projections}} |N_AM_{N_A}\rangle|S_AM_{S_A}\rangle|N_BM_{N_B}\rangle|S_BM_{S_B}\rangle|lm_l\rangle \\ \times \mathcal{G}(JM|j_Am_AN_AM_{N_A}S_AM_{S_A};j_Bm_BN_BM_{N_B}S_BM_{S_B};lm_l), \tag{9.79}$$

where the coefficients are the products of the Clebsch–Gordan coefficients

$$\mathcal{G}(JM|j_Am_AN_AM_{N_A}S_AM_{S_A};j_Bm_BN_BM_{N_B}S_BM_{S_B};lm_l) \\ = C^{JM}_{jm_jlm_l}C^{jm}_{j_Am_Aj_Bm_B}C^{j_Am_A}_{N_AM_{N_A}S_AM_{S_A}}C^{j_Bm_B}_{N_BM_{N_B}S_BM_{S_B}}. \tag{9.80}$$

To appreciate the usefulness of Eq. (9.79), consider the matrix elements of the interaction potential between two diatomic molecules in a Σ electronic state. Each of the molecules possesses the electron spin $\boldsymbol{S}_A = \boldsymbol{S}_B = 1/2$. These two spins can be combined to form the states of total spin $S = 0$ and 1. These two values of the total spin correspond to the different electronic states of the two-molecule complex. The interaction between the molecules in each of these two spins is described by a different potential energy surface.

The *full* molecule–molecule interaction potential that enters the Hamiltonian can be written as a sum of two terms

$$U(\boldsymbol{R},\boldsymbol{r}_\text{A},\boldsymbol{r}_\text{B}) = V_\text{AB}(\boldsymbol{R},\boldsymbol{r}_\text{A},\boldsymbol{r}_\text{B}) + A_\text{AB}(\boldsymbol{R},\boldsymbol{r}_\text{A},\boldsymbol{r}_\text{B})\boldsymbol{S}_\text{A}\cdot\boldsymbol{S}_\text{B}, \tag{9.81}$$

where $V_\text{AB}(\boldsymbol{R},\boldsymbol{r}_\text{A},\boldsymbol{r}_\text{B})$ is the spin-independent part of the intermolecular potential and the term $A_\text{AB}(\boldsymbol{R},\boldsymbol{r}_\text{A},\boldsymbol{r}_\text{B})\boldsymbol{S}_\text{A}\cdot\boldsymbol{S}_\text{B}$ describes the spin-dependent interaction that splits the interaction potential surfaces of the two-molecule complex corresponding to different values of total spin.

By analogy with Eq. (9.19), both of the terms $V_\text{AB}(\boldsymbol{R},\boldsymbol{r}_\text{A},\boldsymbol{r}_\text{B})$ and $A_\text{AB}(\boldsymbol{R},\boldsymbol{r}_\text{A},\boldsymbol{r}_\text{B})$ can be expanded in products of spherical harmonics as

$$V_\text{AB}(\boldsymbol{R},\boldsymbol{r}_\text{A},\boldsymbol{r}_\text{B}) = \sum_{\lambda_\text{A}\lambda_\text{B}\lambda} V_{\lambda_\text{A}\lambda_\text{B}\lambda}(R,r_\text{A},r_\text{B}) \\ \times \sum_{m_{\lambda_\text{A}}m_{\lambda_\text{B}}m_\lambda} C^{\lambda m_\lambda}_{\lambda_\text{A}m_{\lambda_\text{A}}\lambda_\text{B}m_{\lambda_\text{B}}} Y_{\lambda_\text{A}m_{\lambda_\text{A}}}(\hat{r}_\text{A})Y_{\lambda_\text{B}m_{\lambda_\text{B}}}(\hat{r}_\text{B})Y_{\lambda m_\lambda}(\hat{R}). \tag{9.82}$$

If these expansions are sandwiched between the direct product states of the fully uncoupled representation, the evaluation of the matrix elements becomes a simple application of the Wigner–Eckart theorem, as discussed in previous examples in this chapter. This is a consequence of the fact that the spherical harmonic $Y_{\lambda_\text{A}m_{\lambda_\text{A}}}(\hat{r}_\text{A})$ acts only on the states $|N_AM_{N_A}\rangle$ and not on any of the other states in the direct product basis. Similarly, $Y_{\lambda_\text{B}m_{\lambda_\text{B}}}(\hat{r}_\text{B})$ acts only on the states $|N_BM_{N_B}\rangle$ and $Y_{\lambda m_\lambda}(\hat{R})$ acts only on the states $|lm_l\rangle$.

One can, of course, try to evaluate the matrix elements of the operator V_{AB} directly in the total angular momentum representation using the advanced techniques of angular momentum algebra. To do this, one would represent the full interaction potential as

$$U = \sum_S \sum_{M_S} |SM_S\rangle V^S \langle SM_S|, \tag{9.83}$$

where $V^S(\boldsymbol{R}, \boldsymbol{r}_{\mathrm{A}}, \boldsymbol{r}_{\mathrm{B}})$ are the potential energy surfaces of the molecule - molecule collision system for fixed total spin S. Each of the potential surfaces can then be expanded as in Eq. (9.82)

$$\begin{aligned} V^S(\boldsymbol{R}, \boldsymbol{r}_{\mathrm{A}}, \boldsymbol{r}_{\mathrm{B}}) = & \sum_{\lambda_{\mathrm{A}}\lambda_{\mathrm{B}}\lambda} V^S_{\lambda_A\lambda_B\lambda}(R, r_{\mathrm{A}}, r_{\mathrm{B}})(-1)^{m_\lambda}(2\lambda+1)^{-1/2} \\ & \times \sum_{m_{\lambda_{\mathrm{A}}} m_{\lambda_{\mathrm{B}}} m_\lambda} C^{\lambda - m_\lambda}_{\lambda_A m_{\lambda_{\mathrm{A}}} \lambda_{\mathrm{B}} m_{\lambda_{\mathrm{B}}}} Y_{\lambda_{\mathrm{A}} m_{\lambda_{\mathrm{A}}}}(\hat{r}_{\mathrm{A}}) Y_{\lambda_{\mathrm{B}} m_{\lambda_{\mathrm{B}}}}(\hat{r}_{\mathrm{B}}) Y_{\lambda m_\lambda}(\hat{R}). \end{aligned} \tag{9.84}$$

This approach would ultimately lead to the following equation [321]:

$$\begin{aligned} & \langle JMjl(j_{\mathrm{A}}j_{\mathrm{B}})|\hat{V}|J'M'j'l'(j'_{\mathrm{A}}j'_{\mathrm{B}})\rangle \\ & = \sum_{\lambda_A,\lambda_B,\lambda,S} V^S_{\lambda_A\lambda_B\lambda}(-1)^{\lambda_A-\lambda_B} X^S_{\lambda_A,\lambda_B,\lambda,l,l'}, \end{aligned} \tag{9.85}$$

where X can be expressed as a product of 3j-symbols, a 6j symbol, and an 18j symbol

$$\begin{aligned} X^S_{\lambda_A,\lambda_B,\lambda,l,l'} = & \delta_{M,M'}\delta_{J,J'}(-1)^{N_{\mathrm{A}}+N'_{\mathrm{A}}-S_{\mathrm{A}}-S'_{\mathrm{A}}+j_{\mathrm{A}}+j'_{\mathrm{A}}-j'-l-\lambda-J} \\ & [(j_{\mathrm{A}})(j'_{\mathrm{A}})(j_{\mathrm{B}})(j'_{\mathrm{B}})(j)(j')(\lambda_A)(\lambda_B)(\lambda)]^{\frac{1}{2}}(S) \\ & \times c_\lambda(l,l')c_{\lambda_A}(N_A,N'_A)c_{\lambda_B}(N_B,N'_B)\begin{Bmatrix} j & j' & \lambda \\ l' & l & J \end{Bmatrix} \\ & \times \begin{Bmatrix} \lambda_A & & N_{\mathrm{A}} & & S_{\mathrm{A}} & & S'_{\mathrm{B}} & & N'_{\mathrm{B}} & & \lambda_B \\ & & & & & S & & & & & \\ & N'_{\mathrm{A}} & & j_{\mathrm{A}} & & & & j'_{\mathrm{B}} & & N_{\mathrm{B}} & \\ & & & & & \lambda & & & & & \\ S'_{\mathrm{A}} & & j'_{\mathrm{A}} & & j' & & j & & j_{\mathrm{B}} & & S_{\mathrm{B}} \end{Bmatrix}. \end{aligned}$$

Here, I use the short-hand notation $(j) \equiv 2j+1$ and

$$c_\lambda(l,l') = (-1)^l \left[\frac{(l)(l')}{(4\pi)}\right]^{1/2} \begin{pmatrix} l & \lambda & l' \\ 0 & 0 & 0 \end{pmatrix}. \tag{9.86}$$

While elegant and beautiful, Eq. (9.86) requires the evaluation of the 18j symbol. While this can be done by expressing the 18j-symbol as a sum of products of 9j-symbols, it is less transparent to code than the equation based on simple products of the Clebsch–Gordan coefficients as in Eq. (9.79).

10

Field-Induced Scattering Resonances

This chapter can be considered as a prelude to the next one, where we discuss broadly the external field control of molecular collisions. Here, we consider a specific phenomenon, namely, the effect of scattering resonances on molecular collisions.

What is a scattering resonance?

From the point of view of an experimentalist, a scattering resonance is a sharp variation in the dependence of the scattering cross section on the collision energy or on the external field strength. The characteristic variation of the scattering cross section near a resonance is shown in Figure 10.1. *Typically*, the cross sections for elastic, inelastic, and chemically reactive collisions of molecules are greatly affected by scattering resonances. I emphasize the word "typically" because not all resonances lead to a detectable variation of the collision cross sections and the shape of the cross section variation may be different from that shown in Figure 10.1. For example, the trough may often be absent. Also, not every structure in the energy or field dependence of the cross sections corresponds to a scattering resonance.

The goal of this chapter is to understand what determines the resonant variation of the cross sections and how it can be induced by tuning the strength of an external field.

10.1 Feshbach vs Shape Resonances

The scattering resonances occur when the energy of the scattering state coincides with the energy of a quasi-bound state of the collision complex; provided the two are coupled by some interaction.

There are two types of scattering resonances: the Feshbach resonances and the shape resonances. In order to understand the mechanism of the scattering

Molecules in Electromagnetic Fields: From Ultracold Physics to Controlled Chemistry, First Edition. Roman V. Krems.

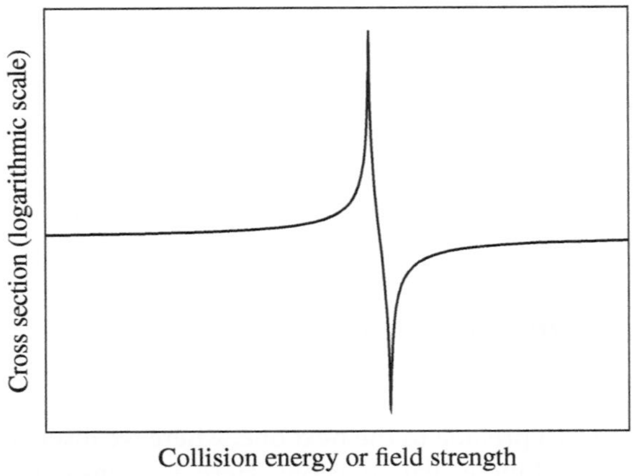

Figure 10.1 The effect of a resonance on a scattering cross section.

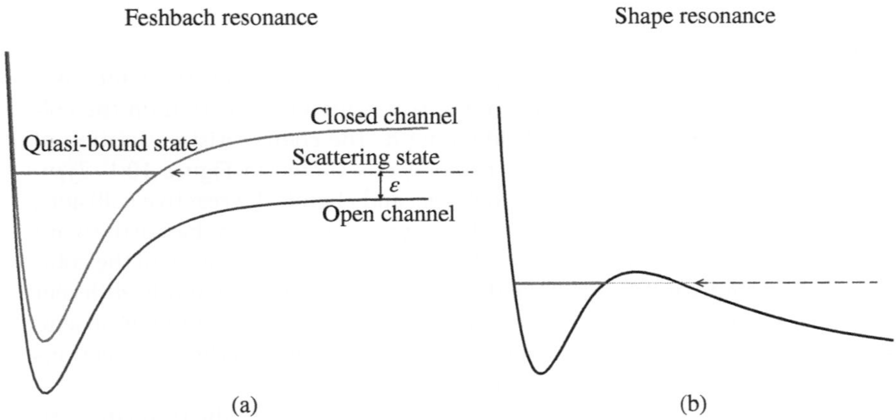

Figure 10.2 Schematic diagrams illustrating the mechanisms of Feshbach and shape scattering resonances. The resonances occur when a scattering state (with the energy shown by the dashed lines) interacts with a quasi-bound state (shown by the full horizontal lines). The nature of the quasi-bound states is different for the two types of resonances. See text for a detailed discussion.

resonances, consider the diagrams shown in Figure 10.2. These diagrams display the typical shape of the intermolecular interaction potentials as functions of the separation between the colliding particles. Going left to right in these diagrams corresponds to increasing the distance R between the colliding particles. The quantum states energetically accessible in the limit $R \to \infty$ are often referred to as open channels and the quantum states that are energetically inaccessible as closed channels.

Consider now the colliding particles prepared in some quantum states corresponding to the $R \to \infty$ limit of the lower curve in the left panel of Figure 10.2. The kinetic energy of the collision is labeled by ε. As the collision energy increases from zero, the total energy of the scattering particles increases and at some point becomes degenerate with a quasi-bound state shown by a solid horizontal line. This is when a Feshbach resonance occurs. The scattering cross section near this energy is typically enhanced as shown in Figure 10.1.

It is important to clarify the concept of a quasi-bound state in this context. Figure 10.2a suggests that the quasi-bound state is a bound state in the interaction potential corresponding to the closed channel. Strictly speaking, this is only true if the open and closed channels are not coupled. Figure 10.2a should thus be interpreted as the limit of zero coupling between the open and closed channels. In this limit, the bound state is infinitely long-lived and the width of the resonant enhancement of the cross section is zero, i.e. there is no scattering resonance. If the coupling is nonzero, the bound state becomes quasi-bound. The two-particle system prepared in this state can *predissociate* into two particles in the open channel. The lifetime of the bound states is inversely proportional to the resonance width (defined more rigorously in Section 10.2). In the presence of a coupling between the two channels, it is impossible to treat the bound and scattering states separately. We represent them separately as the zeroth-order states, assuming that the interaction between the scattering state and the quasi-bound state is a perturbation.

The shape resonances similarly arise because of a coupling between a scattering state and a quasi-bound state. However, the quasi-bound state now has a different nature: it occurs due to the presence of a long-range centrifugal barrier, as shown in Figure 10.2b. The state is *quasi*-bound because it can decay through tunnelling under the centrifugal barrier. The higher-energy closed channels are not necessary for the shape resonances to occur.

The scattering resonances most naturally show up as peaks in the dependence of the cross sections on the collision energy (ε in Figure 10.2). They can also appear in the dependence of the cross sections on the external field strength. When an external field is applied, it shifts the energy levels of the isolated molecules and also the quasi-bound states of the collision system. If these shifts are different, tuning an external field can be used to move the $R \to \infty$ limit of the open channel closer to or further away from the bound state. Tuning an external field with a fixed ε is thus equivalent to varying the collision energy.

In this case, the collision cross section will exhibit the resonant variation as a function of the field strength.

In order to understand the characteristic variation of the cross sections near a scattering resonance, we need to consider the theoretical description of resonances in terms of the strength of the coupling between the scattering and quasi-bound states. This is best done with the projection operator method used by Feshbach [322, 323]. To understand and apply Feshbach's method, we need to reformulate the scattering theory using Green's operators.

10.2 The Green's Operator in Scattering Theory

First, consider a particle moving in a potential V, described by the Hamiltonian

$$\hat{H} = \hat{K} + V, \tag{10.1}$$

where $\hat{K}$ is the kinetic energy operator. The eigenstates of the system $|\phi\rangle$ satisfy the Schrödinger equation

$$(E - \hat{H})|\phi\rangle = 0. \tag{10.2}$$

Given the eigenstates of the kinetic energy operator

$$(E - \hat{K})|\chi\rangle = 0, \tag{10.3}$$

we can write the full state

$$|\phi\rangle = |\chi\rangle + |\psi\rangle, \tag{10.4}$$

where $|\psi\rangle$ is some state to be determined. Using this decomposition, Eq. (10.2) can be written as

$$(E - \hat{H})|\psi\rangle = -(E - \hat{H})|\chi\rangle \tag{10.5}$$

or, using Eq. (10.3), as

$$(E - \hat{H})|\psi\rangle = V|\chi\rangle. \tag{10.6}$$

This equation can be formally inverted by means of the Green's operator

$$\hat{G}(E) = \frac{1}{E - \hat{H}}, \tag{10.7}$$

to yield

$$|\psi\rangle = \frac{1}{E - \hat{H}} V|\chi\rangle. \tag{10.8}$$

If E is an eigenvalue of $\hat{H}$, the matrix elements of the Green's operator exhibit singularities. To avoid such singularities, it is convenient to introduce the complex variables

$$E^{\pm} = E \pm i\eta, \tag{10.9}$$

where η is a small positive real number. We can thus define two Green's operators, one for positive η and one for negative η:

$$\hat{G}^{\pm}(E) = \frac{1}{E \pm i\eta - \hat{H}}. \tag{10.10}$$

All the calculations can be carried out with the operators $\hat{G}^{\pm}$ and the limit of $\eta \to 0^{+}$ should be taken at the end of the calculations.

By analogy with (10.8), we can write

$$|\psi^{\pm}\rangle = \frac{1}{E \pm i\eta - \hat{H}} V|\chi\rangle \tag{10.11}$$

and

$$|\phi^{\pm}\rangle = |\chi\rangle + \frac{1}{E \pm i\eta - \hat{H}} V|\chi\rangle. \tag{10.12}$$

The states $|\phi^{\pm}\rangle$ approach the eigenstates of the full Hamiltonian in the limit of $\eta \to 0^{+}$. The state $|\phi^{+}\rangle$ corresponds to the outgoing scattering state [324]. Since the system is invariant under time reversal, $|\phi^{-}\rangle$ corresponds to the incoming scattering state solution.

10.3 Feshbach Projection Operators

We are now ready to apply Feshbach's formalism [322, 323] to describe resonant scattering. The discussion in the following two sections is based on the work in Ref. [325]. Some of the discussion in Sections 10.4 and 10.5 previously appeared in an early version of Ref. [325] available on arXiv.org. Within Feshbach's approach,

> The Hilbert space of the total Hamiltonian $\hat{H}$ for the collision system is partitioned into two orthogonal Hilbert spaces: P, comprising all open channels of the system; and Q, containing all closed channels.

Consider the eigenstate of the full Hamiltonian $|\Psi^{+}(E)\rangle$ that corresponds to the outgoing wave scattering solution. The state $|\Psi^{+}(E)\rangle$ belongs to the entire Hilbert space and satisfies the Schrödinger equation

$$(E - \hat{H})|\Psi^{+}(E)\rangle = 0, \tag{10.13}$$

where E is the total energy of the colliding particles. Let us define two operators $\hat{P}$ and $\hat{Q}$ that project the full state $|\Psi^{+}\rangle$ onto the P and Q subspaces

$$\hat{P}|\Psi^{+}\rangle \equiv |\Psi_{P}^{+}\rangle \tag{10.14}$$

$$\hat{Q}|\Psi^{+}\rangle \equiv |\Psi_{Q}\rangle. \tag{10.15}$$

By writing the full state as

$$|\Psi^+\rangle = \hat{P}|\Psi_P^+\rangle + \hat{Q}|\Psi_Q\rangle, \tag{10.16}$$

inserting $|\Psi^+\rangle$ into the Schrödinger equation (10.13) and acting on the left with either $\hat{P}$ or $\hat{Q}$, we can obtain the following equations:

$$(E - \hat{H}_{PP})|\Psi_P^+\rangle = \hat{H}_{PQ}|\Psi_Q\rangle \tag{10.17}$$

$$(E - \hat{H}_{QQ})|\Psi_Q\rangle = \hat{H}_{QP}|\Psi_P^+\rangle, \tag{10.18}$$

where $\hat{H}_{PP} \equiv \hat{P}\hat{H}\hat{P}$, $\hat{H}_{QQ} \equiv \hat{Q}\hat{H}\hat{Q}$, $\hat{H}_{PQ} \equiv \hat{P}\hat{H}\hat{Q}$, and $\hat{H}_{QP} \equiv \hat{Q}\hat{H}\hat{P}$. This is a system of coupled equations for the projected states $|\Psi_P^+\rangle$ and $|\Psi_Q\rangle$.

What we now want is to write an effective equation for $|\Psi_P^+\rangle$ (the scattering state of interest), which does not involve $|\Psi_Q\rangle$ (the quasi-bound states in the subspace of closed channels). In order to do that, let us formally invert Eq. (10.18) by acting on both sides with the Green's operator $(E - \hat{H}_{QQ})^{-1}$, giving

$$|\Psi_Q\rangle = \frac{1}{E - \hat{H}_{QQ}}\hat{H}_{QP}|\Psi_P^+\rangle, \tag{10.19}$$

which, when substituted into Eq. (10.17), yields

$$(E - \hat{H}_{PP})|\Psi_P^+\rangle = \hat{H}_{PQ}\frac{1}{E - \hat{H}_{QQ}}\hat{H}_{QP}|\Psi_P^+\rangle. \tag{10.20}$$

This is an effective Schrödinger equation for $|\Psi_P^+\rangle$, which can be written as

$$(E - \hat{H}_{\text{eff}}(E))|\Psi_P^+\rangle = 0, \tag{10.21}$$

with an energy-dependent pseudo-Hamiltonian

$$\hat{H}_{\text{eff}}(E) = \hat{H}_{PP} + \hat{H}_{PQ}(E - \hat{H}_{QQ})^{-1}\hat{H}_{QP}. \tag{10.22}$$

In the absence of coupling to the closed channels, the Hamiltonian in the open-channel subspace can be written as a sum of the kinetic energy operator $\hat{K}_{PP}$ and the potential energy term V_{PP}:

$$\hat{H}_{PP} = \hat{K}_{PP} + V_{PP}. \tag{10.23}$$

The term $\hat{H}_{PQ}(E - \hat{H}_{QQ})^{-1}\hat{H}_{QP}$ can be viewed as an effective addition to the potential energy operator, thus yielding an effective potential acting in the P-subspace

$$V_{\text{eff}}(E) = V_{PP} + \hat{H}_{PQ}\frac{1}{E - \hat{H}_{QQ}}\hat{H}_{QP}. \tag{10.24}$$

This potential accounts for the coupling to the closed channel subspace without including $|\Psi_Q\rangle$ explicitly. The scattering problem can now be described by the Hamiltonian

$$\hat{H}_{\text{eff}}(E) = \hat{K}_{PP} + V_{\text{eff}}(E). \tag{10.25}$$

We need to remember to include E as the arguments of $\hat{H}_{\text{eff}}$ and V_{eff} as they are energy-dependent operators.

For reference, consider the solutions to the Schrödinger equation in the P-subspace in the absence of couplings to the closed channel. A scattering state solution for the isolated P-subspace $|\phi_P^+(E)\rangle$ satisfies

$$(E - \hat{H}_{PP})|\phi_P^+\rangle = 0. \tag{10.26}$$

We can also define the free states $|\chi_P(E)\rangle$ as the eigenstates of the kinetic energy operator, which satisfy the following equation:

$$(E - \hat{K}_{PP})|\chi_P\rangle = 0. \tag{10.27}$$

The functions $|\chi_P\rangle$ and $|\phi_P^+\rangle$ are related by (cf., Eq. (10.12))

$$|\phi_P^+\rangle = |\chi_P\rangle + \frac{1}{E^+ - \hat{H}_{PP}} V_{PP}|\chi_P\rangle. \tag{10.28}$$

Applying the Green's operator $(E^+ - \hat{H}_{PP})^{-1}$ to Eq. (10.20), we obtain an implicit equation for $|\Psi_P^+\rangle$,

$$|\Psi_P^+\rangle = |\phi_P^+\rangle + \frac{1}{E^+ - \hat{H}_{PP}}\hat{H}_{PQ}\frac{1}{E - \hat{H}_{QQ}}\hat{H}_{QP}|\Psi_P^+\rangle. \tag{10.29}$$

We see that the scattering state in Eq. (10.29) has two contributions: the state $|\phi_P^+\rangle$ that describes the scattering event in the subspace of open channels without the effect of the closed channel (the background scattering) and the term accounting for the coupling to the Q-subspace of closed channels (that will lead to the resonant scattering). The background scattering state (10.28) is in turn a sum of a free state and a state perturbed by the interaction potential in the subspace of open channels.

We use P and P' to label different channels within the P-subspace. The probability amplitude for a transition from the incoming free state $|\chi_P(E)\rangle$ to the outgoing state is given by the on-shell[1] T-matrix element $T_{P'P}(E)$. This T-matrix element is given by [324]

$$T_{P'P} = \langle\chi_{P'}|V_{\text{eff}}|\Psi_P^+\rangle. \tag{10.30}$$

Using the expression for the effective potential (10.24) and Eq. (10.29), we can write this matrix element as a sum of three terms

$$T_{P'P} = \langle\chi_{P'}|V_{PP}|\phi_P^+\rangle + \langle\chi_{P'}|V_{PP}\frac{1}{E^+ - \hat{H}_{PP}}\hat{H}_{PQ}\frac{1}{E - \hat{H}_{QQ}}\hat{H}_{QP}|\Psi_P^+\rangle$$
$$+ \langle\chi_{P'}|\hat{H}_{PQ}\frac{1}{E - \hat{H}_{QQ}}\hat{H}_{QP}|\Psi_P^+\rangle, \tag{10.31}$$

1 i.e. between states of the same energy.

or in a more suggestive form

$$T_{P'P} = \langle \chi_{P'} | V_{PP} | \phi_P^+ \rangle + \left[\langle \chi_{P'} | V_{PP} \frac{1}{E^+ - \hat{H}_{PP}} + \langle \chi_{P'} | \right] \hat{H}_{PQ} \frac{1}{E - \hat{H}_{QQ}} \hat{H}_{QP} | \Psi_P^+ \rangle. \tag{10.32}$$

Now consider Eq. (10.28). If we flip the sign from + to − and complex conjugate the resulting state, we will obtain

$$\langle \phi_{P'}^- | = \langle \chi_{P'} | + \langle \chi_{P'} | V_{PP} \frac{1}{E^+ - \hat{H}_{PP}}, \tag{10.33}$$

so that the T-matrix element can be written as

$$\begin{aligned} T_{P'P} &= \langle \chi_{P'} | V_{PP} | \phi_P^+ \rangle + \langle \phi_{P'}^- | H_{PQ} \frac{1}{E - H_{QQ}} H_{QP} | \Psi_P^+ \rangle. \\ &\equiv T_{P'P}^b + T_{P'P}^r. \end{aligned} \tag{10.34}$$

The transition amplitude can thus be written as a sum of two terms: the transition amplitude for background scattering in the isolated P-subspace

$$T_{P'P}^b = \langle \chi_{P'} | V_{PP} | \phi_P^+ \rangle \tag{10.35}$$

and the resonant contribution due to coupling to the Q-channel

$$T_{P'P}^r = \langle \phi_{P'}^- | H_{PQ} \frac{1}{E - H_{QQ}} H_{QP} | \Psi_P^+ \rangle. \tag{10.36}$$

10.4 Resonant Scattering

To simplify matters, let us now assume that the Q-subspace contains an isolated bound state $|\phi_1\rangle$ giving rise to the resonance. In this case, we can ignore all other states in the Q-subspace and write

$$\frac{1}{E - H_{QQ}} \approx \frac{|\phi_1\rangle\langle\phi_1|}{E - \epsilon_1}, \tag{10.37}$$

where ϵ_1 is the energy of the bound state. Using this expression, we can write for the resonant part of the T-matrix element

$$T_{P'P}^r = \frac{\langle \phi_{P'}^- | H_{PQ} | \phi_1 \rangle \langle \phi_1 | H_{QP} | \Psi_P^+ \rangle}{E - \epsilon_1}. \tag{10.38}$$

The resonant scattering thus couples the incoming scattering state to the outgoing state via the bound state in the closed channel subspace.

To evaluate the matrix element $\langle \phi_1 | H_{QP} | \Psi_P^+ \rangle$, we multipy Eq. (10.29) on the left by $\langle \phi_1 | H_{QP}$ and solve to obtain

$$\langle \phi_1 | H_{QP} | \Psi_P^+ \rangle = \frac{\langle \phi_1 | H_{QP} | \phi_P^+ \rangle}{1 - \frac{\langle \phi_1 | H_{QP} \frac{1}{E^+ - H_{PP}} H_{PQ} | \phi_1 \rangle}{E - \epsilon_1}}. \tag{10.39}$$

Substituting this into Eq. (10.38), we obtain

$$T^r_{P'P} = \frac{\langle \phi^-_{P'} | H_{PQ} | \phi_Q \rangle \langle \phi_Q | H_{QP} | \phi^+_P \rangle}{E - \epsilon_1 - \langle \phi_Q | H_{QP} \frac{1}{E^+ - H_{PP}} H_{PQ} | \phi_Q \rangle}. \tag{10.40}$$

Let us make the following definitions to simplify the expression for this T-matrix element:

$$\Gamma_{P'P} \equiv 2\pi \langle \phi^-_{P'} | H_{PQ} | \phi_Q \rangle \langle \phi_Q | H_{QP} | \phi^+_P \rangle, \tag{10.41}$$

$$\Delta(E) \equiv \text{Re}\ \langle \phi_Q | H_{QP} \frac{1}{E^+ - H_{PP}} H_{PQ} | \phi_Q \rangle, \tag{10.42}$$

and

$$-\frac{\Gamma}{2} \equiv \text{Im}\ \langle \phi_Q | H_{QP} \frac{1}{E^+ - H_{PP}} H_{PQ} | \phi_Q \rangle. \tag{10.43}$$

Using these new parameters, we can rewrite the expression (10.40) for the T-matrix element as

$$T^r_{P'P} = \frac{\Gamma_{P'P}(E)}{2\pi \left(E - \epsilon_1 - \Delta(E) + \frac{i\Gamma(E)}{2} \right)}. \tag{10.44}$$

In order to evaluate the matrix elements defining Γ and Δ, let us represent the Green's operator in Eq. (10.40) by an expansion in terms of the projectors on states $|\phi^+_{P''}(E')\rangle$, yielding

$$\frac{1}{E^+ - H_{PP}} = \int_0^\infty \mathrm{d}E' \sum_{P''} \frac{|\phi^+_{P''}(E')\rangle \langle \phi^+_{P''}(E')|}{E^+ - E'} \tag{10.45}$$

and

$$\Delta(E) = \mathcal{P} \int_0^\infty \mathrm{d}E' \frac{|\langle \phi_Q | H_{QP} | \phi^+_P(E') \rangle|^2}{E - E'}, \tag{10.46}$$

where $\mathcal{P}$ indicates the Cauchy principal value integral, in which the symmetric limit approaching the singularity at $E' = E$ is taken. Note that, as a result of taking the real part of the expression, the complex-valued E^+ in the energy denominator of Eq. (10.45) is replaced by the real-valued E.

To obtain an expression for Γ, we use

$$\lim_{\eta \to 0^+} \frac{1}{E - i\eta - E'} - \frac{1}{E + i\eta - E'} = 2\pi i \delta(E - E'), \tag{10.47}$$

which gives

$$\frac{\Gamma}{2} = \pi \sum_{P''} |\langle \phi_Q | H_{QP} | \phi^+_{P''}(E) \rangle|^2. \tag{10.48}$$

In Chapter 8 we defined the scattering amplitudes and cross sections in terms of the elements of the scattering S matrix. The S operator is related to the T operator defined above by

$$S = \mathbb{1} - 2\pi i T. \tag{10.49}$$

The matrix elements of S in the basis of free states are

$$\langle \chi_{P'}(E')|S|\chi_P(E)\rangle = \delta(E - E')s_{P'P}(E) = \delta(E - E')(\delta_{P'P} - 2\pi i T_{P'P}(E)). \tag{10.50}$$

As in the case of the T-operator, the elements of the scattering matrix may be expressed as the sum of background and resonant parts: $s_{P'P} = s^b_{P'P} + s^r_{P'P}$. The background S operator $S^b = \mathbb{1} - 2\pi i T^b$ describes scattering in the open channels in the absence of coupling to the closed channels. The resonant part of $s_{P'P}$ is

$$s^r_{P'P} = \frac{-i\Gamma_{P'P}(E)}{E - \epsilon_1 - \Delta(E) + \dfrac{i\Gamma(E)}{2}}. \tag{10.51}$$

Using Eq. (10.33), the adjoint of Eq. (10.28), and Eq. (10.47), one may show that

$$\langle \phi^-_{P'}| = \sum_{P''} (\delta_{P'P''} - 2\pi i \langle \chi_{P'}|V_{PP}|\phi^+_{P''}\rangle)\langle \phi_{P''}|. \tag{10.52}$$

The quantity in parentheses is $s^b_{P'P''}$, hence from Eq. (10.41), we find

$$\Gamma_{P'P} = \sum_{P''} 2\pi s^b_{P'P''} \langle \phi^+_{P''}|H_{PQ}|\phi_Q\rangle\langle \phi_Q|H_{QP}|\phi^+_P\rangle. \tag{10.53}$$

Since the matrix of S^b is unitary,

$$|\Gamma_{PP}| = 2\pi|\langle \phi_Q|H_{QP}|\phi^+_P\rangle|^2 \tag{10.54}$$

and $\Gamma = \sum_P |\Gamma_{PP}|$.

Equation (10.51) shows that the resonance gives rise to a peak in all open channels, with the magnitude of the resonant scattering in channel P determined by $|\Gamma_{PP}|$. The position and width of the resonance are the same in all channels. The position E_r of the resonance satisfies the equation

$$E_r - \epsilon_1 - \Delta(E_r) = 0. \tag{10.55}$$

If the resonance is sufficiently narrow and Γ does not vary significantly with energy, then its width is approximately $\Gamma(E_r)$.

10.5 Calculation of Resonance Locations and Widths

10.5.1 Single Open Channel

In this section, we discuss the numerical procedure for computing the positions and widths of the scattering resonances. First, consider the simple case of a single open channel. In this case, there is only one element of the *S*-matrix. It may be written as

$$s_P(E) = e^{2i\delta}, \tag{10.56}$$

where $\delta(E)$ is the scattering phase shift. In the case of a single open channel, there is only elastic scattering and the cross section for elastic collisions can be written in terms of the *S*-matrix elements, and hence in terms of the scattering shift, as was described in Chapter 8.

Here, our interest is to decompose the scattering shift into the background part and the resonant part $\delta = \delta_b + \delta_r$. The background part of s_P is $s_P^b = e^{2i\delta_b}$. Equation (10.53) takes the form $2\pi\langle\phi_P^-|H_{PQ}|\phi_Q\rangle\langle\phi_Q|H_{QP}|\phi_P^+\rangle = e^{2i\delta_b}\Gamma$. Using Eqs. (10.41) and (10.44), we obtain

$$s_P(E) = e^{2i\delta_b}\left(\frac{E - \epsilon_1 - \Delta - \dfrac{i\Gamma}{2}}{E - \epsilon_1 - \Delta + \dfrac{i\Gamma}{2}}\right) \tag{10.57}$$

and find that the resonant phase shift is given by

$$\delta_r(E) = \tan^{-1}\frac{-\Gamma(E)}{2(E - \epsilon_1 - \Delta(E))}. \tag{10.58}$$

As the energy E increases through a resonance, where $E - \epsilon_1 - \Delta(E) = 0$, the scattering shift (10.58) increases by a factor $\sim \pi$. The resonance locations and widths can thus be identified by fitting the energy dependence of the scattering shift to

$$\delta(E) = \delta_b + \tan^{-1}\frac{-\Gamma(E)}{2(E - \epsilon_1 - \Delta(E))}. \tag{10.59}$$

10.5.2 Multiple Open Channels

In the case of multiple open channels, we must deal with the *S*-matrix. As derived in Chapter 8, the scattering matrix **S** is a unitary symmetric matrix of complex elements.

The derivation in this section follows the work in Ref. [326]. In general, if **A** is a unitary symmetric matrix of complex elements, it can be diagonalized to yield

$$\mathbf{C}^T\mathbf{A}\mathbf{C} = \mathbf{\Upsilon}^2, \tag{10.60}$$

where **C** is a real orthogonal matrix and $\mathbf{\Upsilon}$ is a diagonal matrix with the elements

$$\mathbf{\Upsilon}_{nn} = e^{i\upsilon_n}, \tag{10.61}$$

where the coefficients υ_n are the *eigenphases* of the matrix **A**.

For the S-matrix, we have

$$\mathbf{S}(E) = \mathbf{B}(E)\mathbf{\Lambda}(E)^2\mathbf{B}^T(E) \tag{10.62}$$

with

$$\mathbf{\Lambda}_{nn} = e^{i\lambda_n}. \tag{10.63}$$

The locations and widths of the scattering resonances can be obtained from computing the energy dependence of the eigenphase sum [326, 327]

$$\eta(E) = \sum_n^N \lambda_n(E). \tag{10.64}$$

In order to see what happens to the eigenphase sum near a resonance, it is convenient to write the scattering matrix as [326]

$$\mathbf{S} = \mathbf{S}_b - i\frac{\mathbf{g}(E)\mathbf{g}^T(E)}{E - \epsilon_1 - \Delta(E) + \dfrac{i\Gamma(E)}{2}}, \tag{10.65}$$

where **S** and $\mathbf{S}_b$ are the square matrices of dimension $N \times N$ with N being the number of open channels and $\mathbf{g}(E)$ is a vector of dimension N. The elements of the vector **g** are

$$|\mathbf{g}_P|^2 = \Gamma_{PP} \tag{10.66}$$

so that (cf., Eq. (10.54))

$$\Gamma = \sum_P |\mathbf{g}_P|^2. \tag{10.67}$$

The matrix $\mathbf{S}_b$ – just like the full scattering matrix – is unitary and symmetric. So we can write for the background scattering matrix

$$\mathbf{S}_b(E) = \mathbf{B}_b(E)\mathbf{\Lambda}_b(E)^2\mathbf{B}_b^T(E). \tag{10.68}$$

with

$$[\mathbf{\Lambda}_b]_{nn} = e^{i\lambda_n^b}. \tag{10.69}$$

Following [326, 327], let us introduce the matrix

$$\mathbf{X} = \mathbf{B}_b(E)\mathbf{\Lambda}_b(E)\mathbf{B}_b^T(E) \tag{10.70}$$

and note that

$$\mathbf{X}\mathbf{X} = \mathbf{S}_b. \tag{10.71}$$

This can be used to rewrite Eq. (10.65) in matrix form

$$\mathbf{S} = \mathbf{X}\mathbf{S}_r\mathbf{X}, \tag{10.72}$$

where

$$\mathbf{S}_r = 1 - i\Gamma \frac{\mathbf{w}(E)\mathbf{w}^T(E)}{E - \epsilon_1 - \Delta(E) + \dfrac{i\Gamma(E)}{2}} \tag{10.73}$$

with the vector $\mathbf{w}$ defined as follows:

$$\mathbf{w} = \Gamma^{-1/2}\mathbf{X}^\dagger \mathbf{g} \tag{10.74}$$

Because $\mathbf{g}^T\mathbf{g} = \Gamma$,

$$\mathbf{w}^\dagger\mathbf{w} = 1. \tag{10.75}$$

Note that

$$\mathbf{X}^\dagger\mathbf{X} = \mathbf{B}_b\mathbf{\Lambda}_b^*\mathbf{B}_b^T\mathbf{B}_b\mathbf{\Lambda}_b\mathbf{B}_b^T = \mathbb{1}, \tag{10.76}$$

that is the matrix $\mathbf{X}$ is unitary. If both $\mathbf{S}$ and $\mathbf{X}$ are unitary, so must be the matrix $\mathbf{S}_r = \mathbf{X}^\dagger\mathbf{S}\mathbf{X}^\dagger$. The matrix $\mathbf{S}_r$ as defined in Eq. (10.73) can only be unitary if all of the elements of the vector $\mathbf{w}$ are real so that $\mathbf{w}$ is a unit column vector.

Note that because $\mathbf{w}$ is real and because $\mathbf{w}^T\mathbf{w} = 1$, $\mathbf{w}$ is an eigenvector of the matrix $\mathbf{S}_r$. This is easy to see by writing

$$\mathbf{S}_r\mathbf{w} = \left[1 - i\Gamma\frac{\mathbf{w}\mathbf{w}^T}{E - \epsilon_1 - \Delta + \dfrac{i\Gamma}{2}}\right]\mathbf{w} = \left[1 - \frac{i\Gamma}{E - \epsilon_1 - \Delta + \dfrac{i\Gamma}{2}}\right]\mathbf{w}. \tag{10.77}$$

We thus immediately obtain that one of the eigenvalues of the matrix $\mathbf{S}_r$ is

$$1 - \frac{i\Gamma}{E - \epsilon_1 - \Delta + \dfrac{i\Gamma}{2}} = \mathrm{e}^{2i\lambda}, \tag{10.78}$$

where

$$\lambda = \tan^{-1}\left[\frac{\Gamma}{2(E - \epsilon_1 - \Delta)}\right]. \tag{10.79}$$

Since all other eigenvectors must be orthogonal to $\mathbf{w}$, all other eigenvalues of the matrix $\mathbf{S}_r$ are simply 1.

For the unitary matrix $\mathbf{S}_r$, we can write

$$\mathbf{S}_r = \mathbf{B}_r\mathbf{\Lambda}_r{}^2\mathbf{B}_r^T. \tag{10.80}$$

We can now write Eq. (10.72) as

$$\mathbf{B}\mathbf{\Lambda}^2\mathbf{B}^T = \mathbf{B}_b\mathbf{\Lambda}_b\mathbf{B}_b^T\mathbf{B}_r\mathbf{\Lambda}_r{}^2\mathbf{B}_r^T\mathbf{B}_b\mathbf{\Lambda}_b\mathbf{B}_b^T. \tag{10.81}$$

Since the determinant of a product of two matrices is equal to the product of the determinant

$$|\mathbf{AB}| = |\mathbf{A}|\,|\mathbf{B}|, \tag{10.82}$$

and the determinants of the orthogonal transformation matrices are all $|\mathbf{B}| = |\mathbf{B}_b| = |\mathbf{B}_r| = 1$, we obtain

$$|\mathbf{\Lambda}^2| = |\mathbf{\Lambda}_b{}^2|\,|\mathbf{\Lambda}_r{}^2|. \tag{10.83}$$

Given that $|\mathbf{\Lambda}_r{}^2| = e^{2i\eta}$, we obtain the Breit–Wigner formula

$$\eta(E) = \eta_b(E) + \tan^{-1}\left[\frac{\Gamma}{2(E - \epsilon_1 - \Delta)}\right], \tag{10.84}$$

where $\eta(E)$ is the eigenphase sum (10.64) of the full S-matrix and $\eta_r(E) = \sum_n \lambda_n^b$ is the eigenphase sum of $\mathbf{S}_b$. The background scattering eigenphase sum is typically a slowly varying function of energy so most of the variation of the full scattering eigenphase sum near a resonance is due to the inverse tangent. The locations and widths of resonances can be obtained by fitting the energy dependence of the eigenphase sum to Eq. (10.84).

In practice, the eigenphase sum can be computed by diagonalyzing the K-matrix

$$\mathbf{K} = -i[\mathbf{S} + 1]^{-1}[\mathbf{S} - 1] \tag{10.85}$$

instead of the S-matrix. The K-matrix is real so it is easier to diagonalize than the S-matrix that is complex. It is easy to see that

$$\mathbf{B}^T\mathbf{K}\mathbf{B} = \tan \mathbf{\Lambda}, \tag{10.86}$$

where $\tan \mathbf{\Lambda}$ is a diagonal matrix of tangents of the eigenphases of the S-matrix. The eigenphase sum $\eta(E)$ is thus a sum of the inverse tangents of the eigenvalues of the K-matrix.

10.6 Locating Field-Induced Resonances

Consider again the Figure 10.2a. As shown in the previous sections, the scattering resonances manifest themselves as peaks in the dependence of the scattering cross sections on the collision energy ε. Consider now the cross sections at fixed collision energy ε as functions of an external field strength. If – and that is an important *if* – the external field moves the quasi-bound state depicted in Figure 10.2 closer to or further away from the open channel threshold, the scattering resonances can appear in the dependence of the cross sections on the external field strength at a fixed collision energy.

The field-induced resonances are most important for molecular collisions at very low energies, i.e. in the limit $\epsilon \to 0$. In this limit, the velocity distribution of molecules is very narrow so the collision energy can be considered fixed. The field-induced resonances may also appear in the field dependence of cross sections at higher energies. However, the scattering events at higher energies must be averaged over the corresponding velocity distribution and this averaging may wash out the resonant enhancement of the cross sections.

The external fields can be used to induce both the Feshbach and shape resonances. In general, the interaction of molecules with external fields modifies the Hamiltonian matrix, which leads to modifications of the interaction potentials depicted in Figure 10.2. These modifications may result in shifts of the quasi-bound states.

In general, the positions and widths of the field-induced resonances can be obtained by computing the eigenphase sum – as discussed in Section 10.5 – as a function of the external field strength. This procedure involves fitting the eigenphase sum to the Breit–Wigner formula and is quite tedious. If only the positions of the resonances are needed, for field-induced resonances in the limit of zero collision energy ϵ, there are two alternative approaches.

The first approach is based on computing the energies of the bound states of the collision complex as functions of the external field strength [328]. The $\epsilon \to 0$ resonance occurs when a bound state of the collision complex crosses the open channel threshold and becomes quasi-bound. The external field strengths, at which this happens, can be identified by computing the field dependence of the bound states at energies below the open channel threshold.

The bound-state energies of a collision complex can be computed by solving the coupled channel equations (8.11). In order to understand how to do this, consider first the procedure for calculating the bound state in an isolated potential depicted in the left part of Figure 10.3.

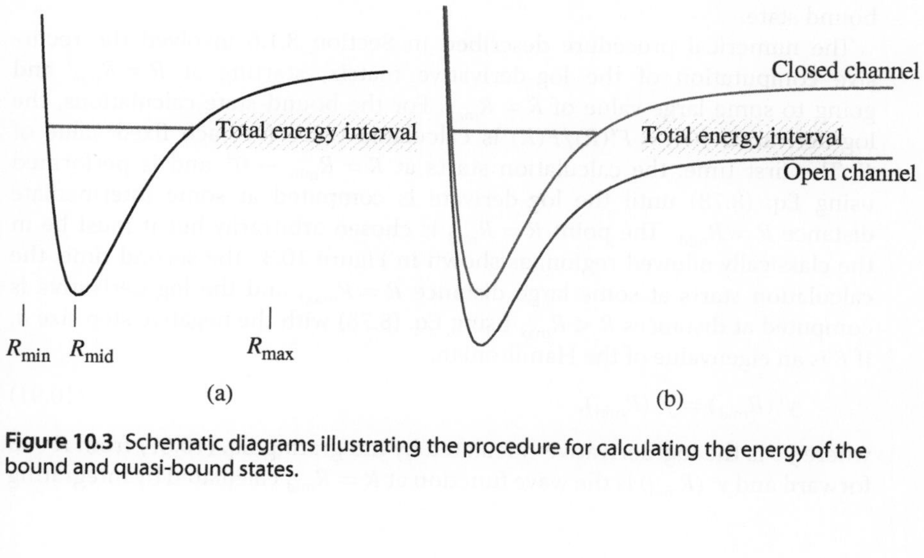

Figure 10.3 Schematic diagrams illustrating the procedure for calculating the energy of the bound and quasi-bound states.

The single-channel Schrödinger equation for a particle moving in an isolated potential is given by Eq. (8.75) and is reproduced here in a simplified form:

$$\left[\frac{\mathrm{d}^2}{\mathrm{d}R^2} + W(R)\right] F(R) = 0, \tag{10.87}$$

where

$$W(R) = \frac{2\mu}{\hbar^2}[E - V(R)] - \frac{l(l+1)}{R^2}, \tag{10.88}$$

and $V(R)$ is the interaction potential depicted in the left part of Figure 10.3. For clarity, let us set the zero of energy so that $W(R) \to 0$ when $R \to \infty$.

In Section 8.1.6, we were concerned with computing the solutions of Eq. (10.87) at positive energies. Here, we would like to calculate the solutions of the Schrödinger equation at negative energies. As in Section 8.1.6, we can use the log-derivative method. However, the boundary conditions for the bound states are different. In particular, we note that at negative energies,

$$F(R) \to 0, R \to 0_+, \tag{10.89}$$

$$F(R) \to 0, R \to +\infty, \tag{10.90}$$

that is the wave function vanishes both in the limit of zero R and at large R.

The differential equation (10.87) is parametrized by the value of E in Eq. (10.88). It is important to remember that at negative energies, where the eigenspectrum of the Hamiltonian is discrete, the solutions of Eq. (10.87) are physical only for certain values of E. The procedure of computing the bound-state energy involves integrating Eq. (10.87) with fixed values of E in an interval indicated by the shaded area in Figure 10.3. The goal is then to select one of the solutions in this interval, the one that corresponds to the bound state.

The numerical procedure described in Section 8.1.6 involved the recursive computation of the log-derivative matrix, starting at $R = R_{\text{min}}$ and going to some large value of $R = R_{\text{max}}$. For the bound-state calculations, the log-derivative $y(R) = F'(R)/F(R)$ is calculated twice for each fixed value of E. The first time, the calculation starts at $R = R_{\text{min}} \to 0^+$ and is performed using Eq. (8.78) until the log-derivate is computed at some intermediate distance $R = R_{\text{mid}}$. The point $R = R_{\text{mid}}$ is chosen arbitrarily but it must be in the classically allowed region, as shown in Figure 10.3. The second time, the calculation starts at some large distance $R = R_{\text{max}}$, and the log-derivative is computed at distances $R < R_{\text{max}}$ using Eq. (8.78) with the negative step size h. If E is an eigenvalue of the Hamiltonian,

$$y^+(R_{\text{mid}}) = y^-(R_{\text{min}}), \tag{10.91}$$

where y^+ is the log-derivative calculated by integrating Eq. (10.87) from $R = 0$ forward and $y^-(R_{\text{mid}})$ is the wave function at $R = R_{\text{mid}}$ calculated by integrating

Eq. (10.87) from $R = +\infty$ backward. The last equation follows from the fact that, if E is the eigenvalue of the Hamiltonian, the corresponding wave function must satisfy

$$F(R_{\text{mid}}) = F^{+}(R_{\text{mid}}) = F^{-}(R_{\text{mid}}) \tag{10.92}$$

and

$$\frac{dF^{+}(R_{\text{mid}})}{dR} = \frac{dF^{-}(R_{\text{mid}})}{dR}, \tag{10.93}$$

where $F^{+}(R_{\text{mid}})$ is the wave function at $R = R_{\text{mid}}$ calculated by integrating Eq. (10.87) from $R = 0$ forward and $F^{-}(R_{\text{mid}})$ is the wave function at $R = R_{\text{mid}}$ calculated by integrating Eq. (10.87) from $R = +\infty$ backward [329–332].

In the case of more than one channels, both y and F become matrices (see Section 8.1.6), but Eqs. (10.92) and (10.93) remain valid. Combining Eqs. (10.92)

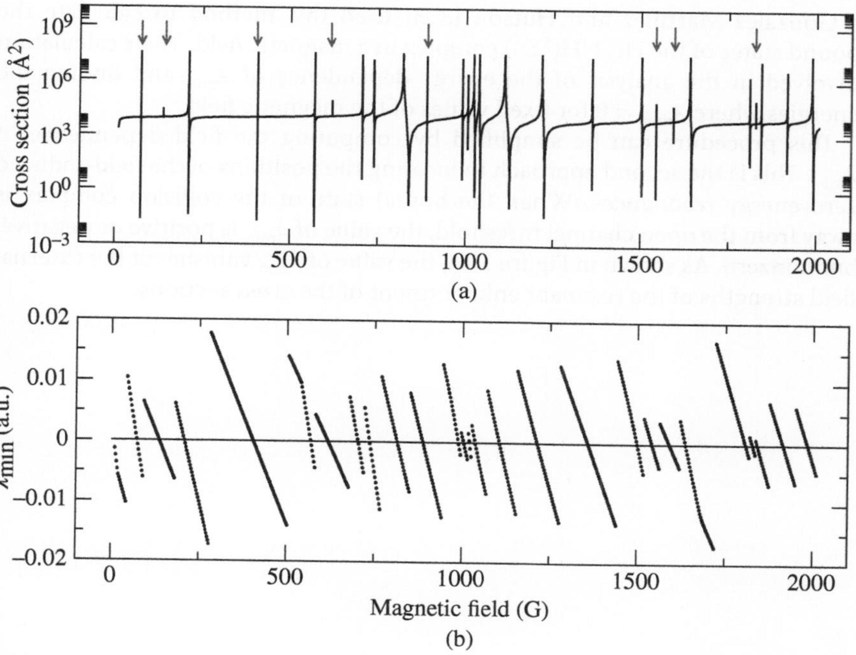

Figure 10.4 (a) The s-wave elastic scattering cross section for collisions of $O_2(^3\Sigma_g^-)$ molecules in the lowest high-field-seeking state $|M_{S_A} = -1\rangle|M_{S_B} = -1\rangle$ as a function of the magnetic field. (b) The minimal eigenvalue of the matching matrix λ_{min} as a function of the magnetic field. The collision energy is 10^{-6} K. The projection of the total angular momentum of the system is −2. New resonances found using the analysis of the magnetic field dependence of λ_{min} are marked by arrows. *Source*: Reproduced with permission from Suleimanov and Krems 2011 [334]. © 2011, American Institute of Physics.

and (10.93), we can write [332]

$$[\mathbf{y}^{+}(R_{\mathrm{mid}}) - \mathbf{y}^{-}(R_{\mathrm{mid}})]\mathbf{F}(R_{\mathrm{mid}}) = 0, \tag{10.94}$$

where $\mathbf{y}^{+}(R_{\mathrm{mid}})$ is the log-derivative matrix at $R = R_{\mathrm{mid}}$ calculated by integrating the close coupling equations from $R = 0$ forward with the boundary condition (10.89) and $\mathbf{y}^{-}(R_{\mathrm{mid}})$ is the log-derivative matrix at $R = R_{\mathrm{mid}}$ calculated by integrating the close coupling equations from $R = +\infty$ backward with the boundary conditions (10.90).

Equation (10.94) shows that $\mathbf{F}(R_{\mathrm{mid}})$ is an eigenvector of the matrix $|\mathbf{y}^{+} - \mathbf{y}^{-}|$ with eigenvalue 0. Several algorithms have been suggested for locating the bound-state energies using Eq. (10.94) [329–332]. The most stable approach for systems with a large number of channels is based on the analysis of the eigenvalue λ_{min} of $|\mathbf{y}^{+} - \mathbf{y}^{-}|$ with the smallest absolute magnitude [332]. At least one eigenvalue of the matching matrix must be zero if the total energy of the system is the same as the energy of a bound state.

González-Martínez and Hutson [333] used this method to calculate the bound states of the He-NH($^3\Sigma^-$) complex in a magnetic field. Their calculation involved in the analysis of the energy dependence of λ_{min} and finding the energies where $\lambda_{\mathrm{min}} = 0$ for fixed values of the magnetic field.

This procedure can be simplified by computing the field dependence of λ_{min}. This is the second approach to locating the positions of the field-induced zero-energy resonances. When the bound state of the collision complex is away from the open channel threshold, the value of λ_{min} is positive or negative, but nonzero. As shown in Figure 10.4, the value of λ_{min} vanishes at the external field strengths of the resonant enhancement of the cross sections.

11

Field Control of Molecular Collisions

The purpose of this chapter is to discuss various strategies for external field control of molecular collisions, reactive and nonreactive. The emphasis is made on the computational and experimental results that have demonstrated the possibility of external field control of molecular collisions. Most of the cases discussed in this chapter are rather peculiar. Despite the work presented, the question of whether microscopic chemical reactions and inelastic collisions of molecules in a gas can be *generally* and *effectively* controlled by external fields remains quite open.

As discussed in Section 11.1, there are two ingredients that we can identify for the effective control of molecular collisions in a gas:

- The molecules need to be prepared in a narrow distribution of internal and translational states;
- The kinetic energy of the molecules should be generally smaller than the perturbations induced by external fields.

Both of these have become possible with the development of cooling and trapping techniques for molecular radicals. The experiments on external field control of molecular collisions can also be performed with molecular beams, allowing for the preparation of bunches of molecules with narrow velocity distributions. The collision energy of molecules traveling in different molecular beams can be controlled by merging beams at different angles.

11.1 Why to Control Molecular Collisions

Controlled chemistry has been a goal of chemical dynamics research for many decades. This goal is often stated without further justification. Yet, the question of why to seek the control of molecular collisions is a valid one.

There is a big gap between the research fields of chemical dynamics and synthetic chemistry. It is thus premature to discuss the experiments on controlled collisions of molecules in the gas phase as a way to controlled synthesis of

Molecules in Electromagnetic Fields: From Ultracold Physics to Controlled Chemistry, First Edition. Roman V. Krems.
© 2019 John Wiley & Sons, Inc. Published 2019 by John Wiley & Sons, Inc.

new molecular species. I believe that the greatest value of controlled collision experiments will be in improving the mechanistic understanding of molecular interactions and reaction processes and in providing a new platform for fundamental studies with controlled molecules as the basic element of measurements.

What can we learn from experiments on controlled collisions? Here is a partial list of ideas:

- Since experiments on controlled collisions are most often carried out with molecular radicals, tuning the collision dynamics can be used as a new tool to examine the role of nonadiabatic interactions. For example, shifting the position of nonadiabatic couplings by applying an external field can be used to compare the collision dynamics with and without conical intersections. The effect of conical intersections on chemical reactions is currently a major outstanding question in chemical reaction dynamics [335].
- By separating the Zeeman or Stark levels, external fields can be used to uncouple angular momenta, which can be exploited to explore the role of fine-structure interactions in determining the reactivity of certain species.
- By shifting the molecular energy levels, external fields can induce scattering resonances. The effect of scattering resonances on reactive collisions of molecules has long been an important question in chemical reaction dynamics [336].
- Since external fields break the symmetry of the collision problem, the experiments with molecules in fields can be used to explore the role of symmetries in reaction dynamics.
- Since external fields hybridize states of different energy (for example, the electric field-induced interactions couple different rotational states), measuring the effects of external fields on low-temperature chemical reactions can be used as a probe of the topology of the reaction barrier.
- The possibility of assembling polyatomic molecules by the photoassociation of controlled molecules at ultralow temperatures seems to be particularly enticing.

The purpose of the list above is to seed the discussion of controlled molecular dynamics in a broad context of physical chemistry. There are also multiple applications of controlled molecular dynamics for fundamental physics research, including

- The preparation of molecular gases with suppressed collisional decoherence for applications requiring long-lived coherent states.
- The suppression of undesirable collision processes in molecular cooling experiments. This is critical for the success of evaporative cooling of molecular ensembles to ultracold temperatures.

Many of the examples of controlled collisions discussed in this chapter are presented with the abovementioned applications in mind.

11.2 Molecular Collisions are Difficult to Control

The 2007 DAMOP[1] abstract by Paul Brumer, Professor at the University of Toronto and one of the pioneers of the research field of coherent control of molecular processes, opens with the following two sentences:

> Experimental and theoretical studies of the Coherent Control of unimolecular processes have seen spectacular growth over the last two decades. By contrast, Coherent Control of collisional processes remains a significant challenge.

Molecular collisions are difficult to control. There are two fundamental reasons for this: the stochastic nature of molecular collisions and the weak effect of external field perturbations compared to the magnitude of intermolecular forces.

As molecules are whizzing in a gas, they encounter each other at random times and at random angles. The collision events are largely independent. The observables in a typical scattering experiment are averages over a random sequence of collision events, a Maxwell–Boltzmann distribution of collision energies, a Boltzmann distribution of internal energies, and the relative angles of approach. This averaging diminishes the effects of external fields on molecular collisions and makes the external field control of collisions hard to achieve.

Let us consider, for example, how the randomness of molecular collisions defies the control schemes based on quantum interferences. One could – in principle – devise a scheme of controlling the outcome of molecular collisions by preparing the collision partners in coherent superpositions of two eigenstates of the full Hamiltonian of the two-particle system,

$$|\Psi_i\rangle = a_1|\psi_1\rangle + a_2|\psi_2\rangle. \tag{11.1}$$

Since the probability for scattering into some final state $|\Psi_f\rangle$ is proportional to the square of the S-matrix element,

$$\begin{aligned} P_{i\to f} &= |\langle\Psi_i|\hat{S}|\Psi_f\rangle|^2 = a_1^2|\langle\psi_1|\hat{S}|\Psi_f\rangle|^2 + a_2^2|\langle\psi_2|\hat{S}|\Psi_f\rangle|^2 \\ &\quad + a_1^* a_2\langle\psi_1|\hat{S}|\Psi_f\rangle\langle\Psi_f|\hat{S}|\psi_2\rangle + a_2^* a_1\langle\psi_2|\hat{S}|\Psi_f\rangle\langle\Psi_f|\hat{S}|\psi_1\rangle, \end{aligned} \tag{11.2}$$

one could control the outcome of collisions by tuning the relative phases of the states $|\psi_1\rangle$ and $|\psi_2\rangle$, thereby changing the interference terms in Eq. (11.2). However, the conditions for the interference terms to be nonzero are very stringent. First of all, because the collision events conserve total energy, the states $|\psi_1\rangle$ and $|\psi_2\rangle$ must correspond to the same eigenvalue of the full Hamiltonian. This is half the problem. More importantly, because the collision events do not affect

1 Division of atomic, molecular and optical physics of the American Physical Society.

the center-of-mass motion of the collision complex, the states $|\Psi_i\rangle$ and $|\Psi_f\rangle$ must correspond to the same eigenstate of the kinetic energy operator for the center-of-mass motion. This means that in order for the interference terms in Eq. (11.2) to be nonzero, the states $|\psi_1\rangle$ and $|\psi_2\rangle$ must correspond to the same state of the center-of-mass motion. Due to the randomness of the molecular motion in a gas, it is very difficult to ensure that multiple collisions occur with the same center-of-mass motion of the collision complex. Therefore, different collision events will generally not interfere.

The averaging over the collision events with different initial conditions also diminishes the effects of molecular shifts due to AC and DC fields on molecular scattering. This happens because the response of the scattering cross sections to external fields is generally different depending on the collision energy and the internal states of the colliding particles. Often, the external fields significantly affect collisions of molecules only in specific internal states and only in a limited range of collision energies. In this case, averaging over a large number of internal states and a wide distribution of collision energies will negate the effects of external fields.

It is clear that the first step toward external field control of molecular collisions must be to reduce the randomness of molecular encounters. This can be achieved by simply cooling the gas, thereby reducing the number of populated internal states and narrowing the collision energy distribution.

But then, there is another problem. The perturbations induced by external fields are on the order of a few wave numbers, at best. When molecules come together for a chemical exchange or an inelastic scattering event, they experience the attraction energy on the order of thousands of wave numbers or more. Therefore, when molecules approach each other to react, they should be expected to largely ignore the presence of an external field and tend to follow the minimum energy path set by the intermolecular interactions.

This problem can be eliminated by preventing the molecules from coming together. In fact, a very powerful mechanism of controlling molecular reactions, demonstrated experimentally [265], relies on modifying the long-range interactions between molecules to create long-range barriers. This control of molecular reactions boils down to allowing or not allowing molecules to collide. While very useful for practical purposes (for example, for eliminating the unwanted reaction processes leading to the destruction of molecules), this method can only be used for molecules with extremely low kinetic energy and does not yield much information about the mechanisms of reactions.

Highly desirable is the development of control mechanisms that would steer molecular collisions to produce particular products in particular states, in a wide range of collision energies and for a wide range of internal states. Whether this is possible remains an open question.

11.3 General Mechanisms for External Field Control

The interactions with external fields, whether magnetic or electric, static or oscillating, have three effects in common:

- They break the spherical symmetry of the collision problem by inducing couplings between states of different total angular momenta.
- They shift the molecular states due to the Zeeman or Stark effect.
- They split the molecular states with nonzero angular momentum into manifolds of the Zeeman or Stark levels.

These effects are at the core of the main mechanisms that can be exploited to control molecular collisions. As will be illustrated by the examples in the subsequent sections, we can identify the following mechanisms for external field control:

- Field-induced scattering resonances (see Chapter 10).
- Field-induced variation of the energy gap between the internal energy of the reactants and products of a scattering event.
- Field-induced hybridization of molecular states leading to variations in the intermolecular interactions.
- Collision-induced absorption (or emission) of radiation.

11.4 Resonant Scattering

As discussed in Chapter 10, scattering resonances may have a dramatic effect on molecular collisions. Because the scattering resonances are often sensitive to external fields – and some may be induced by external fields – they can be used as a versatile tool for controlling the collision cross sections of molecules. There is one important limitation of the control schemes based on tuning scattering resonances. Resonances usually affect the scattering cross sections in a narrow range of collision energies. Therefore, the control schemes based on shifting or inducing scattering resonances with external fields are most effective when applied to molecules prepared in a narrow distribution of collision energies. Field-tuneable scattering resonances are particularly useful for controlling the collision properties of particles in the limit of zero collision energy, that is in an ultracold gas.

For example, magnetic scattering resonances play a pivotal role in the experimental studies of ultracold atomic gases. They are used as a tool to control the scattering length of ultracold atoms. The scattering length parametrizies the microscopic interactions of ultracold atoms and determines the macroscopic

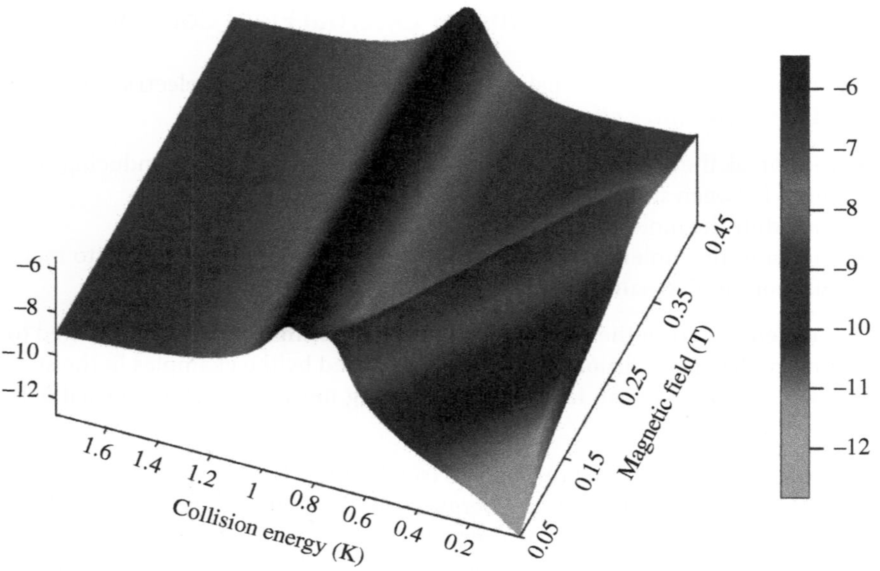

Figure 11.1 The logarithm of the cross section for the $|M_S = 1\rangle \rightarrow |M_S = -1\rangle$ transition in collisions of NH molecules with ^{3}He atoms as a function of the magnetic field and collision energy. *Source*: Adapted with permission from Campbell et al. 2009 [345]. © 2009, the American Physical Society.

behavior of dilute ultracold gases. Magnetic-field-induced resonances can thus be used to tune both the microscopic and macroscopic properties of ultracold atomic gases, which has been used to study such important phenomena as Bose–Einstein condensation, bosonic superfluidity, quantum magnetism, many-body spin dynamics, Efimov states, Bardeen–Cooper–Schrieffer (BCS) superfluidity, and the BEC–BCS crossover.

The experiments with ultracold molecular gases are still in their infancy. However, it is reasonable to expect that the field-induced scattering resonances will play a similarly important role providing a control knob for tuning the microscopic and macroscopic properties of ultracold molecular gases. While, at present, there are no direct experimental observations of field-induced scattering resonances in molecular collisions, the literature abounds with predictions of field-induced scattering resonances in both atom–molecule and molecule–molecule collisions [293, 316, 328, 333, 337–353].

Figures 11.1–11.6 provide a few examples of the predicted and measured effects of scattering resonances on the elementary collisions of molecules at low and ultralow temperatures. These figures need no discussion, they speak for themselves.

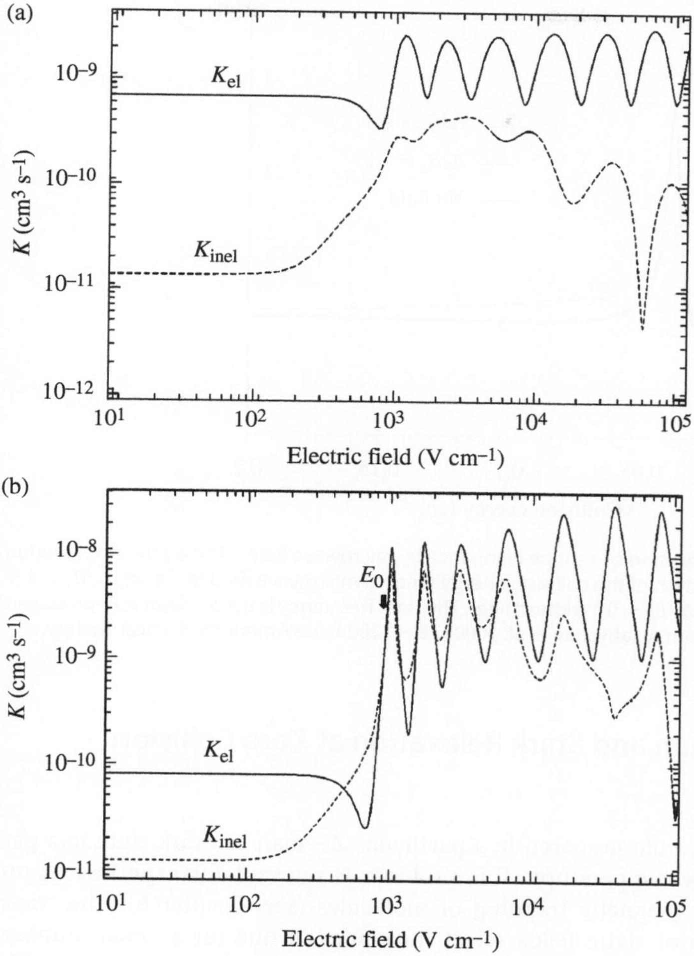

Figure 11.2 Rate constants versus electric field for OH–OH collisions with molecules initially in a particular Stark state. Shown are the collision energies 100 mK (Panel a) and 1 mK (Panel b). Solid lines denote elastic-scattering rates, while dashed lines denote rates for inelastic collisions, in which one or both molecules change their internal state. These rate constants exhibit characteristic oscillations in field when the field exceeds a critical field of about 1000 V cm^{-1}. *Source*: Reproduced with permission from Avdeenkov and Bohn 2002 [338]. © 2002, the American Physical Society.

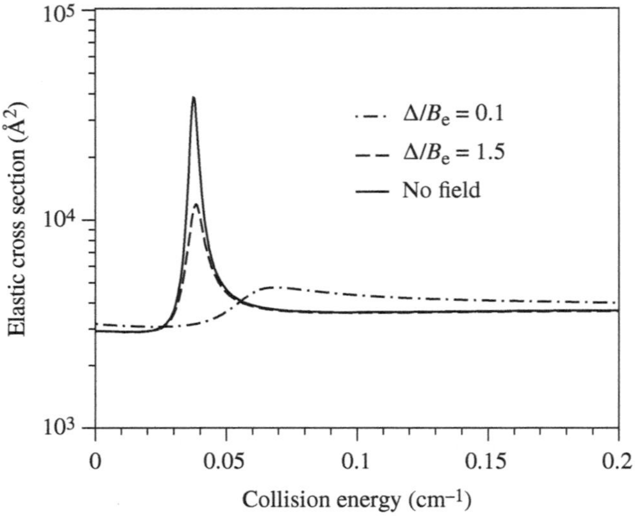

Figure 11.3 Modification of a shape resonance by microwave fields. The elastic cross section is plotted as a function of the collision energy for zero microwave field (full line), $\Delta/B_e = 1.5$ (dashed line), and $\Delta/B_e = 0.1$ (dotted line). The Rabi frequency is 0.5 B_e. *Source*: Reproduced with permission from Alyabyshev et al. 2009 [38]. © 2009, the American Physical Society.

11.5 Zeeman and Stark Relaxation at Zero Collision Energy

Consider a molecule prepared in a particular Zeeman or Stark state in a gas with a very low temperature. This problem is relevant for experiments on electrostatic or magnetic trapping of molecules (see Chapter 6). The traps using gradients of static fields select molecules in one (or a small number of) low-field-seeking Zeeman or Stark state(s). When molecules collide with the particles of some background gas or with each other, they may undergo collision-induced Zeeman or Stark relaxation to high-field-seeking states. As shown in Figure 11.7, when the field is present, the high-field-seeking states are always lower in energy than the low-field-seeking states. Once in a high-field-seeking state, the molecules become untrappable and leave the trap. This is an undesirable process that most often happens in the most dense region of the trap that corresponds to the lowest translational energy. The Zeeman or Stark relaxation thus removes the coldest particles from the trapped gas and leads to heating. This raises the question, can something be done to suppress the undesirable inelastic collisions? The answer is, yes! – make the trap shallower.

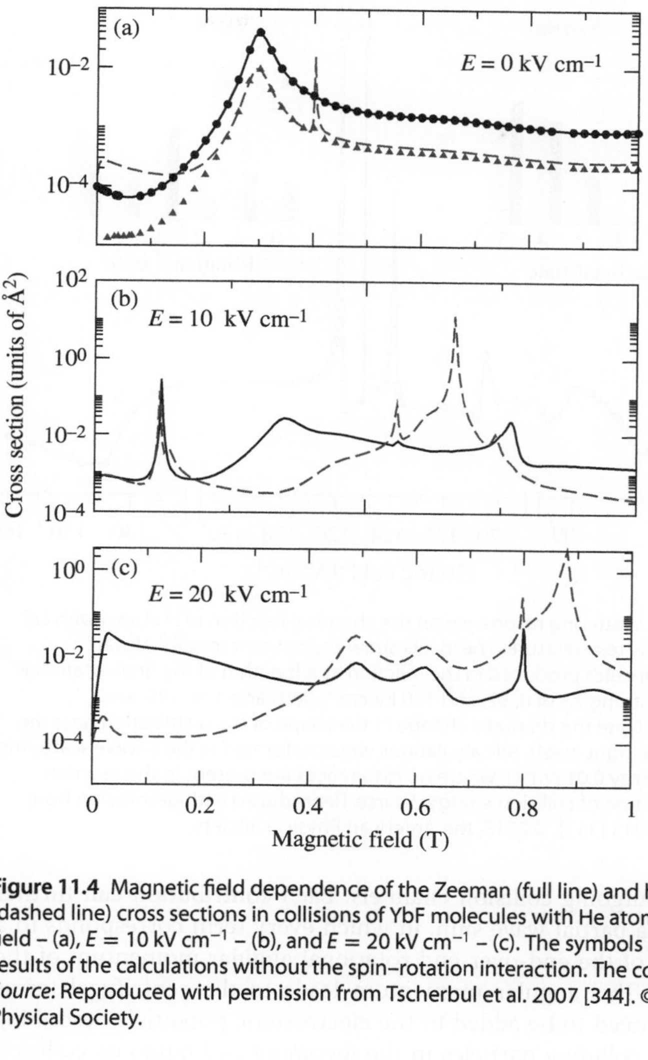

Figure 11.4 Magnetic field dependence of the Zeeman (full line) and hyperfine relaxation (dashed line) cross sections in collisions of YbF molecules with He atoms at zero electric field – (a), $E = 10$ kV cm^{-1} – (b), and $E = 20$ kV cm^{-1} – (c). The symbols in the (a) are the results of the calculations without the spin–rotation interaction. The collision energy is 0.1 K. *Source*: Reproduced with permission from Tscherbul et al. 2007 [344]. © 2007, the American Physical Society.

Figure 11.8 is an example of the typical dependence of the rate constants and cross sections for the Zeeman or Stark relaxation on the external field strength for collisions with very low energy. Note that the rate constant vanishes in the limit of zero field and increases very fast with the field strength at finite field. This phenomenon was first observed and explained by Volpi and Bohn [337] and can be interpreted with the help of Figure 11.9.

As discusses in Chapter 8, the eigenstate of the full Hamiltonian for a pair of colliding particles can be represented by a sum of contributions corresponding

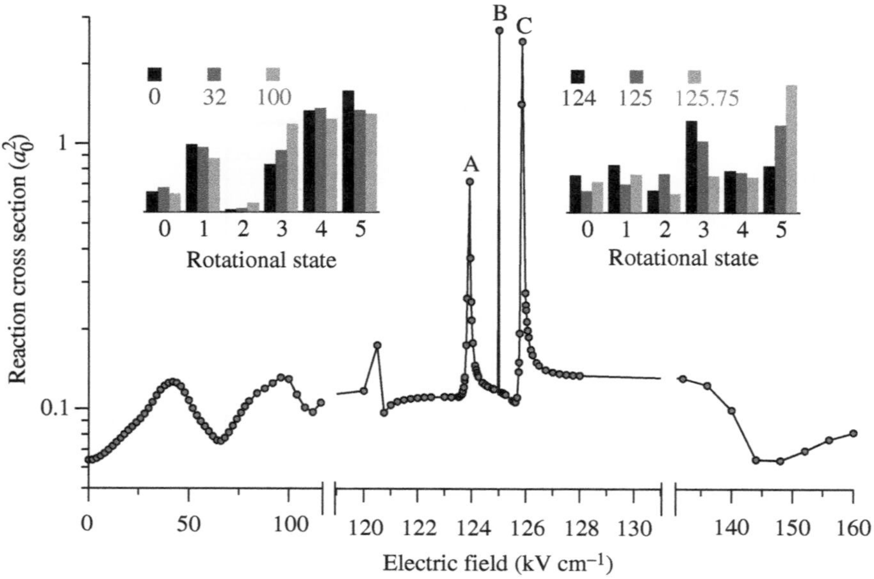

Figure 11.5 Effect of a scattering resonance on the chemical reaction of H atoms with LiF molecules at an ultralow temperature. The insets show the nascent rotational state distributions of HF molecules produced in the reaction as a function of the final rotational state j' at electric field strengths of 0, 32, and 100 kV cm^{-1} (left) and 124, 125, and 125.75 kV cm^{-1} (right). Note the dramatic change in the shape of the distribution near the resonance electric field (right inset). All calculations were performed in the s-wave scattering regime (at collision energy 0.01 cm^{-1}), where no resonances are present in the reaction cross sections as a function of collision energy. *Source*: Reproduced with permission from Tscherbul and Krems 2015 [353]. © 2015, the American Physical Society.

to incoming and outgoing collision channels. Each contribution can further be represented by a partial wave sum, in which every term corresponds to a well-defined value of the end-over-end rotational angular momentum of the collision complex. This angular momentum leads to the centrifugal terms $\propto l(l+1)/R^2$ than need to be added to the electrostatic potentials describing the interactions of colliding particles in the incoming and outgoing collision channels. The centrifugal terms with nonzero angular momenta lead to the centrifugal barriers at large inter-particle separations.

If the collision energy is very small – as is the case for molecules in an ultracold gas – the colliding particles cannot surmount the centrifugal barriers in the incoming collision channel. The scattering wave function for a pair of ultracold particles is therefore dominated by the s-wave (i.e. zero angular momentum) contribution in the partial wave expansion. The contribution from the p-wave, i.e. the term with $l = 1$ is negligible by comparison with the $l = 0$ contribution.

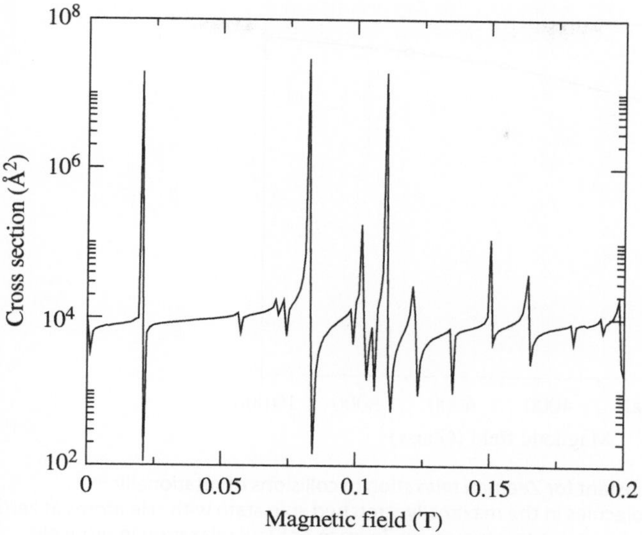

Figure 11.6 The elastic scattering cross section for collisions of O_2 molecules in the lowest energy Zeeman state (in which each molecule has the spin angular momentum projection $M_S = -1$) as a function of the magnetic field. The collision energy is 10^{-6} K $\times k_B$. *Source*: Reproduced with permission from Tscherbul et al. 2009 [347]. © 2009, the Institute of Physics.

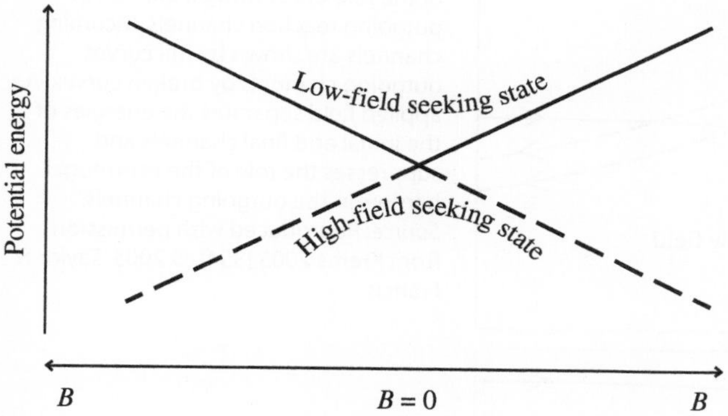

Figure 11.7 Potential energy of a molecule in the low-field-seeking and high-field-seeking Zeeman states in a magnetic trap. The strength of the trapping field *B* increases in all directions away from the middle of the trap. Since molecules in a high-field-seeking state are untrappable, collision-induced relaxation from the low-field-seeking state to the high-field-seeking state leads to trap loss.

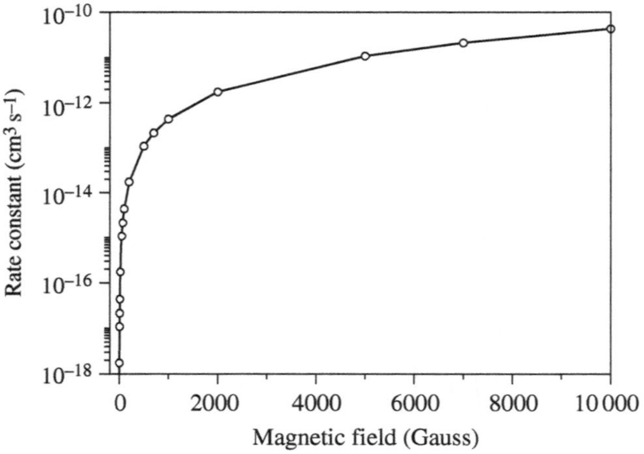

Figure 11.8 The rate constant for Zeeman relaxation in collisions of rotationally ground-state NH($^3\Sigma$) molecules in the maximally stretched spin state with ^{3}He atoms at zero temperature. Such field dependence is typical for Zeeman or Stark relaxation in ultracold collisions of atoms and molecules. The rate for the Zeeman relaxation vanishes in the limit of zero field. The variation of the relaxation rates with the field is stronger and extends to larger field values for systems with smaller reduced mass. *Source*: Reproduced with permission from Krems 2005 [354]. © 2005, Taylor & Francis.

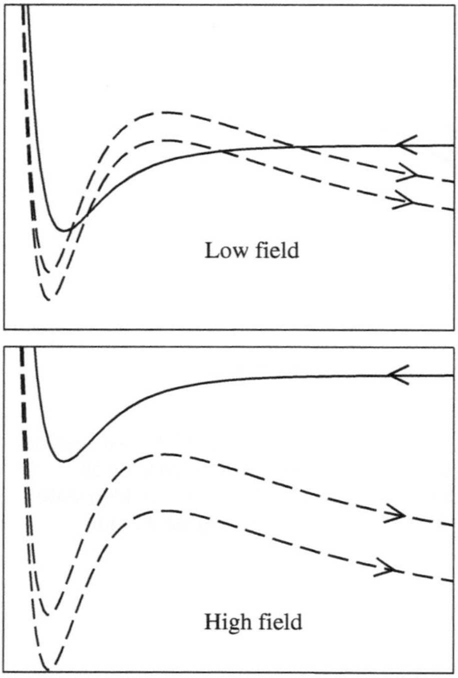

Figure 11.9 External field suppression of the role of centrifugal barriers in outgoing reaction channels. Incoming channels are shown by full curves; outgoing channels by broken curves. An applied field separates the energies of the initial and final channels and suppresses the role of the centrifugal barriers in the outgoing channels. *Source*: Reproduced with permission from Krems 2005 [354]. © 2005, Taylor & Francis.

Each Zeeman or Stark state corresponds to a well-defined projection of the total angular momentum of the molecule on the field axis. The process of the Zeeman or Stark relaxation changes the value of this projection. Because the projection of the total angular momentum on the field axis must be conserved, this change of the projection quantum numbers must be compensated by the change of the end-over-end angular momentum. The Zeeman or Stark relaxation at low collision energies thus begins on an incoming channel potential with zero end-over-end angular momentum (zero centrifugal barrier) and ends on outgoing channel potentials with finite centrifugal barriers.

Mathematically, if m_A and m_B are the projections of the total angular momenta of the molecules A and B on the field axis, the projection of the total angular momentum of the two-molecule system is $M = m_A + m_B + m_l$, where m_l is the projection of $\boldsymbol{l}$. If the external field is directed along the z-axis, there are no matrix elements of the total Hamiltonian between states of different M. Since for s-wave collisions $l = 0$, the projection of the end-over-end angular momentum must be $m_l = 0$ before the collision. If either or both of m_A and m_B decrease, m_l must increase to preserve the value of M. Nonzero values of m_l correspond to nonzero l. Thus, l must be nonzero in the outgoing Zeeman or Stark states of lower energy.

The energy separation between the incoming and outgoing states is given by the amount of the Zeeman or Stark splitting. In the limit of zero field, it is zero. Thus, in the limit of zero field, the process of changing the angular momentum projections is forbidden. At nonzero fields, the outgoing states are separated from the incoming states, as schematically shown in Figure 11.9. This energy separation equal to the Zeeman or Stark splitting is the amount of the kinetic energy acquired by the products of the collision event. As the kinetic energy increases, it becomes easier for the collision products to surmount the centrifugal barriers in the outgoing collision channels, which results in the enhancement of the collision cross section depicted in Figure 11.8.

11.6 Effect of Parity Breaking in Combined Fields

Section 7.1.1 illustrates the effect of combined magnetic and electric fields on molecular radicals in a $^2\Sigma$ electronic state. In particular, Figure 7.1 shows that the Zeeman states of the rotationally ground and first rotationally excited states become degenerate at a particular value of an applied magnetic field. Since these states have different parity, the interaction with an electric field lifts the degeneracy and transforms the real crossing into an avoided crossing. The collision properties of molecules placed in superimposed magnetic and electric fields near these avoided crossings are very sensitive to the field strengths and orientations. This is illustrated in Figure 11.10.

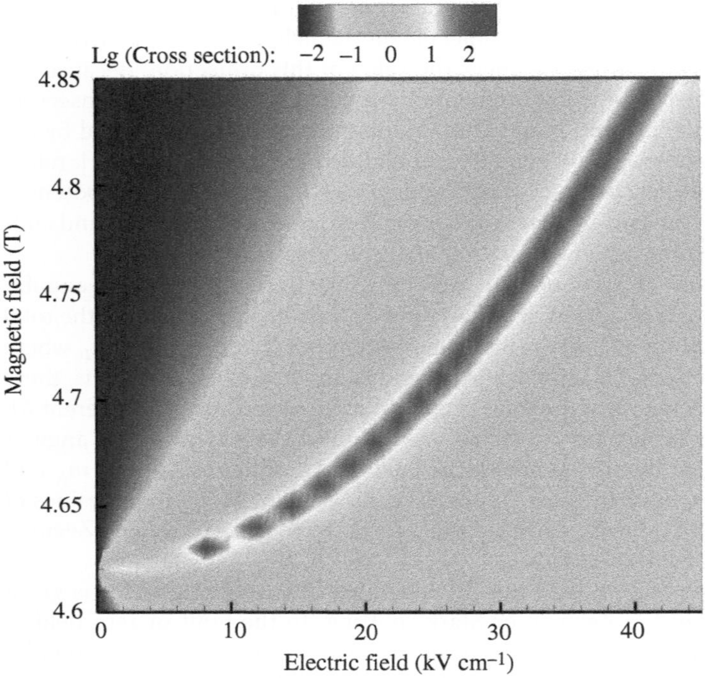

Figure 11.10 Decimal logarithm of the cross section for spin relaxation in collisions of $CaD(^2\Sigma)$ molecules in the rotationally ground state with He atoms as a function of electric and magnetic fields. The fields are parallel. The collision energy is 0.5 K. The cross section increases exponentially near the avoided crossings. *Source*: Reproduced with permission from Abrahamsson et al. 2007 [343]. © 2007, American Institute of Physics.

Figure 11.10 shows the cross section for collisions of molecules prepared in the magnetic low-field-seeking ("spin-up") state of the rotationally ground state, producing molecules in the magnetic high-field-seeking state ("spin-down"). The collision-induced spin-up-to-spin-down conversion is very sensitive to small variations of either of the field, with the cross sections exhibiting formidable peaks at certain combinations of the fields.

Since the collision occurs in a combination of two fields, DC electric and DC magnetic, it is natural to ask if the collision outcome is sensitive not only to the strength of the fields but, perhaps, also the relative orientation of the two fields. If one field vector is rotated with respect to the other, the structure of the Hamiltonian matrix changes. It follows from symmetry considerations. When the two fields are parallel, the cylindrical symmetry of the collision problem is preserved so that the projection of the total angular momentum on the direction of the fields is conserved. This is no longer the case when the angle between the field vectors is nonzero. The change of the Hamiltonian with the

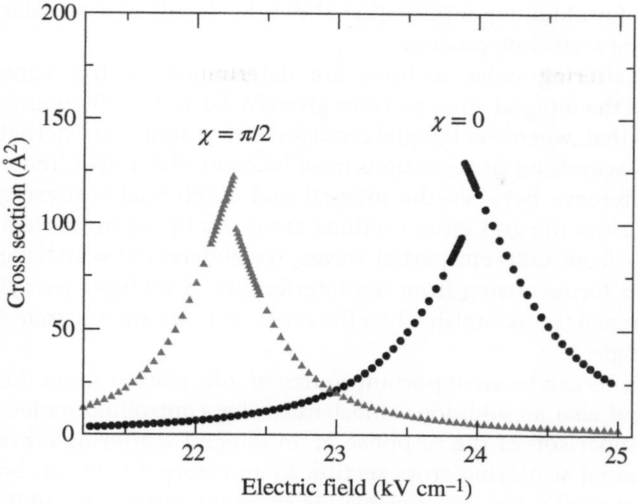

Figure 11.11 Cross sections for spin-up to spin-down transitions in collisions of CaD molecules in the rotationally ground state with He atoms at two different angles χ between the DC electric and DC magnetic fields. The magnetic field is 4.7 T. The positions of the maxima correspond to the locations of the avoided crossings depicted in Figure 7.1 that move as the relative orientation of the fields is changed. *Source*: Reproduced with permission from Abrahamsson et al. 2007 [343]. © 2007, American Institute of Physics.

angle between the fields changes the eigenstates of the molecule and hence the positions of the avoided crossings responsible for the enhancement of the cross sections depicted in Figure 11.10.

Figure 11.11 illustrates how the cross sections for the spin-up to spin-down conversion in collisions of $^2\Sigma$ molecules with atoms respond to the variation of the angle between the electric and magnetic fields. The relative orientation of the combined fields can thus be used as an additional knob to control certain molecular collision processes.

11.7 Differential Scattering in Electromagnetic Fields

The development of the experiments for the production of slow and controlled molecular beams described in Chapter 6 offers the unique possibility to study state- and angle-resolved *differential* scattering of molecules in the presence of external electromagnetic fields. One can imagine an experiment where two beams are brought together at a fixed angle in an area subjected to a magnetic or electric field and the collision products scattered in a particular direction are imaged. This could yield information on state-to-state differential scattering cross sections described by Eq. (8.43). Measuring such cross sections could be an invaluable tool for obtaining information about the details of molecular interactions during the scattering process.

The differential scattering cross sections are determined by the same T-matrix elements as the integral cross sections given by Eq. (8.50). Therefore, it should be expected that, whenever integral cross sections respond to external fields, the differential scattering cross sections must be controllable also. There is one important difference between the integral and differential scattering cross sections. While the integral cross sections are given by an incoherent sum of contributions from different partial waves, the differential scattering cross section involves terms arising from the interference of multiple partial waves. These interference terms vanish when the cross sections are integrated over the scattering angle.

The interference terms can be an important source of information about the scattering process and also an additional mechanism for controlling molecular collisions. A comparison of the response of an integral scattering cross section and a differential scattering cross section to an external field can be used to identify the partial waves responsible for resonant scattering. Additionally, the information on the difference of forward and backward scattering may shed light on whether a collision process occurs via the formation of a quasi-long-lived intermediate state.

Figure 11.12 illustrates how the differential scattering cross sections for spin-up to spin-down transitions in collisions of CaD($^2\Sigma$) molecules with He atoms change with the collision energy in the limit of ultralow energy scattering. This collision system exhibits a shape resonance at the collision energy ≈ 0.5 cm^{-1}. As the energy rises, higher partial waves contribute to the scattering process and the dependence of the differential scattering cross sections on the scattering angle becomes more complex. A similar effect can, in principle, be achieved by tuning a system in or out of a scattering resonances corresponding to a specific partial wave or a narrow range of partial waves. The scattering resonance will enhance the contribution of the corresponding partial waves and thereby affect the angular dependence of the differential scattering cross sections.

The effect of an electric field on the magnetic spin relaxation in CaD($^2\Sigma$)–He collisions is illustrated in Figure 11.13. It can be clearly seen in the figure how the peaks arising from the shape resonance at $\approx$0.5 cm^{-1} become modified by an electric field. It is intriguing to see that the presence of an electric fields forces these particular collisions to be predominantly backward-scattering.

11.8 Collisions in Restricted Geometries

As described in Chapter 9, counterpropagating laser beams create standing waves of electromagnetic fields called as optical lattice. These optical lattices

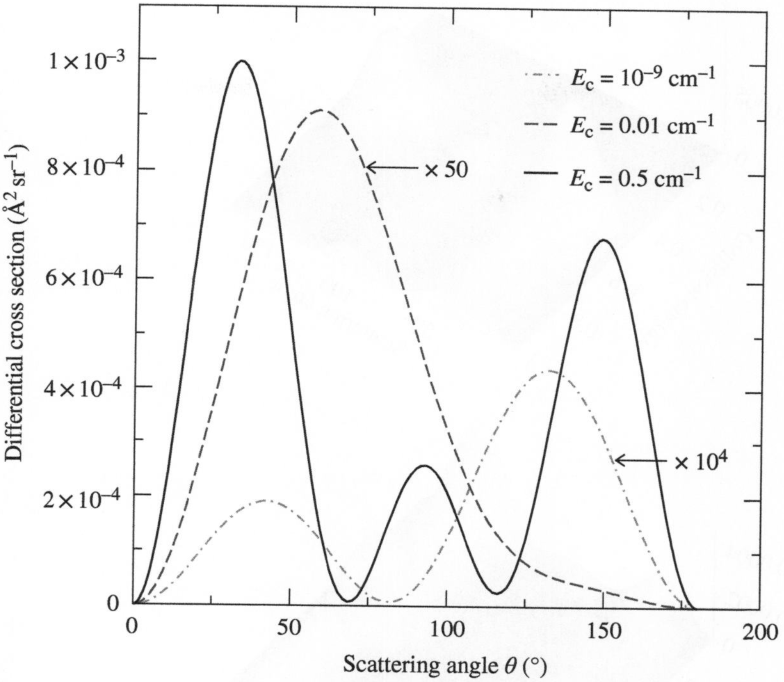

Figure 11.12 Differential scattering cross sections for spin-up to spin-down inelastic transitions in collisions of CaD($^2\Sigma$) radicals in the rotationally ground state with He atoms in a magnetic field $B = 0.5$ T at three different collision energies. The collision energy $\approx 0.5\text{cm}^{-1}$ corresponds to a shape resonance arising from the $l = 3$ partial wave. *Source*: Reproduced with permission from Tscherbul 2008 [355]. © 2008, American Institute of Physics.

can be used to trap atoms due to the AC Stark effect. If the kinetic energy of the atoms is low enough, as in an ultracold gas, the AC Stark shifts push the atoms toward the regions of the lattice where their potential energy is the lowest. Once trapped in an optical lattice, ultracold atoms can be associated with molecules. This makes a lattice of molecules suspended in three dimensions by laser fields. The intensity of the laser field determines the forces that keep the individual molecules in individual lattice sites. If the intensity is decreased, the molecules can tunnel between lattice sites and collide. An ensemble of molecules in a three-dimensional optical lattice is thus an interesting playground for studying controlled molecular interactions.

If the molecules are instead confined by laser forces only in one dimension, one realizes a quasi-two-dimensional gas. In such a trap – often referred to as a pancake-shape trap – molecules are free to move in two dimensions, while the motion in the third dimension is quantized. Similarly, one can realize ensembles

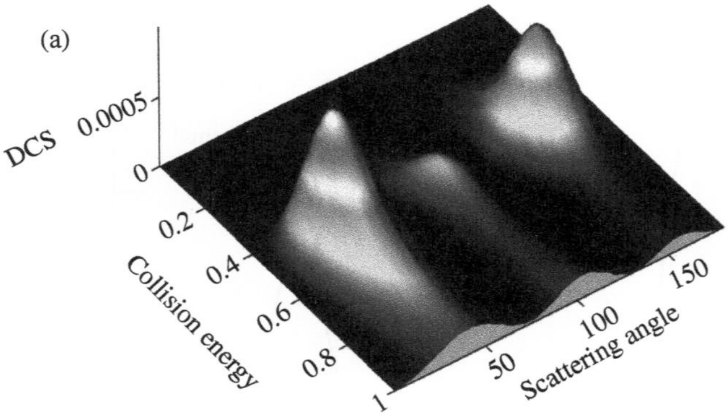

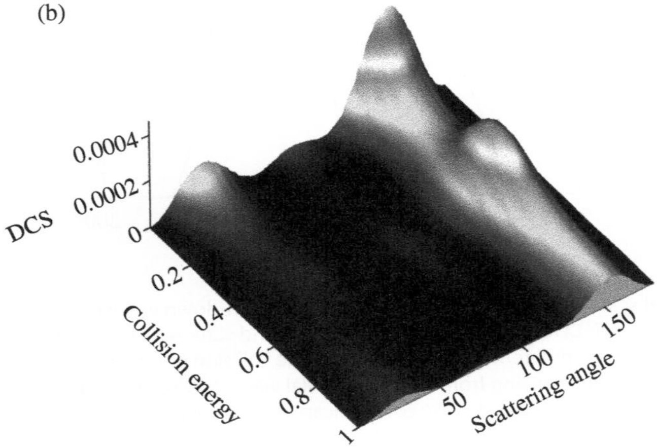

Figure 11.13 Differential scattering cross sections for spin-up to spin-down inelastic transitions in collisions of CaD($^2\Sigma$) radicals in the rotationally ground state with He atoms in a magnetic field $B = 0.5$ T. The graphs show the cross sections for collisions in the absence of an electric field (a) and in a DC electric field with magnitude $E = 100$ kV cm^{-1} (b). *Source*: Reproduced with permission from Tscherbul 2008 [355]. © 2008, American Institute of Physics.

of ultracold particles confined in two dimensions but free to move in a third dimension – a cigar-shaped trap.

These experiments raise the question: Does restricting the molecular motion to reduced geometries alter the dynamics of elastic and inelastic collisions? What about chemical reactions of molecules? The answer can be found in the papers tracing back to the work of Olshanii [356] for cigar-shaped traps and Petrov and Shlyapnikov [357] for pancake-shaped traps.

To understand how the interaction dynamics of molecules changes with the geometry of space, let us first consider the (somewhat unrealistic) limit of extremely tight confinement in one dimension and assume that the molecules are restricted to move in two dimensions. Restricting molecules to two dimensions renormalizes their scattering states, which results in the change of the energy dependence of the elastic and inelastic scattering cross sections at very low energies. This is well explained by Sadeghpour et al. in their paper titled "Collisions near threshold in atomic and molecular physics" [358]. For the theory of inelastic and chemically reactive scattering in reduced dimensions, see also the work by Naidon et al. [359] and Li et al. [360, 361]. To illustrate the difference between ultracold scattering in two and three dimensions here I adopt the original approach of Wigner [317].

As we discussed in Section 8.4, the energy dependence of the cross sections and scattering rates in the limit of very low kinetic energy of the colliding particles is described by simple analytical forms derived by Wigner [317]. In Section 8.4, we assumed that at very low energies, the scattering cross sections are completely determined by the contribution of a single partial wave corresponding to zero angular momentum for the end-over-end rotation of the collision complex. This is usually the case because any nonzero angular momentum will introduce a long-range centrifugal barrier. If the collision energy is low enough, the collision partners cannot overcome this barrier and the collision does not happen. In some cases, however, one needs to consider ultracold scattering involving nonzero partial waves. For example, as discussed in Section 8.1.5, Pauli exclusion principle prohibits the *s*-wave scattering state of identical fermions in identical quantum states. The zero-partial-wave state must thus be excluded and ultracold scattering of identical fermions is determined by *p*-wave scattering, i.e. the scattering state corresponding to one unit of angular momentum in the end-over-end rotational motion. Another example is the scattering process that involves a transition from the *s*-wave scattering state to a state of higher angular momentum without the change of internal energy.

To account for these situations, we will write the Wigner threshold laws more generally, as follows:

$$\sigma \propto v^{2l+2l'} \quad \text{for elastic scattering} \tag{11.3}$$

$$\sigma \propto v^{2l-1} \quad \text{for inelastic or reactive scattering.} \tag{11.4}$$

Here, σ is the collision cross section, v is the collision velocity, l is the value of the orbital (end-over-end rotation) angular momentum before the collision, and l' is the orbital angular momentum of the outgoing scattering wave.

Now the question is, how do Eqs. (11.3) and (11.4) change if the molecules are restricted to move in two, instead of three, dimensions? There must be a difference. For one, the quantum number l is not defined for collisions in two dimensions. To answer this question, let us formulate the scattering problem for particles colliding in a two-dimensional plane.

11.8.1 Threshold Scattering of Molecules in Two Dimensions

To solve the collision problem, we will repeat the main steps of Chapter 8 but for molecules A and B now moving in two dimensions. The following discussion is based on the work in Ref. [360]. We will direct the quantization axis along the normal to the confinement plane so the molecules are confined to the (xy)-plane before the collision. In this case, the Hamiltonian of the two particles A and B can be written as

$$\hat{H} = -\frac{\hbar^2}{2\mu\rho}\frac{d}{d\rho}\rho\frac{d}{d\rho} + \frac{\hbar^2 \boldsymbol{l}_Z^2(\phi)}{2\mu\rho^2} + \hat{H}_A + \hat{H}_B + U, \tag{11.5}$$

where ρ is the center-of-mass separation between the colliding particles, μ is the reduced mass of the collision complex, $\boldsymbol{l}_Z$ is the operator describing the rotation of the collision complex about the quantization axis, ϕ is the angle specifying the orientation of the vector ρ in the space-fixed coordinate system, $\hat{H}_A$ and $\hat{H}_B$ are the asymptotic Hamiltonians describing the separated molecules A and B, and U is the interaction potential between the colliding particles. The eigenstates of $\boldsymbol{l}_Z$ are

$$|m\rangle \equiv \frac{1}{\sqrt{(2\pi)}} e^{im\phi}. \tag{11.6}$$

Similarly to before, we will represent the eigenstates of the full Hamiltonian by an expansion over the products of states $|m\rangle$, the eigenstates $|\alpha\rangle$ of $\hat{H}_\mathrm{A}$, and the eigenstates $|\beta\rangle$ of $\hat{H}_\mathrm{B}$, as following:

$$|\Psi\rangle = \sum_\alpha \sum_\beta \sum_m |m\rangle|\alpha\beta\rangle F_{\alpha\beta m}(\rho). \tag{11.7}$$

The insertion of this expansion into the Schrödinger equation with the Hamiltonian (11.5) leads to the following system of the coupled differential equations (see Chapter 8 for details):

$$\left(\frac{1}{\rho}\frac{\mathrm{d}}{\mathrm{d}\rho}\rho\frac{\mathrm{d}}{\mathrm{d}\rho} - \frac{m^2}{\rho^2} + k_{\alpha\beta}^2\right) F_{\alpha\beta m}(\rho) = 2\mu \sum_{\alpha'\beta'm'} U_{\alpha\beta m;\alpha'\beta'm'} F_{\alpha'\beta'm'}(\rho), \tag{11.8}$$

where $k_{\alpha\beta}^2 = \frac{2\mu}{\hbar^2}(\epsilon - \epsilon_\alpha - \epsilon_\beta)$, ϵ is the total energy of the system, ϵ_α and ϵ_β are the eigenvalues of $\hat{H}_\mathrm{A}$ and $\hat{H}_\mathrm{B}$, and $U_{\alpha\beta m;\alpha'\beta'm'}$ denotes the matrix elements of the interaction potential U.

We will assume that the interaction potential U vanishes faster than $\propto 1/\rho^3$ as $\rho \to \infty$. This is always going to be the case if A and B are neutral molecules. If this condition is satisfied, U – and hence the matrix elements of U – can be neglected at large ρ, where the term m^2/ρ^2 dominates. The solutions to Eqs. (11.8) with $U_{\alpha\beta m;\alpha'\beta'm'} = 0$ can be written as superpositions of Hankel functions of the first $H_m^{(1)}(k_{\alpha\beta}\rho)$ and second $H_m^{(2)}(k_{\alpha\beta}\rho)$ kind [42]. These are the asymptotic ($\rho \to \infty$) solutions of Eq. (11.8), and we will seek to match the full solutions to these asymptotic forms.

As in Chapter 8, we can write the incoming $|\alpha\beta m\rangle_{\text{inc}}$ and outgoing $|\alpha\beta m\rangle_{\text{out}}$ parts of the full collision state in terms of these asymptotic solutions as follows:

$$|\alpha\beta m\rangle_{\text{inc}} = I_{\alpha\beta m}(k_{\alpha\beta}\rho)|m\rangle|\alpha\beta\rangle \tag{11.9}$$

$$|\alpha\beta m\rangle_{\text{out}} = O_{\alpha\beta m}(k_{\alpha\beta}\rho)|m\rangle|\alpha\beta\rangle \tag{11.10}$$

with

$$I_{\alpha\beta m}(k_{\alpha\beta}\rho) = \sqrt{\frac{\pi\mu}{2}}(-i)^m e^{-i\frac{\pi}{4}} H_m^{(2)}(k_{\alpha\beta}\rho)$$

$$O_{\alpha\beta m}(k_{\alpha\beta}\rho) = \sqrt{\frac{\pi\mu}{2}}(i)^m e^{i\frac{\pi}{4}} H_m^{(1)}(k_{\alpha\beta}\rho). \tag{11.11}$$

The states $|m\rangle$ are the two-dimensional analogs of the partial wave states in three dimensions. If the colliding particles are prepared with specific energies ϵ_α and ϵ_β and in a specific rotational state $|m\rangle$, the collision state must be written as

$$I_{\alpha\beta m}|m\rangle|\alpha\beta\rangle - \sum_{\alpha'\beta'm'} W_{\alpha\beta m;\alpha'\beta'm'} O_{\alpha'\beta'm'}|m'\rangle|\alpha'\beta'\rangle, \tag{11.12}$$

which defines the collision W matrix [317], analogous to the S-matrix we defined before for collisions in three dimensions.

To write the scattering cross sections in terms of the matrix elements of $\boldsymbol{W}$, we will assume that the incoming particles can be described by a plane wave propagating along the x-direction. By analogy with the partial wave expansion of the plane wave in Chapter 8, we can represent this plane wave by the Jacobi–Anger expansion in terms of the incident and outgoing scattering waves (11.11)

$$\left(\frac{k_{\alpha\beta}}{\mu}\right)^{-\frac{1}{2}} e^{ik_{\alpha\beta}x} = \sum_m (-1)^m e^{i\frac{\pi}{4}} k_{\alpha\beta}^{-\frac{1}{2}} [I_{\alpha\beta m}(k_{\alpha\beta}\rho) + (-1)^m e^{-i\frac{\pi}{2}} O_{\alpha\beta m}(k_{\alpha\beta}\rho)]. \tag{11.13}$$

Using this expansion, we can then repeat the steps in Chapter 8 to obtain the following expression for the integral scattering cross section:

$$\sigma_{\alpha\beta m,\alpha'\beta'm'} = \frac{1}{k_{\alpha\beta}}\left|(-1)^{m+1} e^{i\frac{\pi}{2}} W_{\alpha\beta m;\alpha'\beta'm'} - \delta_{\alpha,\alpha'}\delta_{\beta\beta'}\delta_{m,m'}\right|^2. \tag{11.14}$$

We can now determine the dependence of the two-dimensional (2D) scattering cross sections on the collision energy in the limit of very low collision energies using the approach of Wigner [317, 360]. Wigner wrote the $\boldsymbol{W}$-matrix in

Eq. (11.12) as follows:

$$\boldsymbol{W} = \boldsymbol{\omega}(1 + i\boldsymbol{j}(\boldsymbol{q} - \boldsymbol{R})^{-1}\boldsymbol{j})\boldsymbol{\omega}, \tag{11.15}$$

where $\boldsymbol{R}$ is the Wigner's R-matrix, $\boldsymbol{q}$ is a matrix of the reciprocal logarithmic derivatives of $O_{\alpha\beta m}$ from Eq. (11.11), i.e. the matrix with the elements

$$q_{\alpha\beta m} = O_{\alpha\beta m}/e_{\alpha\beta m}, \tag{11.16}$$

where

$$e_{\alpha\beta m} = (\hbar/2\mu)dO_{\alpha\beta m}/d\rho. \tag{11.17}$$

The matrix $\boldsymbol{\omega}$ is a unitary diagonal matrix with the elements defined by

$$e_{\alpha\beta m} = |e_{\alpha\beta m}|\omega^*_{\alpha\beta m}, \tag{11.18}$$

and $\boldsymbol{j}$ is a diagonal matrix of the elements defined by

$$j^2_{\alpha\beta m} = i(q_{\alpha\beta m} - q^*_{\alpha\beta m}). \tag{11.19}$$

We will now focus on finding the dependence of the W matrix elements, and hence the cross sections, on the collision energy at very low energies. The limit of very low collision energies corresponds to the limit $k_{\alpha\beta} \to 0$, and hence $k_{\alpha\beta}\rho \to 0$ for any finite ρ.

S-wave scattering in two dimensions corresponds to the value of $m = 0$, so we will begin by considering the A–B collision complex initially in state with $m = 0$. For convenience, we will define a new variable $\eta = k_{\alpha\beta}\rho$. As $\eta \to 0$, the Hankel function $H_0^{(1)}$ approaches the following function of η [42]

$$H_0^{(1)}(\eta) \to 1 - \frac{\eta^2}{4} + \frac{2i}{\pi}\left\{\ln\left(\frac{\eta}{2}\right) + \gamma\right\}. \tag{11.20}$$

The derivative of the Hankel function with respect to η is

$$(H_0^{(1)}(\eta))' \to -\frac{\eta}{2} + \frac{2i}{\pi\eta}. \tag{11.21}$$

Using these expressions and Eqs. (11.18) and (11.19), we see that $j^2_{\alpha\beta 0} \propto$ const and the matrix elements $[(q - R)^{-1}]_{\alpha\beta m,\alpha\beta m'}$ with one or both of m and m' equal to zero vanish as $1/\ln(\eta)$ as $\eta \to 0$.

Using Eqs. (11.14) and (11.12), the elastic scattering cross section for s-wave scattering in two dimensions can be written as

$$\sigma_{\alpha\beta 0,\alpha\beta 0} = \frac{1}{k_{\alpha\beta}}|1 + i\omega^2_{\alpha\beta 0} + i\omega^2_{\alpha\beta 0}j^2_{\alpha\beta 0}[(q - R)^{-1}]_{\alpha\beta 0,\alpha\beta 0}|^2. \tag{11.22}$$

The term $(1 + i\omega^2_{\alpha\beta 0})$ is proportional to $k^2_{\alpha\beta}$ so it can be omitted at small $k_{\alpha\beta}$, where the third term dominates. We thus find that the energy dependence of the elastic cross section for s-wave scattering in two dimensions is given by

$$\sigma_{\alpha\beta 0,\alpha\beta 0} \propto \frac{1}{k_{\alpha\beta}\ln^2 k_{\alpha\beta}}. \tag{11.23}$$

For collision processes that change the projection of the orbital angular momentum ($\alpha\beta 0 \rightarrow \alpha\beta m$), the cross section can be written as [317]

$$\sigma_{\alpha\beta m,\alpha\beta m'} = \frac{1}{k_{\alpha\beta}} |j_{\alpha\beta m} j_{\alpha\beta m'} [(q-R)^{-1}]_{\alpha\beta m,\alpha\beta m'}|^2. \tag{11.24}$$

To determine the threshold behavior of the cross sections for scattering in collision channels with $|m| > 0$, we express the Hankel functions in Eq. (11.11) in terms of the Bessel and Neuman functions. Using the asymptotic expansions of the Bessel and Neuman functions [42], we find that $j^2_{\alpha\beta m} \propto k^{2|m|}_{\alpha\beta}$ as $k_{\alpha\beta} \rightarrow 0$. This yields the following energy dependence of the cross section near threshold:

$$\sigma_{\alpha\beta 0;\alpha\beta m} \propto k^{2|m|-1}_{\alpha\beta} \frac{1}{\ln^2 k_{\alpha\beta}}. \tag{11.25}$$

The third important case we need to consider is transitions between states of nonzero angular momentum m. As mentioned earlier, such transitions determine the collision outcome of fermionic atoms and molecules confined in 2D. Using the above result $j^2_{\alpha\beta m} \propto k^{2|m|}_{\alpha\beta}$ and noting that $[(q-R)^{-1}]_{\alpha\beta m,\alpha\beta m'} \propto$ const as $k_{\alpha\beta} \rightarrow 0$, we find the following energy dependence of the scattering cross sections

$$\sigma_{\alpha\beta m;\alpha\beta m'} \propto k^{2|m|+2|m'|-1}_{\alpha\beta}. \tag{11.26}$$

Finally, we must consider the case of inelastic scattering that changes the internal states (α and/or β) of the colliding partners and reactive scattering that changes the chemical identity of the colliding species. When collisions release energy, the energy dependence of the scattering cross sections near threshold does not depend on the angular momentum in the final collision channel. For example, as can be seen from Eq. (11.4), the cross section for inelastic energy transfer in collisions in three dimensions is proportional to $k^{(2l-1)}_{\alpha\beta}$ and does not depend on the angular momentum l' in the outgoing collision channel. For inelastic collisions changing one or both of the quantum states (α, β), Eq. (11.24) changes into

$$\sigma_{\alpha\beta m,\alpha'\beta' m'} = \frac{1}{k_{\alpha\beta}} |j_{\alpha\beta m} j_{\alpha'\beta' m'} [(q-R)^{-1}]_{\alpha\beta m,\alpha'\beta' m'}|^2. \tag{11.27}$$

According to Wigner [317], the off-diagonal matrix elements $[(q-R)^{-1}]_{\alpha\beta m,\alpha'\beta' m'}$ are then independent of energy at small collision energies if $m \neq 0$. For $m = 0$, we obtain

$$[(q-R)^{-1}]_{\alpha\beta m,\alpha'\beta' m'} \propto \frac{1}{\ln k_{\alpha\beta}} \tag{11.28}$$

so the energy dependence of the inelastic s-wave scattering cross section is the same as that of the elastic cross section given by Eq. (11.23):

$$\sigma_{\alpha\beta 0,\alpha'\beta' m'} \propto \frac{1}{k_{\alpha\beta} \ln^2 k_{\alpha\beta}}. \tag{11.29}$$

Note again that this is not the case for collisions in three dimensions.

When $|m| > 0$, the energy dependence of the scattering amplitude is determined by the term $j^2_{\alpha\beta m} \propto k^{2|m|}_{\alpha\beta}$ so the inelastic scattering cross section for collisions with angular momentum m is given by

$$\sigma_{\alpha\beta m;\alpha'\beta' m'} \propto k^{2|m|-1}_{\alpha\beta}. \tag{11.30}$$

As can be seen from Eqs. (11.29) and (11.30) and the arguments leading to these equations, the energy dependence of the cross sections for collisions releasing energy is completely determined by the form of the scattering state in the incoming collision channel. As such, these energy dependencies of the cross sections apply to both inelastic scattering and chemically reactive scattering of molecules in two dimensions.

11.8.2 Collisions in a Quasi-Two-Dimensional Geometry

The above analysis can be generalized to describe collisions of molecules on an optical lattice with harmonic confinement in one dimension [357, 359, 361]. In this case, molecules are free to move and collide in two dimensions, while the motion in a third dimension is severely restricted. If the confinement is strong, the geometry is referred to as "quasi-two-dimensional".

The problem of quasi-two-dimensional scattering is different from the one in the previous subsection because the Hamiltonian we must employ describes the motion in three dimensions. Yet, the general features of two-dimensional scattering must somehow appear in the solutions for the quasi-two-dimensional case.

The Hamiltonian of the collision system in a quasi-two-dimensional geometry can be written as

$$\hat{H} = -\frac{\hbar^2}{2\mu}\Delta + \hat{H}_A + \hat{H}_B + U + V_z, \tag{11.31}$$

where the confining potential $V_z = az^2$ acts only on the colliding particles in the initial state $|\alpha\beta\rangle$. As the particles approach each other, the inter-particle interaction potential U becomes strong and the harmonic confinement V_z can be ignored [357, 361]. Therefore, at short to medium inter-particle separations, we can formulate the collision problem as in Chapter 8. However, the three-dimensional solutions of the coupled-channel equations (8.11) written in Chapter 8 must be matched to the asymptotic eigenstates of the Hamiltonian (11.31), which must depend on the harmonic confinement.

At very large inter-particle separation, the interaction potential U vanishes, and Eq. (11.31) can be written as a sum of Eq. (11.5) and a Hamiltonian describing the motion along the z-coordinate. The eigenstate of such Hamiltonian can be represented as products of the solutions to the Schrödinger equation with the Hamiltonian (11.5) describing scattering in two dimensions and the eigenstates of the harmonic oscillator problem describing the motion along the z-direction.

In the case of elastic scattering, the particles feel the harmonic confinement both before and after the collision. In the case of inelastic or chemically reactive scattering, the scattering event produces so much energy that the scattered particles do not feel the confinement. Therefore, the scattering states are affected by the confinement before the collision but not after the collision.

In general, we can represent the total wave function of the collision complex by an expansion in products of the scattering states $F_{\alpha\beta}(x, y, z)$ and the eigenstates $|\alpha\beta\rangle$. For the initial collision channel $\alpha\beta$, the functions $F_{\alpha\beta}(x, y, z)$ can be represented as products $R_{\alpha\beta m}(\rho)|m\rangle|z\rangle$, where $|z\rangle$ are the harmonic oscillator states. The ρ-dependent part $R_{sm}(\rho)$ can be expressed as a superposition of the functions defined in Eq. (11.11).

For all other collision channels with $\alpha' \neq \alpha$ and/or $\beta' \neq \beta$, the functions $F_{\alpha'\beta'}(x, y, z)$ must be represented as in Chapter 8, i.e. as

$$F_{\alpha'\beta'}(x, y, z) \Rightarrow F_{\alpha'\beta' l' m'}(R) Y_{l'm'}(\boldsymbol{R}/R), \tag{11.32}$$

where R is the center-of-mass separation between the colliding particles and Y_{lm} are spherical harmonics. The functions $F_{\alpha'\beta' lm_l}$ can be written as superpositions of *spherical* Hankel functions.

For elastic scattering, we have to consider only the initial channel and Eq. (11.14) applies, leading to the result (11.23). For reactive scattering, Eq. (11.24) must be modified to include $j_{\alpha'\beta' lm'}$ and $[q - R)^{-1}]_{\alpha\beta m, \alpha'\beta' l' m'}$ expressed in terms of the functions $F_{\alpha'\beta' l' m'}$ at large interatomic separation. This modification, however, does not change the energy dependence of the cross section (11.30) entirely determined by $j_{\alpha\beta m}$ and the leading term in the expansion of $[q - R)^{-1}]_{\alpha\beta m, \alpha'\beta' l' m'}$ from Eqs. (11.11). The results (11.23), (11.26), and (11.30) thus apply to scattering in quasi-2D geometry accompanied with the loss of confinement.

Table 11.1 summarizes the comparison of the energy dependence of the elastic and inelastic cross sections on the collision velocity for unconstrained collisions in three dimensions and collisions in the presence of strong confinement. While Table 11.1 does not provide information on the absolute values of the cross sections, it shows that inelastic collisions of molecules in quasi-2D scattering at small energies must be suppressed. External confinement also changes the symmetry of long-range intermolecular interactions. For example, the dipole–dipole interaction averaged over the scattering wave function of polar molecules vanishes in the limit of ultracold s-wave scattering in three dimensions (see Chapter 2), but remains significant in two dimensions. Measurements of molecular collisions in confined geometries may thus provide a sensitive probe of long-range intermolecular interactions and quantum

Table 11.1 Dependence of the scattering cross sections on the collision velocity v in the limit $v \to 0$ for collisions in three dimensions (3D), two dimensions (2D), and in a quasi-two-dimensional geometry.

Elastic collisions	**3D**	**2D or quasi-2D**
s-wave	$\sigma = \text{const}$	$\sigma \propto \frac{1}{v \ln^2 v}$
s-wave to non-s-wave	$\sigma \propto v^{2l'}$	$\sigma \propto v^{2\vert m\vert -1} \frac{1}{\ln^2 v}$
Non-s-wave to non-s-wave	$\sigma \propto v^{2l+2l'}$	$\sigma \propto v^{2\vert m\vert +2\vert m'\vert -1}$
Inelastic or chemically reactive collisions		
Initially s-wave	$\sigma \propto 1/v$	$\sigma \propto \frac{1}{v \ln^2 v}$
Initially non-s-wave	$\sigma \propto v^{2l-1}$	$\sigma \propto v^{2\vert m\vert -1}$

Source: Reproduced with permission from Chapter 4 of Krems et al. [153]. Copyright 2009, Taylor & Francis.

phenomena in collision physics. Confining molecules in low dimensions may be a practical tool for increasing the stability of ultracold molecular gases. The symmetry of the collision problem can be completely destroyed by a combination of confining laser fields and external static electromagnetic fields [360]. Measurements of chemical reactions in confined geometries may thus be a novel approach to study stereodynamics and differential scattering at ultracold temperatures.

12

Ultracold Controlled Chemistry

As discussed in Chapter 11, molecules at very low temperatures can be controlled by external fields through a variety of mechanisms. The development of experiments on cooling molecules to ultracold temperatures has thus laid the ground for a new regime of chemistry research: ultracold controlled chemistry.

Bound by the Coulomb interaction forces alone, atomic nuclei and electrons combine to form an incredible variety of molecular systems. When molecules react, they undergo chemical transformations leading to the rearrangement of atoms within molecules or transfer of atoms between molecules. This process may absorb or release energy; however, the energy change in a chemical reaction is many orders of magnitude smaller than the total energy of the Coulomb interaction in a molecule. This makes chemistry a game of small numbers and a chemical reaction a very complex process to study.

Most of our current understanding of chemical reaction dynamics is due to the development of the technology for producing and colliding molecular beams. The molecules in a molecular beam are often prepared with a narrow distribution of internal energies (up to a few Kelvin) and a low density. When two molecular beams are collided, molecules react under single-collision conditions and the reaction products scatter in a particular direction, where they can be detected by a variety of techniques. By varying the angle between the crossed beams, it is possible to tune the collision energy of the molecules. Molecular beam experiments, however, have two significant limitations: they only probe the outcome of a chemical reaction providing no direct information about the actual process of bond-breaking and bond-making; and, because the density of molecules is usually undetermined, it is difficult to measure the absolute rate of a chemical reaction. What are usually measured are relative rates for different reaction processes. Knowing the absolute magnitudes of reaction rates is crucial for calibrating theories of elementary chemical reactions.

Ultracold chemistry is a completely new approach to the study chemical transformations of molecules. The current experiments usually start with an ensemble of atoms cooled to an ultralow temperature (<10 μK) and confined

Molecules in Electromagnetic Fields: From Ultracold Physics to Controlled Chemistry, First Edition. Roman V. Krems.

in an optical potential of a focused laser beam. The atoms are associated into diatomic molecules with the same (ultralow) temperature as that of the precursor atoms. The atoms and the molecules are trapped in the laboratory frame. The trapped molecules are then allowed to collide with the trapped atoms or with each other and undergo chemical reactions. The collision energy is determined by the temperature of the atom–molecule mixture so these experiments probe chemical reactions at unprecedentedly low temperatures.

Ultracold chemical reactions always release energy and expunge the reaction products from the trap. The reaction rates can be measured by monitoring the trap loss. Because the number of molecules in the trap is known, these measurements yield the absolute rates of the elementary reaction processes. The extremely low temperature of the molecular gas allows for an extremely high resolution of the experiment. For example, it is possible to measure the reactions of molecules prepared in different magnetic sublevels of different hyperfine energy states. This can be exploited for tuning the reaction probabilities by transferring molecules from one hyperfine state to another and to study a variety of phenomena such as the role of quantum statistics in chemical reactivity.

The goal of this chapter is to highlight the recent developments of this new frontier of chemistry research and pause to ask questions about where this field might go next. We will begin with the discussion of the oft-asked question (see Section 12.1).

12.1 Can Chemistry Happen at Zero Kelvin?

Yes!

As discussed in Section 8.4, the rate coefficient for an inelastic scattering or chemically reactive process can be computed by integrating the energy dependence of the scattering cross sections with the Maxwell–Boltzmann distribution

$$R_{\text{reaction}}(T) = \left(\frac{8k_{\text{B}}T}{\pi\mu}\right)^{1/2} \int_0^{\infty} \sigma_{\text{reaction}}(\varepsilon) e^{-\varepsilon/k_{\text{B}}T} \frac{\varepsilon \, d\varepsilon}{(k_{\text{B}}T)^2}, \tag{12.1}$$

where ε is the kinetic energy of the collision, k_{B} is the Boltzmann constant and T is the temperature. Here, "reaction" means an inelastic collision releasing energy $\Delta E \gg \varepsilon$ or a chemical reaction leading to a change in the molecules' identity and release of energy $\Delta E \gg \varepsilon$.

In an ultracold gas, the collision energy is ultralow: $\varepsilon \to 0$. In this limit, the reaction cross section is determined by s-wave scattering for distinguishable molecules and for indistinguishable bosons. If the reacting molecules are fermions prepared in identical quantum states, s-wave scattering is prohibited by symmetry, and the reaction cross section is determined by p-wave scattering.

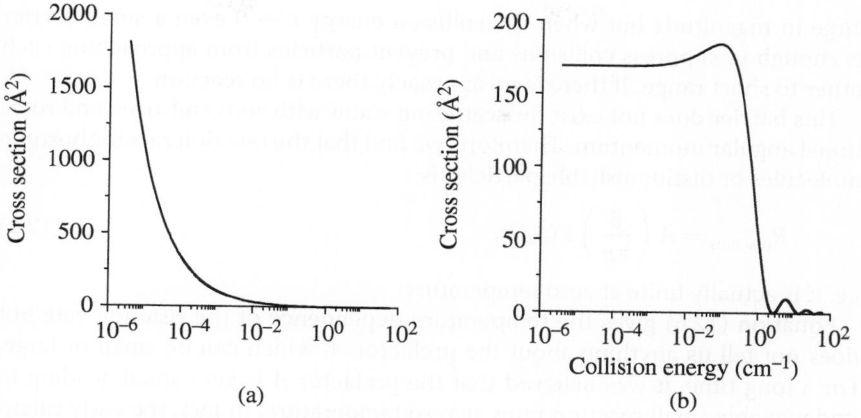

Figure 12.1 Inelastic or reactive (a) and elastic (b) cross sections typical for atomic or molecular scattering near thresholds. *Source*: Reproduced with permission from Krems and Dalgarno 2002 [362]. © 2002, the American Physical Society.

As discussed in Section 11.8, the reaction cross section in the limit $\varepsilon \to 0$ has the following energy dependence:

$$\sigma_{\text{reaction}} \sim \frac{1}{\varepsilon^{l-1/2}}, \tag{12.2}$$

where l is the partial wave of the incoming collision channel, i.e. $l = 0$ for s-wave collisions and $l = 1$ for p-wave collisions. Figure 12.1(a) shows the typical behavior of the low-energy cross section for inelastic or reactive scattering of (bosonic or distinguishable) molecules. It is drastically different from the behavior of the cross section for elastic scattering shown in panel (b) of the same figure.

If the scattering cross section has the form

$$\sigma_{\text{reaction}} = A\varepsilon^{a}, \tag{12.3}$$

where A is some energy-independent constant, then the integral (12.1) is

$$R_{\text{reaction}} = A\left(\frac{8}{\pi\mu}\right)(k_{\text{B}}T)^{a+1/2}(a+1)! \tag{12.4}$$

This shows that the reaction rate for fermionic molecules in identical quantum states

$$R_{\text{reaction}} = A(k_{\text{B}}T)\left(\frac{8}{\pi\mu}\right)\Gamma(5/2), \tag{12.5}$$

where Γ is the gamma function. This rate vanishes as $T \to 0$ so there is no reaction for fermions in identical states at zero temperature. We should have expected this. The angular momentum of p-wave scattering states leads to a centrifugal barrier at large inter-particle separations. This barrier may not be large in magnitude but when the collision energy $\varepsilon \to 0$ even a small barrier is enough to suppress collisions and prevent particles from approaching each other to short range. If there is no approach, there is no reaction.

This barrier does not exist for scattering states with zero end-over-end rotational angular momentum. Therefore, we find that the reaction rate for bosonic molecules or distinguishable particles is

$$R_{\text{reaction}} = A\left(\frac{8}{\pi\mu}\right)\Gamma(3/2), \tag{12.6}$$

i.e. it is actually finite at zero temperature!

Equation (12.6) gives the temperature dependence of the reaction rate but does not tell us anything about the prefactor A, which can be small or large. For a long time, it was believed that the prefactor A is very small, leading to undetectably small reaction rates at zero temperature. In fact, the early calculations for inelastic collisions of molecules yielded very small values of the cross sections in the limit of vanishing collision energy [363].

More recently came a series of calculations of cross sections for inelastic (ro-vibrational) relaxation in atom–molecule collisions,[1] chemically reactive scattering,[2] and inelastic Zeeman relaxation in collisions of molecular radicals in a magnetic field.[3] These papers showed that the rate constants for inelastic energy transfer and chemical reactions may have significant magnitudes at zero Kelvin. In particular, Balakrishnan and Dalgarno [379] found that the chemical reaction $F + H_2 \to HF + H$ occurs very rapidly at ultracold temperatures despite a large activation barrier of about 1.5 kcal mol^{-1}. The rate constant for this reaction was calculated to be as large as 1.25×10^{-12} cm^3 s^{-1} at zero Kelvin. Soldán et al. [381] and Cvitaš et al. [385] showed that chemical reactions without activation barriers (insertion reactions) are even more efficient at ultralow energies. The results of their extensive calculations based on a hyperspherical coordinate representation of the wave function yielded the zero temperature rate constant 5×10^{-10} cm^3 s^{-1} for the $Na + Na_2(v > 0)$ reaction and 4.1×10^{-12} cm^3 s^{-1} for the $^7Li + {}^6Li^7Li(v = 0)$ reaction. These are very large rates!

Still more recently came the experimental verification of these predictions [263, 393]. Figure 12.2 borrowed from Ref. [263] illustrates the measurements of the trap loss of KRb molecule produced at an ultracold temperature below 10^6 K. The molecules are prepared in the absolute ground state and

1 The literature has exploded with studies of inelastic scattering of molecules at ultracold temperatures following the original papers stemming from the collaboration of Balakrishnan, Forrey, and Dalgarno. It is impossible to list all the relevant articles here. The representative references include [319, 364–377]. A recent review of this work can be found in Ref. [378].

2 Similarly, the literature has exploded with studies of ultracold chemical reactions following the original publication by Balakrishnan and Dalgarno [379]. A few representative references include [379–386].

3 Again, there are too many papers to cite here. A few representative references include [315, 316, 351, 352, 387–392].

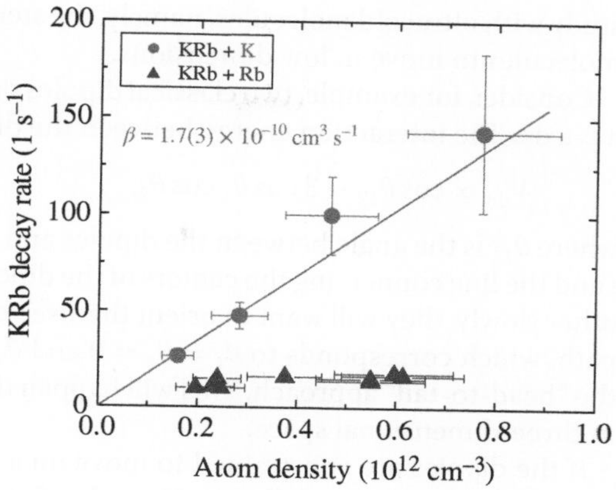

Figure 12.2 Collisions of ultracold molecules and atoms prepared in the lowest-energy quantum state. *Source*: Adapted with permission from Ospelkaus et al. 2010 [263]. © 2010, American Association for the Advancement of Science.

can only decay by undergoing chemical transformations in collisions with either the K atoms or the Rb atoms. The chemical reactions of ultracold KRb molecules with ultracold K atoms occur at a staggering rate of $>10^{-10}$ cm^3 s^{-1}.

12.2 Ultracold Stereodynamics

The groundbreaking advances in molecular beam methods in the early 1970s enabled the study of chemical reactions between field-oriented molecules under single-collision conditions, eventually leading to the development of stereochemistry [47, 48, 100, 101, 104, 394]. In these studies of reaction stereodynamics, molecules are oriented by external electric fields and collided by crossing molecular beams traveling with high speeds. Owing to the high collision energy and consequently short interaction times, molecules interact without changing their orientation, which can be used to study the chemical abstraction of atoms from different sides of a molecule [90]. Unfortunately, the high collision energy of molecules in these experiments restricts the observations to probing a limited range of the intermolecular interaction potential and washes out the sensitivity to subtle effects such as nonadiabatic interactions or field-induced couplings. In order to study the effects of the global topology of inter-molecular interactions and weak intra-molecular interactions on chemical reactions, chemical dynamics should be taken to the limit of low collision energies. What happens to stereodynamics in the limit of zero collision energy?

In a way, "stereodynamics" and "ultracold scattering" are antonyms. At low collision energies, molecules have a large de Broglie wavelength so, as argued by Herschbach [395], the reactants engulf rather than hit each other, which defies stereodynamics. However, there is a different kind of stereodynamics one can study with ultracold molecules, namely, the stereodynamics based on confining molecules to move in low dimensions.

Consider, for example, two classical dipoles oriented by an external field along a z-axis. The interaction energy between the dipoles is

$$V_{dd} \propto \cos\theta_{12} - 3\cos\theta_1 \cos\theta_2, \tag{12.7}$$

where θ_{12} is the angle between the dipoles and θ_i are the angles between dipole i and the line connecting the centers of the dipoles. If the dipoles approach each other slowly, they will want to orient themselves to follow the minimum energy path, which corresponds to $\theta_1 = \theta_2 = 0$ and $\theta_{12} = \pi$. This can be thought of as the "head-to-tail" approach. This will happen if the dipoles are allowed to move in three-dimensional space.

If the dipoles are constrained to move on a two-dimensional "pancake," the situation is drastically different. Consider, for example, a pancake of dipoles in the plane perpendicular to the z-axis of alignment. In this case, $\theta_1 = \theta_2 = \pi/2$ and $\theta_{12} = 0$ so the dipole–dipole interaction is repulsive.

Molecules are not the classical dipoles. But when oriented along an electric field axis and confined in a quasi-2D geometry by an optical lattice, they can be manipulated just like the classical dipoles described earlier. This was illustrated by an experiment of de Miranda et al. [396] (Figure 12.3). The essential goal of the experiment of de Miranda et al. was to create long-range repulsive interactions that would prohibit molecules to approach to short intermolecular distances, where chemical reactions can occur. They have succeeded by squeezing the motion of the ultracold molecules in one dimension by laser forces. An experiment like this was beyond imagination just a decade ago.

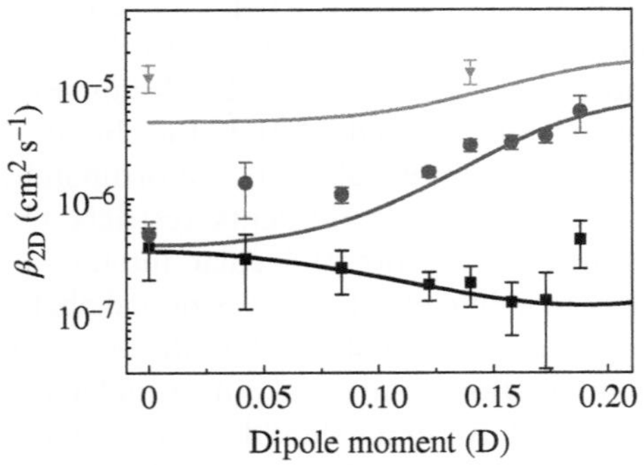

Figure 12.3 Collisions of ultracold molecules in a quasi-2D geometry. The extracted loss-rate constants for collisions of molecules in the same lattice vibrational level (squares) and from different lattice vibrational levels (circles) plotted for several dipole moments. Measured loss-rate constants for molecules prepared in different internal states are shown as triangles. *Source*: Adapted with permission from de Miranda et al. 2011 [265]. © 2010, Rights Managed by Nature Publishing Group.

12.3 Molecular Beams Under Control

As discussed in Section 6.2, the past 10 years have witnessed huge progress of techniques for controlling and manipulating molecular beams. As the protagonists of this research area like to say, “molecular beams have been tamed.” This was made possible hugely thanks to the work of Meijer and coworkers on Stark deceleration of molecules described in Section 6.2. Once a beam of polar molecules is slowed, one can do wonderful things with it. One can load molecules in a trap [121, 163, 165, 167, 212, 254], reflect them off an electrostatic or magnetic mirror [163, 183, 184, 397, 398], inject them into a molecular synchrotron [221]. The molecular synchrotron can be thought of as the ultimate molecule collider [224]. Built by analogy with charged particle accelerators, it is a system of electrodes bent in a ring shape. Molecules injected in the synchrotron travel around the ring so it can be viewed as a trap and used as a storage ring. One can inject into the ring one or more packets of molecules produced by a pulsed molecular beam. These packets can then be manipulated by adjusting the field applied to the ring electrodes. The spread of the packets can be decreased, and the packets can be accelerated or decelerated. They can be sent in the same or opposite directions and collided. The collisions can be repeated again and again, with the collision energy and the interval between the collisions precisely controlled.

Such unprecedented control over molecular collisions has opened a new era in the exploration of molecular interactions. Controlled beams have been used to map out the behavior of the cross sections near the excitation thresholds [399], imaging the diffraction patterns in atom–molecule collisions [400, 401], and detailed imaging of scattering resonances in atom–molecule collisions [402]. The differential scattering cross sections can now be measured with the angle resolution well under one degree [403]. One can measure the absolute magnitudes of cross sections for state-to-state transitions in molecule–molecule collisions [404]. These spectacular experiments serve as a detailed reference for theories of molecular collisions and as a rich source of information about the mechanisms of energy exchange in molecular encounters.

12.4 Reactions in Magnetic Traps

The experiments on collisions and chemical reactions of trapped molecules represent yet another frontier of exciting chemistry research. The alignment of molecular dipole moments by the trapping field restricts the symmetry of the interaction between the reactants in the entrance reaction channel and limits the number of adiabatically accessible states. This can be exploited as a new way to control and study chemical reactions.

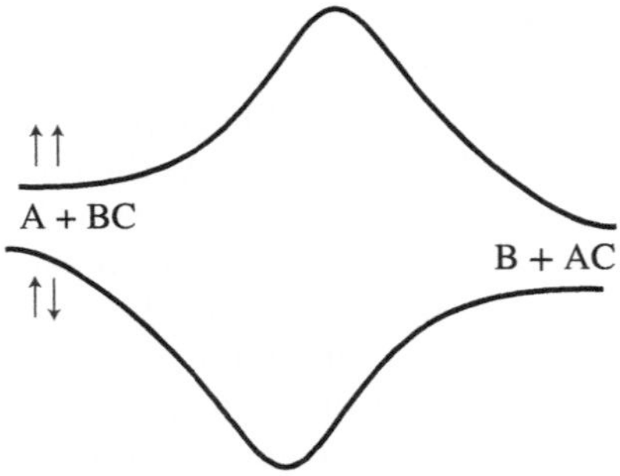

Figure 12.4 Schematic illustration of minimum energy profiles for an A(^2S) + BC($^2\Sigma$) chemical reaction in the singlet-spin (lower curve) and triplet-spin (upper curve) electronic states. Electric fields may induce nonadiabatic transitions between the different spin states and modify the reaction mechanism. *Source*: Reproduced by permission of the PCCP Owner Societies from Krems 2008 [269]. © 2008, the PCCP Owner Societies.

Consider, for example, a chemical reaction between a $^2\Sigma$ diatomic molecule in the rotationally ground state (e.g. CaH) and an atom with one unpaired electron (e.g. Na) in a ^2S electronic state. If the atom and the molecule are both confined in a magnetic trap, their magnetic moments are coaligned. In general, the interaction between two spin-1/2 species gives rise to two electronic states, which correspond to the total spin values $S = 1$ (triplet state) and $S = 0$ (singlet state). If the magnetic moments of the two species point in the same direction, they must be in the triplet state.

The interactions of spin-1/2 atoms and molecules in the triplet state are usually much more repulsive at short range than the interactions of the same species in the singlet state. Chemical reactions of $^2\Sigma$ with ^2S atoms in the triplet state are thus expected to proceed through a significant chemical barrier. The reaction barrier is usually much lower for the chemical reaction of the same species in the singlet state. Often, this reaction barrier is absent altogether, leading to insertion (as opposed to abstraction) chemical reactions. This is schematically illustrated by Figure 12.4. Moreover, the reaction products in the triplet state typically have much higher internal energy than the reaction products of the same species in the singlet state. As a result, reactions of spin-1/2 species in the triplet state may often be endoergic (and energetically forbidden), while reactions of the same species in the singlet state can at the same time be exoergic, i.e. proceeding with the release of energy.

Imagine now a reaction of a $^2\Sigma$ molecule with a ^2S atom in a magnetic trap. If the reaction is endoergic, it will not proceed unless the reaction complex undergoes a transition from the triplet state to the singlet state. Such transitions are induced by nonadiabatic couplings so reactions in magnetic traps can be used as a sensitive probe of nonadiabatic interactions. Such transitions can also be controlled through a mechanism that controls the spin orientation of the molecule in the space-fixed coordinate frame. For example, as we saw in Section 7.1.1, the orientation of the molecule's magnetic moment becomes very unstable under certain conditions in the presence of combined electric and magnetic fields. Combinations of electric and magnetic fields can thus be used to induce nonadiabatic $S = 1 \rightarrow S = 0$ chemical reactions of molecules in a magnetic trap. One can also imagine an experiment where the spin orientation of the molecule is flipped by an rf field. Such a spin flip would again induce a nonadiabatic transition leading to chemical reactions in the singlet state.

12.5 Ultracold Chemistry – The Why and What's Next?

The excitement generated by the advent of ultracold molecules has been hard to contain. Ultracold molecules offer many wonderful research avenues in physics. Some of the most important examples include

- Precision spectroscopy for tests of fundamental symmetries of nature;
- Studies of new regimes of many-body physics;
- Quantum simulation of complex quantum systems;
- Quantum computing based on controlled molecules;

Each of these areas of research has experienced an explosion in the past 10–15 years, after the technology for producing ultracold molecules was developed.

Ultracold chemistry was originally a byproduct of the experiments with ultracold molecules. In some of the experiments, ultracold chemical reactions were considered as a loss mechanism and a lot of effort was expended to prevent chemistry from happening. It was important to understand chemistry at zero Kelvin in order to avoid it. Still, for other experiments, ultracold chemistry has been a goal.

In the latter case, ultracold chemistry has often been portrayed as a new regime of chemistry research that needs to be explored because it is new. While lofty, this justification is not very satisfying. Therefore, the purpose of this section is to pose, and attempt to answer, the following questions: Can ultracold chemistry be of practical importance? Does ultracold chemistry research have the potential to stimulate new areas of research? Can ultracold chemistry research help to answer fundamental questions?

Often, controlled chemistry is used as a motivation for ultracold chemistry. As discussed in this book, ultracold molecules and their interactions can be controlled by external fields through a variety of mechanisms. Therefore, ultracold chemistry is almost always controlled. But then again, why do we need ultracold controlled chemistry? What does one gain by achieving external control over microscopic reactions of molecules in this unique temperature regime?

12.5.1 Practical Importance of Ultracold Chemistry?

Chemistry is a highly utilitarian science. Much of chemistry is about the design and synthesis of new molecular species and materials. Since ultracold chemistry is highly controllable, can it be used to synthesize new molecular species in a controlled way? What does it take to make ultracold chemistry useful as a practical tool for general chemical synthesis?

The current technology to produce ultracold molecules is very complex, molecule selective, and limited to a rather small set of molecular species. Most of the molecules studied in the ultracold temperature regime (see Table 6.9) are exotic by traditional chemistry standards. Yet, one can envision this area of research to grow into a tool for material's design.

As recently mentioned by Herschbach [153], "attaining ultracold temperatures requires liberation from thermodynamics." This suggests that ultracold molecules and ultracold chemistry offer a route to assembly of materials that are thermodynamically unstable. This also suggests that molecules can be brought together at angles of approach that would not be possible when multiple internal degrees of molecules are agitated.

In order to achieve this, one must find a general way to cool complex molecules and confine them in traps. The work on cooling organic molecules to ultracold temperature is already underway at multiple laboratories [405–407]. For example, a number of recent experiments have shown that molecules as complex as naphthalene [408] or *trans*-Stilbene [409] can be cooled to a temperature of a few Kelvin by immersing them in a cold He gas. The problem that needs to be solved is trapping those complex molecules in an external field in order to isolate them from the environment.

Large polyatomic molecules do not have low-field-seeking Stark states so they cannot be confined in a DC electric trap. However, trapping complex molecules can be achieved by means of an optical trap. In fact, optical tweezers have been used, albeit at higher temperatures, for many years of research in chemistry. One can also envision trapping organometallic molecules in a magnetic trap [406]. If the metal has a magnetic moment, it can be used as an anchor to park the molecule in a magnetic trap. A molecule with two or more spatially separated magnetic anchors could be confined in a magnetic trap to assume a particular orientation, making magnetic traps an interesting new platform for assembling complex molecules into aggregates.

Ultracold chemistry can also be envisioned as a new way to synthesize new organometallic compounds. Most, if not all, of the experiments in the field of ultracold atoms are performed with vapors of metals. Much of recent research in the field of ultarcold atoms has been focussed on cooling and trapping transition metal and lanthanide atoms. This work has culminated in the creation of Bose–Einstein condensates of ultracold lanthanide atoms (Cr, Dy, Er) in multiple laboratories [410–414]. Combining a Bose–Einstein condensate of species like Dy with organic molecules must lead to reaction pathways drastically different from those at elevated temperatures. To achieve this, one must bring the temperature of organic molecules to the ultracold regime of trapped atoms.

12.5.2 Fundamental Importance of Ultracold Controlled Chemistry

The experiments with ultracold molecules allow one to examine the process of bond-making and bond-breaking in ways previously unfeasible. The preparation of molecules in individual ro-vibrational-fine-and-hyperfine states combined with the control over molecular association and dissociation by external fields allows one to ask new fundamental questions, such as: What is the difference between a pair of atoms and a diatomic molecule [415]? What is the role of relativistic interaction in the dynamics of chemical bond formation?

The experiments with molecules in confined geometries and in the presence of symmetry-breaking fields lead to the question of whether microscopic chemical reactions are affected by external space symmetry. And if yes, can the confinement be engineered to ensure a desired reaction outcome?

The possibility of tuning nonadiabatic couplings by external fields puts the tests of adiabaticity of chemical reactions to a new level of scrutiny. Of particular interest appears the possibility of studying the effects of the geometric phase, which have been shown to be magnified at ultracold temperatures [416]. The geometric phase arises when a quantum system evolves in the presence of a conical intersection between the potential energy of two (or more) states. With molecules placed in a combination of electric and magnetic fields, such conical intersections can be engineered in the laboratory frame [417]. If the molecular ensemble is trapped, the conical intersection can then be moved around the molecular ensemble by tuning the electric or magnetic field magnitude, leading to the appearance of the geometric phase. The geometric phase can thus be engineered and, consequently, probed in experiments with trapped molecules.

Precise measurements of the ro-vibrational energy levels possible with ultracold molecules put quantum chemistry to the test like never before. The experimental measurements have reached the level allowing us to question the limitations of the Schrödinger equation itself.

Chemistry of molecules in a quantum degenerate gas remains a rather unexplored area. As was proposed by Moore and Vardi [418], reaction dynamics of molecules in a Bose–Einstein condensate may be dramatically altered by the Bose-enhancement effects. Similar phenomena must lead to Pauli blocking in a Fermi degenerate gas. If true, chemical reactions in quantum degenerate gases may offer a new way to test the Pauli exclusion principle.

The ability to link ultracold molecules together by means of Feshbach resonances or by tuning the trapping fields of an optical lattice offers the possibility to compare the dynamics of molecular reactions in two- and three-body collisions. This may uncover new reaction mechanisms, leading to the research of few-body chemistry.

The above examples illustrate that ultracold controlled chemistry is indeed a new regime of fundamental chemistry research poised to push the boundaries of our current understanding of molecular structure and molecular reactions. The research of ultracold chemistry has already stimulated us to reexamine conventional approaches and concepts. The scattering theory approach, as described in Chapter 8, has had to be reformulated in response to experiments on molecular collisions in external field traps. In the process of the work on ultracold collisions in fields, the Arthurs–Dalgarno representation, which was used in molecular scattering theory since 1960 [314], was abandoned and then reintroduced. The experiments with molecules in optical lattices or in controlled beams have made us bolder in our pursuit of understanding intermolecular interactions. In short, – and this is perhaps the most important contribution of ultracold chemistry thus far, – the experiments with ultracold molecules have taught us that there is still a lot to learn and that there is no fundamental obstacle to how much we can learn about molecules, and ways they interact.

12.5.3 A Brief Outlook

Where do the fields of ultracold molecules and ultracold chemistry go from here? There are a few major problems that need to be tackled and several enticing research directions that can't be missed. The major problems include:

- Bridging the gap between the cold and ultracold temperature regimes. There are many experimental techniques that can be used to prepare cold ensembles of molecules with temperatures $\gtrsim$1 mK. However, many of the above applications require a source of molecules at temperatures below 10^{-6} degrees Kelvin. Bringing the molecules from 10^{-3} to $<10^{-6}$ degrees Kelvin has proven a difficult task. Atoms are cooled to ultracold temperatures by evaporative cooling. However, evaporative cooling requires collisions in external field traps. Unfortunately, molecular collisions at these temperatures often lead to inelastic processes resulting in the loss of trapped molecules. It is therefore very important to find a working mechanism for cooling large ensembles of molecules from milliKelvin temperatures to microKelvin temperatures.
- Cooling complex (organic) molecules to ultracold temperatures. As mentioned earlier, this is needed to make ultracold chemistry of practical interest to chemists. While multiple experiments have demonstrated cooling organic molecules from ambient to few-Kelvin temperatures, the next important step is to create a large ensemble of organic molecules at ultracold temperatures.
- Probing state-to-state chemical reactions of ultracold molecules. At present, ultracold chemical reactions are detected by measuring the loss of trapped molecules, assuming the loss is due to reactions after ruling out other loss mechanisms. However, in order to probe ultracold chemistry in detail, one must develop methods for imaging the products as well as the reactants of an ultracold chemical process. This is challenging because the products

are generally few in numbers and are quickly departing the trap in all directions.

- Measuring the absolute rates of chemical reactions in molecular beam experiments. While this has already been achieved for inelastic collisions of molecules [404], it is necessary to generalize these measurements to chemically reactive scattering.
- The theoretical description of ultracold molecular collisions in general and chemical reactions in particular remains hampered by the problem arising from the sensitivity of the collision cross sections to small variations of the potential energy surfaces. In order to overcome this problem, one needs to develop theoretical methods that would produce the cross sections or rate constants averaged over reasonable variations of the underlying potential surfaces accounting for the inherent inaccuracy of the intermolecular potentials. This is a difficult problem because quantum scattering calculations of molecular collisions in the presence of external fields are computationally demanding, and it is very difficult to perform multiple calculations with multiple potential surfaces to account for the potential uncertainties. As we proposed in our recent work [419, 420], this problem can potentially be solved by combining quantum dynamics computations with established machine learning techniques.

The work on the above problems requires breakthroughs in the technologies currently used for the production and imaging of molecules at very low temperatures and a close interplay of theoretical and experimental research. It will undoubtedly stimulate new research directions. Within chemistry, a few examples could include:

- A new platform for studying the problem of internal energy redistribution in large polyatomic molecules, its connection to Anderson localization, and many-body localization [421]. With molecules prepared in the lowest energy vibrational state, injecting a single quantum of vibrational excitation and imaging its propagation between the vibrational modes should reveal many intricate details about the quantum dynamics of internal energy flow in polyatomic systems. Recent years have witnessed an explosion of interest in using ultracold atoms and ultracold molecules trapped on an optical lattice as quantum simulators of condensed-matter physics phenomena [422]. With the development of sources of ultracold polyatomic molecules, similar ideas can be explored with individual molecules serving as a platform for quantum simulators, which currently require an array of atoms or molecules.
- The exploration of new reaction paths of ultracold metals with ultracold molecules that cannot happen at ambient temperatures. If found, such reaction paths may pave the way to potentially exotic organometallic compounds with novel properties.

- Controlled assembly of molecules into clusters. Can the external field control over ultracold molecules be exploited for the controlled assembly of two- and few-molecule clusters? If yes, this will open up a path to interesting minituarized quantum materials consisting of a handful of complex molecules brought together with the help of external field forces. This will also make possible the systematic exploration of how the properties of materials change as the complexity builds up from a single molecule, to a few molecules, to a system with macroscopic size.
- Controlled study of bond-breaking and bond-making. This is perhaps the most important fundamental application of ultracold controlled chemistry. If molecules can be reacted under perfectly controlled conditions and the reaction products can be imaged with the single quantum state resolution, one has an unprecedented tool for studying microscopic chemistry in action.
- Quantum-tunnelling-driven chemistry. Trapped molecules allow for a long interrogation time. Combined with the possibility of tuning the initial energy of the reactants by external fields, this may be a perfect platform for studying tunnelling-driven reactions of importance to organic chemistry and biology. One can envision experiments mapping out the width and height of the chemical reaction barriers for such reaction. With the barriers mapped out, one can then make predictions regarding the role of the tunnelling-driven reactions at ambient temperatures of relevance to biological processes.
- As described in this book, the work on the production of ultracold molecules has led to major developments in the molecular beam techniques. These techniques will continue to be perfected and will likely provide the most detailed information about a chemical process one can imagine, including the dependence of the chemical reactivity on the angle of the molecule–molecule approach, full quantum state resolution, and the effects of quantum shape and Feshbach resonances on chemical reactions of various kinds. There is much to look forward to in this area of research.

A

Unit Conversion Factors

This appendix summarizes the conversion factors between different units commonly used for computations of molecular energy levels. The atomic units are defined so that $\hbar = 1$, the electron charge $e = 1$, and the electron mass $m_e = 1$. In atomic units, the speed of light is $c = 1/\alpha$, where α is the fine structure constant $\alpha = 1/137.035999074$. The atomic units are particularly important for calculations in atomic and molecular physics. I always convert all quantities to atomic units before each calculation and convert the result to a desired unit after the calculations. This makes the calculations simpler and helps me to avoid mistakes.

Units of Energy

1 atomic unit of energy (Hartree) = 219474.6314 cm^{-1}
1 Hartree = 27.2117 eV
1 Hartee = 627.5095 kcal mol^{-1}
1 Hartree = $4.3597482 \times 10^{-18}$ J
1 Hartree = $6.579683920721(44) \times 10^{15}$ Hz

1 eV = 23.06037 kcal mol^{-1}
1 eV = 8065.478 cm^{-1}
1 cm^{-1} = 29.9792458 GHz

Units of Length

1 atomic unit of length (Bohr) = 0.5291772083 Å = $0.5291772083 \times 10^{-10}$ m.

Units of Mass

We will define the atomic unit of mass as the rest mass of the electron.
1 atomic unit of mass = 1 $m_e = 9.10938188(72) \times 10^{-31}$ kg.
electron-to-proton mass ratio = $m_e/m_p = 5.446170232(12) \times 10^{-4}$

1 U = mass (^{12}C)/12
1 U = 1823.15406 a.u.
1 m_p = 1.00727646688(13) U

Units of Time

1 atomic unit of time = 2.418885×10^{-17} s

Units of the Electric Field

1 atomic unit of electric field = 5.142×10^{9} V cm^{-1}

Units of the Dipole Moment [35]

1 atomic unit of electric dipole moment $[ea_0]$ = 2.541747 D = 8.47836×10^{-30} C m

Units of the Dipole Polarizability [35]

1 atomic unit of dipole polarizability $[e^2 a_0^2 E_h^{-1} = 4\pi\epsilon_0 a_0^3]$ = 1.64878×10^{41} C^2 m^2 J^{-1}

Units of Laser Intensity [423]

$I_0 = \epsilon_0 c E_0/2$
1 atomic unit of laser intensity = 3.54×10^{16} W cm^{-2}

B

Addition of Angular Momenta

This appendix discusses the basics of angular momentum algebra. There are several excellent books written on the subject, including "Angular Momentum: Understanding Spatial Aspects in Chemistry and Physics" by Zare [9] and "Quantum Theory of Angular Momentum: Irreducible Tensors, Spherical Harmonics, Vector Coupling Coefficients, $3nj$ Symbols" by Varshalovich et al. [10].

The orbital motion of a particle in a spherically symmetric potential (think: electron in a hydrogen atom) is described by its angular momentum $\boldsymbol{l}$. In classical mechanics, the angular momentum is defined as $\boldsymbol{l} = \boldsymbol{r} \times \boldsymbol{p}$, where $\boldsymbol{r}$ is the position vector of the particle and $\boldsymbol{p}$ is its linear momentum. In quantum mechanics, $\boldsymbol{l}$ is a vector operator. For convenience, we will define the angular momentum operator $\boldsymbol{l}$ by the equation

$$\hbar \boldsymbol{l} = \boldsymbol{r} \times \boldsymbol{p}. \tag{B.1}$$

This will remove $\hbar$ from most of angular momentum algebra. The explicit expressions for the Cartesian components of this operator can be obtained by replacing the Cartesian components of the linear momentum by the corresponding quantum mechanical operators:

$$\boldsymbol{l} = \begin{vmatrix} \hat{x} & \hat{y} & \hat{z} \\ x & y & z \\ -i\dfrac{\partial}{\partial x} & -i\dfrac{\partial}{\partial y} & -i\dfrac{\partial}{\partial x} \end{vmatrix}. \tag{B.2}$$

If expressed in spherical polar coordinates, the operator $\boldsymbol{l}$ is a function of the spherical polar angles θ and ϕ that specify the direction of the vector $\boldsymbol{r}$ in the three-dimensional coordinate frame centered at the origin of the spherically symmetric potential.

The wave function of the particle $\psi(r, \theta, \phi)$ is a function of three coordinates: r, θ, and ϕ. This wave function can be generally represented as $\psi(r, \theta, \phi) = \sum_{lm} f_{lm}(r) Y_{lm}(\theta, \phi)$, where the functions $Y_{lm}(\theta, \phi)$ form a complete basis set in the coordinate subspace (θ, ϕ). There are two quantum numbers l and m necessary to specify these basis functions, one for each dimension. It is convenient to

Molecules in Electromagnetic Fields: From Ultracold Physics to Controlled Chemistry, First Edition. Roman V. Krems.

choose $Y_{lm}(\theta, \phi)$ to be the eigenfunctions of two commuting operators $\boldsymbol{l}^2$ and $\boldsymbol{l}_z$, where $\boldsymbol{l}_z$ is the z-component of the angular momentum operator $\boldsymbol{l}$. Then, $Y_{lm}(\theta, \phi)$ are the spherical harmonics [9] that satisfy the following eigenvalue equations:

$$\boldsymbol{l}^2|lm\rangle = l(l+1)|lm\rangle,$$
$$\boldsymbol{l}_z|lm\rangle = m|lm\rangle, \tag{B.3}$$

where I use the Dirac notation $Y_{lm} \equiv |lm\rangle$ for the basis vectors. The quantum number l can take any nonnegative integer values ($l = 0, 1, 2, \ldots$) and $m = -l, -l+1, \ldots, l$. Thus, there are $2l+1$ functions Y_{lm} for a given value of l. Because the orbital angular momentum for the particle in a spherically symmetric potential is conserved, the Hamiltonian matrix must be diagonal in the basis $|lm\rangle$ so the eigenvalues of the Hamiltonian can be labeled by the quantum numbers l and m.

If there are two particles in a spherically symmetric potential (think: helium atom), there are two angular momenta $\boldsymbol{l}_1$ and $\boldsymbol{l}_2$ required to describe the system. We can construct a set of products $|l_1 m_1\rangle|l_2 m_2\rangle$ that form a complete angular basis set for the two-particle problem. This basis set can be used to construct the matrix of the two-particle Hamiltonian. If the particles are interacting (e.g. through the electron–electron repulsion in He), the Hamiltonian matrix in the basis $|l_1 m_1\rangle|l_2 m_2\rangle$ is a general nondiagonal matrix. The eigenstates of the Hamiltonian are superpositions of states $|l_1 m_1\rangle|l_2 m_2\rangle$ with different values of l_1, m_1, l_2, and m_2.

The problem can be simplified by introducing the total angular momentum $\boldsymbol{L}$ as the vector sum of $\boldsymbol{l}_1$ and $\boldsymbol{l}_2$. The beauty of the angular momentum theory is that we don't need to know the explicit form of the operator $\boldsymbol{L}$. All we need to know is that a sum of angular momenta is also an angular momentum so the rules of angular momentum algebra apply to $\boldsymbol{L}$ just like they apply to $\boldsymbol{l}_1$ and $\boldsymbol{l}_2$. The products $|l_1 m_1\rangle|l_2 m_2\rangle$ are the eigenvectors of four operators ($\boldsymbol{l}_1^2$, $\boldsymbol{l}_{1z}$, $\boldsymbol{l}_2^2$, and $\boldsymbol{l}_{2z}^2$). An alternative basis set representation can be formed by the eigenvectors of $\boldsymbol{L}^2$, $\boldsymbol{L}_z$, $\boldsymbol{l}_1^2$, and $\boldsymbol{l}_2^2$:

$$\boldsymbol{L}^2|L(l_1 l_2)M\rangle = L(L+1)|L(l_1 l_2)M\rangle,$$
$$\boldsymbol{L}_z|L(l_1 l_2)M\rangle = M|L(l_1 l_2)M\rangle,$$
$$\boldsymbol{l}_1^2|L(l_1 l_2)M\rangle = l_1(l_1+1)|L(l_1 l_2)M\rangle,$$
$$\boldsymbol{l}_2^2|L(l_1 l_2)M\rangle = l_2(l_2+1)|L(l_1 l_2)M\rangle. \tag{B.4}$$

We will denote these eigenvectors by $|L(l_1 l_2)M\rangle$, although sometimes we will forget to include the indices l_1 and l_2 in order to simplify the notation. The quantum number L can take any integer value from $|l_1 - l_2|$ to $l_1 + l_2$.

The number of basis states $|L(l_1 l_2)M\rangle$ is the same as the number of states $|l_1 m_1\rangle|l_2 m_2\rangle$. However, the two-particle Hamiltonian is block-diagonal in the representation $|L(l_1 l_2)M\rangle$. There are no matrix elements coupling states with different values of L or different values of M. This reflects the conservation of the *total* angular momentum and its z-component for an ensemble of particles in a spherically symmetric potential. By expressing the Hamiltonian matrix in the basis $|L(l_1 l_2)M\rangle$, we have reduced the eigenvalue problem to a set of smaller problems. The eigenfunctions of the Hamiltonian can be labeled by the quantum numbers L and M.

Example B.1 Let's consider two particles, each with angular momentum $l_1 = l_2 = 1$. The *uncoupled* representation includes the following basis vectors:

$$|l_1 = 1, m_1 = -1\rangle|l_2 = 1, m_2 = -1\rangle$$
$$|l_1 = 1, m_1 = -1\rangle|l_2 = 1, m_2 = 0\rangle$$
$$|l_1 = 1, m_1 = -1\rangle|l_2 = 1, m_2 = +1\rangle$$
$$|l_1 = 1, m_1 = 0\rangle|l_2 = 1, m_2 = -1\rangle$$
$$|l_1 = 1, m_1 = 0\rangle|l_2 = 1, m_2 = 0\rangle$$
$$|l_1 = 1, m_1 = 0\rangle|l_2 = 1, m_2 = +1\rangle$$
$$|l_1 = 1, m_1 = +1\rangle|l_2 = 1, m_2 = -1\rangle$$
$$|l_1 = 1, m_1 = +1\rangle|l_2 = 1, m_2 = 0\rangle$$
$$|l_1 = 1, m_1 = +1\rangle|l_2 = 1, m_2 = +1\rangle$$

The *coupled* representation consists of the following states:

$$|L = 0, M = 0\rangle$$
$$|L = 1, M = -1\rangle$$
$$|L = 1, M = 0\rangle$$
$$|L = 1, M = +1\rangle$$
$$|L = 2, M = -2\rangle$$
$$|L = 2, M = -1\rangle$$
$$|L = 2, M = 0\rangle$$
$$|L = 2, M = 1\rangle$$
$$|L = 2, M = 2\rangle.$$

B.1 The Clebsch–Gordan Coefficients

The eigenvectors $|\psi\rangle$ of the two-particle Hamiltonian $\hat{H}$ can be expanded either as

$$|\psi\rangle = \sum_{n=1}^{N} \sum_{m_1=-l_1}^{l_1} \sum_{m_2=-l_2}^{l_2} a_{nm_1m_2} |n\rangle|l_1 m_1\rangle|l_2 m_2\rangle \qquad \text{(B.5)}$$

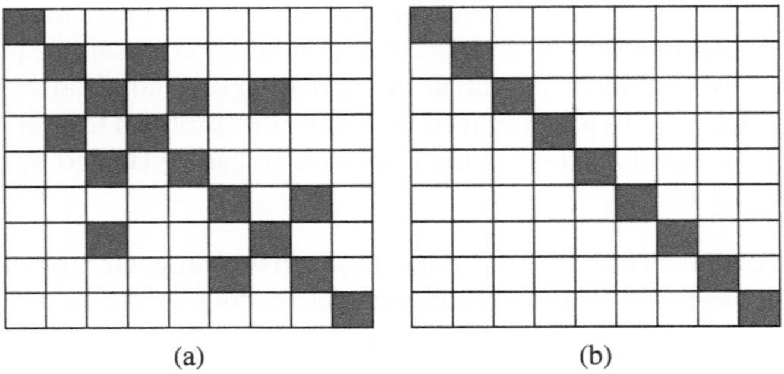

Figure B.1 Schematic structure of the Hamiltonian matrix for two interacting particles in a spherically symmetric potential. Part (a) shows the matrix in the uncoupled angular momentum basis (B.5) and part (b) shows the matrix in the coupled angular momentum basis (B.6). Each square of the tables represents the $N \times N$ block of the radial matrix elements $\langle n|\hat{H}|n'\rangle$. The empty squares show the blocks of the matrices, in which all matrix elements are zero. The shaded squares show the nonvanishing blocks of the matrices.

or as

$$|\psi\rangle = \sum_{n=1}^{N} \sum_{L=|l_1-l_2|}^{l_1+l_2} \sum_{M=-L}^{L} b_{nLM} |n\rangle |L(l_1 l_2)M\rangle, \tag{B.6}$$

where $|n\rangle$ denotes the basis vectors of the coordinate subspace (r_1, r_2). The eigenvectors of $\hat{H}$ can be determined by the diagonalization of the square matrix with the matrix elements $\langle l_2 m_2|\langle l_1 m_1|\langle n|\hat{H}|n'\rangle|l_1 m_1'\rangle|l_2 m_2'\rangle$ or the matrix with the elements $\langle L(l_1 l_2)M|\langle n|\hat{H}|n'\rangle|L'(l_1 l_2)M'\rangle$. For the example considered earlier, the matrices have the dimension $9N \times 9N$. Figure B.1 shows schematically the structure of the Hamiltonian matrix in the uncoupled and coupled angular momentum basis representations.

It is clear from Eqs. (B.5) and (B.6) that the vectors $|L(l_1 l_2)M\rangle$ and $|l_1 m_1\rangle|l_2 m_2\rangle$ are related by a unitary transformation. So we can write

$$|L(l_1 l_2)M\rangle = \sum_{m_1, m_2} C^{LM}_{l_1 m_1 l_2 m_2} |l_1 m_1\rangle |l_2 m_2\rangle$$

or

$$|l_1 m_1\rangle |l_2 m_2\rangle = \sum_{L,M} C^{l_1 m_1 l_2 m_2}_{LM} |L(l_1 l_2)M\rangle.$$

The expansion coefficients $C^{LM}_{l_1 m_1 l_2 m_2} = C^{l_1 m_1 l_2 m_2}_{LM}$ in these equations are the Clebsch–Gordan coefficients. The Clebsch–Gordan coefficients are thus the matrix elements of the unitary transformation between the coupled and uncoupled bases. Alternatively, they are projections of uncoupled states onto coupled states:

$$C^{LM}_{l_1 m_1 l_2 m_2} = \langle l_1 m_1|\langle l_2 m_2|LM\rangle. \tag{B.7}$$

The angular momentum theory offers analytical expressions for calculating the Clebsch–Gordan coefficients for given sets of the six quantum numbers. These expressions are somewhat cumbersome and, in practice, we use computer programs to calculate the Clebsch–Gordan coefficients. These coefficients can also be calculated with Mathematica [424]. Because the Clebsch–Gordan coefficients are the matrix elements of a unitary transformation,

$$\sum_{m_1}\sum_{m_2}\sum_{m_1'}\sum_{m_2'} C^{LM}_{l_1m_1l_2m_2} C^{L'M'}_{l_1m_1'l_2m_2'} = \delta_{LL'}\delta_{MM'}$$

and

$$\sum_{L}\sum_{M}\sum_{L'}\sum_{M'} C^{LM}_{l_1m_1l_2m_2} C^{L'M'}_{l_1m_1'l_2m_2'} = \delta_{m_1m_1'}\delta_{m_2m_2'}.$$

The Clebsch–Gordan coefficients have many other useful properties [9, 10]. For example,

$$C^{LM}_{l_1m_1l_2m_2} = 0 \text{ unless } m_1 + m_2 = M. \tag{B.8}$$

This ensures that the total angular momentum projection is the sum of the projections of the constituent angular momenta. I used this property, and the fact that the total angular momentum projection is conserved, to conclude that the matrix elements in the empty squares in Figure B.1a must be zero. We also need to remember that the Clebsch–Gordan coefficient $C^{LM}_{l_1m_1l_2m_2}$ must vanish unless the three angular momenta satisfy the triangle rule: $|l_1 - l_2| \leq L \leq l_1 + l_2$. Otherwise, it would be unphysical.

The quantum mechanical operators l_x, l_y, and l_z derived from the equation $\hbar \boldsymbol{l} = \boldsymbol{r} \times \boldsymbol{p}$ do not commute. This is a consequence of the uncertainty principle. In quantum mechanics, an angular momentum operator can be defined as any vector operator $\boldsymbol{j}$, whose Cartesian components j_x, j_y, and j_z satisfy the same commutation relations as $[l_x, l_y]$, $[l_y, l_z]$, and $[l_z, l_x]$ with l_x, l_y, and l_z defined by Eq. (B.2). This more general definition extends the angular momentum algebra to apply to spin angular momenta, which do not have an analog in classical mechanics and cannot be defined by an explicit equation such as (B.2). The quantum numbers j and m of a spin angular momentum $\boldsymbol{j}$ can be integer or half-integer.

B.2 The Wigner 3*j*-Symbols

It is often more convenient to use Wigner 3*j* symbols instead of Clebsch–Gordan coefficients. The Wigner 3*j* symbols are related to the Clebsch–Gordan coefficients through the following expressions:

$$\begin{pmatrix} j_1 & j_2 & j_3 \\ m_1 & m_2 & m_3 \end{pmatrix} = (-1)^{j_1 - j_2 - m_3}(2j_3 + 1)^{-1/2} C^{j_3 - m_3}_{j_1m_1j_2m_2}. \tag{B.9}$$

The $3j$ symbols have nice symmetry properties [9, 10]. For example, a permutation of any two columns of a $3j$ symbol introduces a phase factor $(-1)^{j_1+j_2+j_3}$,

$$\begin{pmatrix} j_1 & j_2 & j_3 \\ m_1 & m_2 & m_3 \end{pmatrix} = (-1)^{j_1+j_2+j_3} \begin{pmatrix} j_1 & j_3 & j_2 \\ m_1 & m_3 & m_2 \end{pmatrix} = \begin{pmatrix} j_2 & j_3 & j_1 \\ m_2 & m_3 & m_1 \end{pmatrix}. \tag{B.10}$$

Of course, the $3j$ symbols in Eq. (B.10) must vanish unless $m_1 + m_2 + m_3 = 0$ (because of Eq. (B.8), note the minus sign in Eq. (B.9)) and unless $|j_1 - j_2| \leq j_3 \leq j_1 + j_2$. We will also use the following properties of the $3j$ symbols:

$$\begin{pmatrix} j_1 & j_2 & j_3 \\ m_1 & m_2 & m_3 \end{pmatrix} = (-1)^{j_1+j_2+j_3} \begin{pmatrix} j_1 & j_2 & j_3 \\ -m_1 & -m_2 & -m_3 \end{pmatrix} \tag{B.11}$$

and the fact that

$$\begin{pmatrix} j_1 & j_2 & j_3 \\ 0 & 0 & 0 \end{pmatrix} = 0 \text{ if } j_1 + j_2 + j_3 \text{ is an odd integer.} \tag{B.12}$$

A complete description of the properties of the $3j$ symbols can be found in Ref. [10].

B.3 The Raising and Lowering Operators

Consider an angular momentum $\boldsymbol{J} = \boldsymbol{L} + \boldsymbol{S}$ defined as a sum of two other angular momenta. Often, we need to evaluate the matrix elements of $\boldsymbol{J}^2$ in the uncoupled basis $|LM_L\rangle|SM_S\rangle$. The operator $\boldsymbol{J}^2$ can be written as

$$\boldsymbol{J}^2 = \boldsymbol{L}^2 + \boldsymbol{S}^2 + 2\boldsymbol{L} \cdot \boldsymbol{S} \tag{B.13}$$

and the last term in the last equation can be expanded using

$$\boldsymbol{L} \cdot \boldsymbol{S} = L_x S_x + L_y S_y + L_z S_z. \tag{B.14}$$

We know the matrix elements of $\boldsymbol{L}^2$, $\boldsymbol{S}^2$, and $L_z S_z$ in the uncoupled basis. However, we don't know how the operators L_x or L_y act on the states $|LM_L\rangle$.

In order to evaluate the matrix elements of the x- and y-components of angular momenta, we will introduce the raising/lowering operators as

$$L_\pm = L_x \pm iL_y. \tag{B.15}$$

These operators are also called the ladder operators. It can be proven [9] that

$$L_\pm|LM_L\rangle = |LM_L \pm 1\rangle\sqrt{L(L+1) - M_L(M_L \pm 1)}. \tag{B.16}$$

Note that the operators $L_\pm$ are not Hermitian but this is usually not a problem as they come in combinations that preserve the Hermiticity of the Hamiltonians.

Eq. (B.15) can be inverted to write

$$L_x = (L_+ + L_-)/2 \tag{B.17}$$

$$L_y = (L_+ - L_-)/2i. \tag{B.18}$$

Given these equations and Eq. (B.16), we can evaluate the matrix elements of the x- and y-components of any angular momentum operators.

Exercises

B.1 Using the equation $\hbar \boldsymbol{l} = \boldsymbol{r} \times \boldsymbol{p}$ as a starting point, find the explicit expressions for the Cartesian components l_x, l_y, and l_z of the angular momentum operator in the spherical polar coordinates.

B.2 Using the results of the previous exercise, find the explicit expression for the square of the angular momentum operator $l^2 = l_x^2 + l_y^2 + l_z^2$ in the spherical polar coordinates.

B.3 Find the explicit expression for the Laplace operator

$$\Delta = \frac{\partial^2}{\partial x^2} + \frac{\partial^2}{\partial y^2} + \frac{\partial^2}{\partial z^2}$$

in the spherical polar coordinate system.

B.4 Find the commutators $[l_x, l_y]$, $[l_y, l_z]$, and $[l_z, l_x]$, where l_x, l_y, and l_z are the operators for the Cartesian components of the angular momentum $\boldsymbol{l}$.

B.5 Find the commutators $[l_x, \boldsymbol{l}^2]$, $[l_y, \boldsymbol{l}^2]$, and $[l_z, \boldsymbol{l}^2]$.

B.6 What are the properties of the spherical harmonics $Y_{lm}(\theta, \phi)$ under inversion of the coordinate system (are they even or odd functions)?

B.7 Consider two interacting particles in a spherically symmetric potential. Assuming that the angular momenta of the individual particles can take values 0, 1, and 2, list the basis vectors of the uncoupled and coupled representations. Draw a table (similar to the matrix in Figure B.1a) representing schematically the nonzero blocks of the Hamiltonian matrix in the uncoupled representation. Draw a table representing the matrix of the Clebsch–Gordan coefficients $C^{LM}_{l_1 m_1 l_2 m_2}$. Label the nonzero elements of the Clebsch–Gordan transformation matrix. Using Mathematica [424], compute the values of the nonzero Clebsch–Gordan coefficients.

C

Direction Cosine Matrix

When dealing with molecules in external fields, we need to be able to write operators in both the coordinate system defined by the direction of the external field (space-fixed frame) and the coordinate system defined by the interatomic axis (molecule-fixed frame). To do this, it is first of all necessary to know how to write a position vector in one coordinate system in terms of the same vector in the other coordinate system. The treatment here follows closely that by Hougen [11].

We will denote the coordinates of a particle in the space-fixed coordinate frame by the uppercase X, Y, and Z and in the body-fixed frame by the lowercase x, y, and z. In the presence of an axially symmetric field, it is convenient to choose the Z-axis to point in the direction of the field vector. The body-fixed z-axis is directed along the interatomic axis. The relation between the coordinates of the vector in the space-fixed frame (X, Y, Z) and the coordinates in the body-fixed frame (x, y, z) is given by the direction cosine matrix

$$\begin{pmatrix} X \\ Y \\ Z \end{pmatrix} = \begin{bmatrix} -\sin\varphi & -\cos\theta\cos\varphi & \sin\theta\cos\varphi \\ -\cos\varphi & -\cos\theta\sin\varphi & \sin\theta\sin\varphi \\ 0 & \sin\theta & \cos\theta \end{bmatrix} \begin{pmatrix} x \\ y \\ z \end{pmatrix}, \tag{C.1}$$

Factor	$J' = J+1$	$J' = J$	$J' = J-1$
$f(J';J)$	$\{4(J+1)$ $[(2J+1)(2J+3)]^{1/2}\}^{-1}$	$[4J(J+1)]^{-1}$	$\{4J[(2J+1)$ $(2J-1)]^{1/2}\}^{-1}$
$g_z(J',\Omega;J,\Omega)$	$2[(J+\Omega+1)(J-\Omega+1)]^{1/2}$	2Ω	$2\Omega[(J+\Omega)(J-\Omega)]^{1/2}$
$g_x(J',\Omega\pm1;J,\Omega)$	$\mp[(J\pm\Omega+1)(J\pm\Omega+2)]^{1/2}$	$[(J\mp\Omega)(J\pm\Omega+1)]^{1/2}$	$\pm[(J\mp\Omega)(J\mp\Omega-1)]^{1/2}$
$\mp ig_y(J',\Omega\pm1;J,\Omega)$	$\mp[(J\pm\Omega+1)(J\pm\Omega+2)]^{1/2}$	$[(J\mp\Omega)(J\pm\Omega+1)]^{1/2}$	$\pm[(J\mp\Omega)(J\mp\Omega-1)]^{1/2}$
$h_z(J',M;J,M)$	$2[(J+M+1)(J-M+1)]^{1/2}$	$2M$	$2[(J+M)(J-M)]$
$h_x(J',M\pm1;J,M)$	$\mp[(J\pm M+1)(J\pm M+2)]^{1/2}$	$[(J\mp M)(J\pm M+1)]^{1/2}$	$\pm[(J\mp M)(J\mp M-1)]^{1/2}$
$\pm ih_y(J',M\pm1;J,M)$	$\mp[(J\pm M+1)(J\pm M+2)]^{1/2}$	$[(J\mp M)(J\pm M+1)]^{1/2}$	$\pm[(J\mp M)(J\mp M-1)]^{1/2}$

Molecules in Electromagnetic Fields: From Ultracold Physics to Controlled Chemistry, First Edition. Roman V. Krems.
© 2019 John Wiley & Sons, Inc. Published 2019 by John Wiley & Sons, Inc.

where the angles θ and φ are the spherical polar angles that specify the orientation of the intermolecular axis in the space-fixed frame. The elements of the direction cosine matrix $\alpha_{R,s}$, with $R = X, Y$, or Z and $s = x, y$, or z, can be thought of as operators that act on the rotational functions $|JM\Omega\rangle$. The integrals $\langle J'M'\Omega'|\alpha_{R,s}|JM\Omega\rangle$ are nonzero only if $|J - J'| = 0$ or 1, $|M - M'| = 0$ or 1 and $|\Omega - \Omega'| = 0$ or 1. These integrals can be represented as a product of three factors

$$\langle J'M'\Omega'|\alpha_{R,s}|JM\Omega\rangle = f(J';J) \times g_s(J',\Omega';J,\Omega) \times h_R(J',M;J,M). \qquad \text{(C.2)}$$

These factors are given in Ref. [11] and reproduced in the table on page 307 for use in the present work.

D

Wigner D-Functions

Consider again the two coordinate systems (x, y, z) and (X, Y, Z) related to each other by a rotation. While in the previous appendix we were concerned with the relation between the coordinates of a vector in these two coordinate systems, here we will consider what happens to the angular momentum states upon the rotation of the coordinate system.

An angular momentum $\boldsymbol{J}$ has the component J_Z in one of the coordinate systems and the component J_z in the other coordinate frame. The eigenstates of $\boldsymbol{J}^2$ and J_Z are the states $|JM\rangle$. The eigenstates of $\boldsymbol{J}^2$ and J_z are the states $|J\Omega\rangle$. The question is, what is the relationship between the states $|JM\rangle$ and $|J\Omega\rangle$?

Since the value of J is the same in both cases, the number of different states $|JM\rangle$ is the same as the number of states $|J\Omega\rangle$. Moreover, the states $|JM\rangle$ form a complete basis set as long as J is fixed. Therefore, the states $|J\Omega\rangle$ can be written as superpositions of states $|JM\rangle$ with different values of M, as follows:

$$|J\Omega\rangle = \sum_M |JM\rangle D^J_{M\Omega}. \tag{D.1}$$

The expansion coefficients

$$D^J_{M\Omega} = \langle JM|J\Omega\rangle \tag{D.2}$$

can be viewed as the amplitudes of the probability that the quantum system prepared in the state with the Z-projection of angular momentum $\hbar M$ will have the z-projection equal to $\hbar\Omega$.

Obviously the coefficients $D^J_{M\Omega}$ depend on the relative orientation of the two coordinate frames. A three-dimensional coordinate frame (X, Y, Z) can be rotated to coincide with the three-dimensional coordinate frame (x, y, z) by a succession of three rotations about three axes. For example, the transformation of (X, Y, Z) into (x, y, z) can by achieved by the following three rotations [9]:

- Counterclockwise rotation of the coordinate frame (X, Y, Z) by an angle α about the Z-axis in order to bring the axis Y to lie along the line of the intersection of the original xy- and XY-planes;

Molecules in Electromagnetic Fields: From Ultracold Physics to Controlled Chemistry, First Edition. Roman V. Krems.

- Counterclockwise rotation of the resulting coordinate frame by an angle β about the new Y-axis (that is about the intersection axis of xy- and XY-planes) in order to bring the Z-axis in alignment with the z-axis;
- Counterclockwise rotation of the resulting coordinate frame by an angle γ about the new Z-axis in order to bring the Y-axis in alignment with the y-axis;

The angles α, β, γ thus defined are called the Euler angles.

The expansion coefficients $D^J_{M\Omega}$ are functions of the Euler angles. The are called Wigner D-functions and form the matrix elements of a square matrix that transforms the states $|JM\rangle$ into the states $|J\Omega\rangle$.

From Eq. (D.1), we have

$$\langle J\Omega' | J\Omega\rangle = \sum_{M'} \sum_{M} \langle JM' | JM\rangle D^J_{M\Omega} D^{J*}_{M'\Omega'}. \tag{D.3}$$

Since the eigenstates of $\boldsymbol{J}^2$ and J_z are orthonormal, this leads to

$$\sum_M D^J_{M\Omega} D^{J*}_{M\Omega'} = \delta_{\Omega,\Omega'}. \tag{D.4}$$

The last equation can be used to invert Eq. (D.1) and write

$$|JM\rangle = \sum_\Omega |J\Omega\rangle D^{J*}_{M\Omega}. \tag{D.5}$$

This also implies that

$$\sum_\Omega D^J_{M\Omega} D^{J*}_{M'\Omega} = \delta_{M,M'}. \tag{D.6}$$

In other words, the matrix of D-functions is a unitary transformation matrix.

We can define the quantum states

$$|JM\Omega\rangle \equiv \left[\frac{2J+1}{8\pi^2}\right]^{1/2} D^{J*}_{M\Omega}. \tag{D.7}$$

These states satisfy the following eigenvalue equations:

$$\boldsymbol{J}^2 |JM\Omega\rangle = J(J+1)|JM\Omega\rangle, \tag{D.8}$$

$$J_Z |JM\Omega\rangle = M|JM\Omega\rangle, \tag{D.9}$$

$$J_z |JM\Omega\rangle = \Omega|JM\Omega\rangle. \tag{D.10}$$

As discussed in Section 1.3, these states are the rotational states of a Hund's case (a) molecule.

It is also useful to know that if one of the quantum numbers M or Ω is zero, the D-function becomes a spherical harmonic:

$$D^J_{M0}(\alpha, \beta, \gamma) = \left[\frac{4\pi}{2J+1}\right]^{1/2} Y^*_{JM}(\beta, \alpha), \tag{D.11}$$

$$D^J_{0\Omega}(\alpha, \beta, \gamma) = (-1)^\Omega \left[\frac{4\pi}{2J+1}\right]^{1/2} Y^*_{J\Omega}(\beta, \gamma). \tag{D.12}$$

This is why the rotational states of molecules in a $^1\Sigma$ electronic state (for which $\Omega = 0$) are spherical harmonics.

D.1 Matrix Elements Involving *D*-Functions

Since $|JM\Omega\rangle$ are the eigenstates of Hermitian operators, they must be orthogonal to each other. The prefactor on the right-hand side of Eq. (D.7) is introduced to make these states orthonormal so that

$$\langle JM\Omega|J'M'\Omega'\rangle = \delta_{JJ'}\delta_{MM'}\delta_{\Omega\Omega'}. \tag{D.13}$$

As discussed in the following appendix, many operators can be generally written as products of *D*-functions so it is useful to consider the matrix elements of the following kind:

$$\langle JM\Omega|D^k_{q\omega}|J'M'\Omega'\rangle = ? \tag{D.14}$$

In order to evaluate these matrix elements, consider first an angular momentum $\boldsymbol{j}$ formed by a vector sum of two other angular momenta $\boldsymbol{j}_1$ and $\boldsymbol{j}_2$. The eigenstates $|j\omega\rangle$ of $\boldsymbol{j}^2$ and j_z can then be written in terms of the angular momentum states in the uncoupled representation (see Appendix B) as follows:

$$|j\omega\rangle = \sum_{\omega_1}\sum_{\omega_2} |j_1\omega_1\rangle|j_2\omega_2\rangle C^{j\omega}_{j_1\omega_1 j_2\omega_2}. \tag{D.15}$$

Let us now apply Eq. (D.1) to each of the angular momentum states in the above equation, yielding

$$\sum_m D^j_{m\omega}|jm\rangle = \sum_{\omega_1}\sum_{\omega_2}\sum_{m_1}\sum_{m_2} |j_1 m_1\rangle|j_2 m_2\rangle D^{j_1}_{m_1\omega_1} D^{j_2}_{m_2\omega_2} C^{j\omega}_{j_1\omega_1 j_2\omega_2}. \tag{D.16}$$

Multiplying both sides of this equation by $\langle jm|$ and noting that $\langle jm|j_1 m_1\rangle|j_2 m_2\rangle = C^{jm}_{j_1 m_1 j_2 m_2}$, we obtain

$$D^j_{m\omega} = \sum_{\omega_1}\sum_{\omega_2}\sum_{m_1}\sum_{m_2} D^{j_1}_{m_1\omega_1} D^{j_2}_{m_2\omega_2} C^{j\omega}_{j_1\omega_1 j_2\omega_2} C^{jm}_{j_1 m_1 j_2 m_2}. \tag{D.17}$$

Using the orthogonality properties of the Clebsch–Gordan coefficients, we can invert this equation. Specifically, if we multiply both sides of the last equation by $C^{j\omega}_{j_1\omega_1 j_2\omega_2} C^{jm}_{j_1 m_1 j_2 m_2}$ and sum over j, m, and ω, we will obtain

$$D^{j_1}_{m_1\omega_1} D^{j_2}_{m_2\omega_2} = \sum_j\sum_m\sum_\omega D^j_{m\omega} C^{j\omega}_{j_1\omega_1 j_2\omega_2} C^{jm}_{j_1 m_1 j_2 m_2}. \tag{D.18}$$

Note that the sum over m and ω can be dropped because the Clebsch–Gordan coefficients vanish unless $m = m_1 + m_2$ and $\omega = \omega_1 + \omega_2$.

The last equation can help us to evaluate the matrix elements in Eq. (D.14). Let us write

$$\begin{aligned}\langle JM\Omega|D^k_{q\omega} &= \left[\frac{2J+1}{8\pi^2}\right]^{1/2} D^k_{q\omega} D^J_{M\Omega}\\ &= \left[\frac{2J+1}{8\pi^2}\right]^{1/2} \sum_p D^p_{q+M,\omega+\Omega} C^{p,q+M}_{kqJM} C^{p,\omega+\Omega}_{k\omega J\Omega}\\ &= \left[\frac{2J+1}{8\pi^2}\right]^{1/2} \sum_p \left[\frac{2p+1}{8\pi^2}\right]^{-1/2} \langle p, q+M, \omega+\Omega| C^{p,q+M}_{kqJM} C^{p,\omega+\Omega}_{k\omega J\Omega}.\end{aligned} \tag{D.19}$$

Inserting this equation back into Eq. (D.14) and using the orthonormality of the states $|JM\Omega\rangle$, we obtain

$$\langle JM\Omega|D^k_{q\omega}|J'M'\Omega'\rangle = \left[\frac{2J+1}{2J'+1}\right]^{1/2} C^{J'M'}_{kqJM} C^{J'\Omega'}_{k\omega J\Omega}. \tag{D.20}$$

Note that these matrix elements vanish unless $q + M = M'$ and $\omega + \Omega = \Omega'$, which follows from the properties of the Clebsch–Gordan coefficients discussed in Appendix B.

This is a very powerful equation as it allows us to reduce the matrix elements of many different operators to simple products of Clebsch–Gordan coefficients. Of particular importance to molecules in external fields are the matrix elements of $\cos\beta$. To evaluate these matrix elements, we note that

$$\cos\beta = \sqrt{\frac{4\pi}{3}} Y_{1,0}(\beta) = D^1_{00}, \tag{D.21}$$

which gives

$$\langle JM\Omega|\cos\beta|J'M'\Omega'\rangle = \left[\frac{2J+1}{2J'+1}\right]^{1/2} C^{J'M'}_{10JM} C^{J'\Omega'}_{10J\Omega}. \tag{D.22}$$

These matrix elements vanish unless $M = M'$, $\Omega = \Omega'$, and $J = J' \pm 1$.

Given the relations (D.11) and (D.12) between the D-functions and the spherical harmonics, we can use Eq. (D.22) to evaluate the integrals over products of three spherical harmonics:

$$\langle JM|Y_{kq}|J'M'\rangle = \left[\frac{(2J+1)(2k+1)}{(2J'+1)4\pi}\right]^{1/2} (-1)^q C^{J'M'}_{k-qJM} C^{J'0}_{k0J0}, \tag{D.23}$$

where

$$|JM\rangle = \sqrt{2\pi}|JM0\rangle = Y_{JM} \tag{D.24}$$

is the spherical harmonic. The factor $(4\pi)^{-1/2}$ appears because the normalization of the spherical harmonics is different from that of D-functions. Using the relation (B.9) between the Clebsch–Gordan coefficients and the $3j$-symbols, the last equation can be written in terms of the $3j$-symbols as follows:

$$\langle JM|Y_{kq}|J'M'\rangle = \left[\frac{(2J+1)(2k+1)(2J'+1)}{4\pi}\right]^{1/2}(-1)^{(2J-2k-M'+q)}$$
$$\times\begin{pmatrix} k & J & J' \\ -q & M & -M' \end{pmatrix}\begin{pmatrix} k & J & J' \\ 0 & 0 & 0 \end{pmatrix}, \tag{D.25}$$

which, using the properties of the $3j$-symbols (B.10), (B.11), and (B.12), can be written in a more convenient form

$$\langle JM|Y_{kq}|J'M'\rangle = \left[\frac{(2J+1)(2k+1)(2J'+1)}{4\pi}\right]^{1/2}(-1)^{-M},$$
$$\times\begin{pmatrix} J & k & J' \\ -M & q & M' \end{pmatrix}\begin{pmatrix} J & k & J' \\ 0 & 0 & 0 \end{pmatrix}, \tag{D.26}$$

which is the same as the result obtained in Section 9.1 with the help of the Wigner–Eckart Theorem.

E

Spherical Tensors

Equation (D.5) can be generalized to define spherical tensors. By analogy with the angular momentum states $|JM\rangle$, imagine a set of operators $\hat{T}^k_q$ with k an integer and $q = -k, -k+1, \dots, k-1, k$. The operators $\hat{T}^k_q$ must generally change when the coordinate system is rotated. Let us again define the coordinate frames (X, Y, Z) and (x, y, z) related by three Euler rotations discussed in appendix D.

The $(2k+1)$ components $\hat{T}^k_q$ define a spherical tensor operator of rank k, if they behave under coordinate frame rotation in the same way as the angular momentum states $|JM\rangle$. In other words, if $\hat{T}^k_q$ are the components of a spherical tensor in the coordinate frame (X, Y, Z) and $\hat{T}^k_\omega$ are the components of the same tensor in the coordinate frame (x, y, z), $\hat{T}^k_q$ and $\hat{T}^k_\omega$ must be related by

$$\hat{T}^k_\omega = \sum_q \hat{T}^k_q D^k_{q\omega}(\alpha, \beta, \gamma), \tag{E.1}$$

where $D^k_{q\omega}$ are the Wigner D-functions discussed in appendix D and (α, β, γ) are the Euler angles. Note that the rotation does not change the rank of the spherical tensor.

To make this more transparent, consider a few examples.

Example E.1 First of all, any scalar operator can be considered as a spherical tensor of rank $k = 0$. In this case, there is only one term in the sum in Eq. (E.1) and since $D^0_{00} = 1$, the space-fixed and body-fixed tensor components are equal: $\hat{T}^0_{q=0} = \hat{T}^0_{\omega=0}$. This shows that scalars do not change under rotations.

Example E.2 Any spherical harmonic Y_{lm} will satisfy Eq. (E.1) so any spherical harmonic is automatically a spherical tensor of rank $k = l$:

$$Y_{lm} \Rightarrow \hat{T}^l_m. \tag{E.2}$$

Example E.3 Position vectors can be written in terms of spherical tensors of rank 1. This follows from the transformation between the Cartesian coordinates and spherical polar coordinates,

$$\begin{aligned} z &= r\cos\theta \\ x &= r\sin\theta\cos\phi \\ y &= r\sin\theta\sin\phi \end{aligned} \tag{E.3}$$

and the mathematical expressions of the spherical harmonics of rank 1

$$\begin{aligned} Y_{1,0} &= \frac{1}{2}\sqrt{\frac{3}{\pi}}\cos\theta \\ Y_{1,-1} &= \frac{1}{2}\sqrt{\frac{3}{2\pi}}\sin\theta e^{-i\phi} \\ Y_{1,+1} &= -\frac{1}{2}\sqrt{\frac{3}{2\pi}}\sin\theta e^{i\phi}. \end{aligned} \tag{E.4}$$

Using Eqs. (E.3) and (E.4), we can write any vector as

$$\begin{aligned} \boldsymbol{r} &= x\hat{x} + y\hat{y} + z\hat{z} \\ &= A^{-1}\left[\hat{x}\frac{1}{\sqrt{2}}(Y_{1,-1} - Y_{1,1}) - i\hat{y}\frac{}{\sqrt{2}}(Y_{1,-1} - Y_{1,1}) + \hat{z}Y_{1,0}\right] \\ &= \hat{\boldsymbol{e}}_0\hat{T}_0^1 + \hat{\boldsymbol{e}}_{-1}\hat{T}_{-1}^1 + \hat{\boldsymbol{e}}_{+1}\hat{T}_{+1}^1, \end{aligned} \tag{E.5}$$

where $A = \sqrt{3/4\pi r^2}$, the spherical tensor components are $\hat{T}_q^1 \equiv A^{-1}Y_{1,q}$, and the unit vectors are

$$\hat{\boldsymbol{e}}_0 = \hat{z}, \tag{E.6}$$

$$\hat{\boldsymbol{e}}_{-1} = \frac{1}{\sqrt{2}}(\hat{x} - i\hat{y}), \tag{E.7}$$

$$\hat{\boldsymbol{e}}_{+1} = -\frac{1}{\sqrt{2}}(\hat{x} + i\hat{y}). \tag{E.8}$$

Note that

$$\hat{\boldsymbol{e}}_0^* \cdot \hat{\boldsymbol{e}}_0 = \hat{\boldsymbol{e}}_{-1}^* \cdot \hat{\boldsymbol{e}}_{-1} = \hat{\boldsymbol{e}}_{+1}^* \cdot \hat{\boldsymbol{e}}_{+1} = 1 \tag{E.9}$$

and

$$\hat{\boldsymbol{e}}_0^* \cdot \hat{\boldsymbol{e}}_{+1} = \hat{\boldsymbol{e}}_0^* \cdot \hat{\boldsymbol{e}}_{-1} = \hat{\boldsymbol{e}}_{-1}^* \cdot \hat{\boldsymbol{e}}_{+1} = 0. \tag{E.10}$$

E.1 Scalar and Vector Products of Vectors in Spherical Basis

From Eq. (E.5), we can see that the scalar product of vector $\boldsymbol{r}$ with itself can be written as

$$\boldsymbol{r}^* \cdot \boldsymbol{r} = \sum_{q=-1}^{+1} (-1)^q \hat{T}_q^1 \hat{T}_{-q}^1. \tag{E.11}$$

Here, we have used the property of the spherical harmonics $Y_{lm_l}^* = (-1)^l Y_{l-m_l}$ and the orthonormality of the vectors $\hat{\boldsymbol{e}}_q$.

For two different vectors, $\boldsymbol{r}_1$ and $\boldsymbol{r}_2$, we can write the scalar product as

$$\boldsymbol{r}_1^* \cdot \boldsymbol{r}_2 = \sum_{q=-1}^{+1} (-1)^q \hat{T}_q^1(r_1) \hat{T}_{-q}^1(r_2), \tag{E.12}$$

where I have added r_1 and r_2 in parentheses because, in our definition, the spherical tensors depend on the length of the vector.

What about a cross product of $\boldsymbol{r}_1$ and $\boldsymbol{r}_2$? We can attempt to write it in terms of the cross products of $\hat{\boldsymbol{e}}_q$, as follows:

$$\boldsymbol{r}_1 \times \boldsymbol{r}_2 = \sum_{q_1} \sum_{q_2} \hat{T}_{q_1}^1(r_1) \hat{T}_{q_2}^1(r_2) (\hat{\boldsymbol{e}}_{q_1} \times \hat{\boldsymbol{e}}_{q_2}). \tag{E.13}$$

By direct inspection, we can see that

$$\hat{\boldsymbol{e}}_{-1} \times \hat{\boldsymbol{e}}_0 = \begin{vmatrix} \hat{x} & \hat{y} & \hat{z} \\ \frac{1}{\sqrt{2}} & -\frac{i}{\sqrt{2}} & 0 \\ 0 & 0 & 1 \end{vmatrix} = \frac{-i}{\sqrt{2}}\hat{x} - \frac{1}{\sqrt{2}}\hat{y} = -i\hat{\boldsymbol{e}}_{-1}, \tag{E.14}$$

$$\hat{\boldsymbol{e}}_0 \times \hat{\boldsymbol{e}}_{+1} = \begin{vmatrix} \hat{x} & \hat{y} & \hat{z} \\ 0 & 0 & 1 \\ -\frac{1}{\sqrt{2}} & -\frac{i}{\sqrt{2}} & 0 \end{vmatrix} = \frac{i}{\sqrt{2}}\hat{x} - \frac{1}{\sqrt{2}}\hat{y} = -i\hat{\boldsymbol{e}}_{+1}, \tag{E.15}$$

$$\hat{\boldsymbol{e}}_{-1} \times \hat{\boldsymbol{e}}_{+1} = \begin{vmatrix} \hat{x} & \hat{y} & \hat{z} \\ \frac{1}{\sqrt{2}} & -\frac{i}{\sqrt{2}} & 0 \\ -\frac{1}{\sqrt{2}} & -\frac{i}{\sqrt{2}} & 0 \end{vmatrix} = i\hat{z} = i\hat{\boldsymbol{e}}_0, \tag{E.16}$$

and we note that, for any two vectors, $\boldsymbol{a} \times \boldsymbol{b} = -\boldsymbol{b} \times \boldsymbol{a}$.

We can thus write the vector product as

$$\boldsymbol{r}_1 \times \boldsymbol{r}_2 = \hat{\epsilon}_0 \hat{T}^1_0 + \hat{\epsilon}_{-1} \hat{T}^1_{-1} + \hat{\epsilon}_{+1} \hat{T}^1_{+1} \tag{E.17}$$

with

$$\hat{T}^1_0 = i(\hat{T}^1_{-1}(r_1)\hat{T}^1_1(r_2) - \hat{T}^1_1(r_1)\hat{T}^1_{-1}(r_2)) \tag{E.18}$$

$$\hat{T}^1_{-1} = i(\hat{T}^1_0(r_1)\hat{T}^1_{-1}(r_2) - \hat{T}^1_{-1}(r_1)\hat{T}^1(r_2)) \tag{E.19}$$

$$\hat{T}^1_{+1} = i(\hat{T}^1_{+1}(r_1)\hat{T}^1_0(r_2) - \hat{T}^1_0(r_1)\hat{T}^1_{+1}(r_2)), \tag{E.20}$$

which shows that a vector product of two vectors is a spherical tensor of rank 1, as expected.

We note that the last three equations can be rewritten in a more compact form using the Clebsch–Gordan coefficient $C^{1q}_{1q_1 1q_2}$ as follows (cf. Ref. [425]):

$$\hat{T}^1_q = i\sqrt{2} \sum_{q_1} \sum_{q_2} C^{1q}_{1q_1 1q_2} \hat{T}^{k_1}_{q_1} \hat{T}^{k_2}_{q_2}. \tag{E.21}$$

Here, we take advantage of the fact that the Clebsch–Gordan coefficient $C^{1q}_{1q_1 1q_2}$ is zero when $q_1 + q_2 \neq q$, has the absolute magnitude $1/\sqrt{2}$ [9] when $q_1 + q_2 = q$ and that $C^{1q}_{1q_1 1q_2} = -C^{1q}_{1q_2 1q_1}$.

E.2 Scalar and Tensor Products of Spherical Tensors

Equations (E.11) and (E.21) illustrate that the components of multiple spherical tensors can be combined to form other spherical tensors.

In general, a product of two spherical tensors with components $\hat{T}^{k_1}_{q_1}$ and $\hat{T}^{k_2}_{q_2}$ forms spherical tensors of rank k with components $\hat{T}^k_q$ defined as

$$\hat{T}^k_q = \sum_{q_1} \sum_{q_2} C^{kq}_{k_1 q_1 k_2 q_2} \hat{T}^{k_1}_{q_1} \hat{T}^{k_2}_{q_2}, \tag{E.22}$$

where $C^{kq}_{k_1 q_1 k_2 q_2}$ is a Clebsch–Gordan coefficient.

Not coincidentally, this equation is analogous to the transformation between the coupled and uncoupled angular momentum representations (see Appendix B):

$$|jm\rangle = \sum_{m_1} \sum_{m_2} |j_1 m_1\rangle |j_2 m_2\rangle C^{jm}_{j_1 m_1 j_2 m_2}. \tag{E.23}$$

We defined spherical tensors as operators that transform under rotations in the same way as the angular momentum states. Coupling the angular momentum states $|j_1 m_1\rangle$ and $|j_2 m_2\rangle$ as in Eq. (E.23) results in states $|jm\rangle$ that have the same transformation properties as the original angular momentum states. Therefore, coupling the spherical tensors $\hat{T}^{k_1}_{q_1}$ and $\hat{T}^{k_2}_{q_2}$ in a similar way results in operators $\hat{T}^k_q$ that have the same transformation properties.

The Clebsch–Gordan coefficient $C^{kq}_{k_1q_1k_2q_2}$ vanishes unless $|k_1 - k_2| \leq k \leq k_1 + k_2$. This immediately shows that scalar products can only be formed by spherical tensors of equal rank.

The tensor product of two spherical tensors of ranks k_1 and k_2 is often denoted by

$$[\hat{T}^{k_1} \otimes \hat{T}^{k_2}]^k_q = \sum_{q_1} \sum_{q_2} C^{kq}_{k_1q_1k_2q_2} \hat{T}^{k_1}_{q_1} \hat{T}^{k_2}_{q_2}. \tag{E.24}$$

References

1 Lemeshko, M., Doyle, J., Krems, R.V., and Kais, S. (2013). Manipulation of molecules with electromagnetic fields. *Eur. Phys. J. D* **111**: 1648.

2 Mizushima, M. (1975). *The Theory of Rotating Diatomic Molecules*. Wiley.

3 Lefebvre-Brion, H. and Field, R.W. (1986). *Perturbations in the Spectra of Diatomic Molecules*. Orlando, FL: Academic Press.

4 Brown, J.M. and Carrington, A. (2003). *Rotational Spectroscopy of Diatomic Molecules*. Cambridge University Press.

5 Fohlisch, A., Feulner, P., Hennies, F. et al. (2005). Direct observation of electron dynamics in the attosecond domain. *Nature* **436**: 373.

6 Parlant, G. and Yarkony, D.R. (1999). A theoretical analysis of the state-specific decomposition of OH (A $^2\Sigma^+$, v', N', F_1/F_2) levels, including the effects of spin–orbit and Coriolis interactions. *J. Chem. Phys.* **110**: 363.

7 Krems, R.V., Jamieson, M.J., and Dalgarno, A. (2006). The 1D-3P transitions in atomic oxygen induced by impact with atomic hydrogen. *Astrophys. J.* **647**: 1531.

8 Landau, L.D. and Lifshitz, E.M. (2003). *Quantum Mechanics (Non-Relativistic Theory)*. Elsevier Science Limited.

9 Zare, R.N. (1988). *Angular Momentum: Understanding Spatial Aspects in Chemistry and Physics*. Wiley.

10 Varshalovich, D.A., Moskalev, A.N., and Khersonskii, V.K. (1988). *Quantum Theory of Angular Momentum: Irreducible Tensors, Spherical Harmonics, Vector Coupling Coefficients, 3nj Symbols*. World Scientific.

11 Hougen, J.T. (1970). *The Calculations of Rotational Energy Levels and Rotational Line Intensities in Diatomic Molecules*, National Bureau of Standards Monograph 115. Washington, DC: National Bureau of Standards.

12 Wigner, E. and Witmer, E.E. (1928). Über die Struktur der zweiatomigen Molekelspektren nach der Quantenmechanik. *Z. Phys.* **51**: 859.

13 Landau, L.D. and LIfshitz, E.M. (2003). *The Classical Theory of Fields*. Elsevier Science Limited.

Molecules in Electromagnetic Fields: From Ultracold Physics to Controlled Chemistry, First Edition. Roman V. Krems.

14 Pavlovic, Z., Tscherbul, T.V., Sadeghpour, H.R. et al. (2009). Cold collisions of OH($^2\Pi$) molecules with He atoms in external fields†. *J. Phys. Chem. A* **113**: 14670.

15 Landau, L.D. and LIfshitz, E.M. (2003). *Mechanics*. Elsevier Science Limited.

16 Bergstrröm, L. and Hansson, H. (1999). *Lecture Notes in Relativistic Quantum Mechanics*. Stockholm: Department of Physics, Stockholm University.

17 Karplus, R. and Kroll, N.M. (1950). Fourth-order corrections in quantum electrodynamics and the magnetic moment of the electron. *Phys. Rev.* **77**: 536.

18 Gabrielse, G., Hanneke, D., Kinoshita, T. et al. (2006). New determination of the fine structure constant from the electron *g* value and QED. *Phys. Rev. Lett.* **97**: 030802.

19 National Institute of Standards and Technology database: www.nist.gov (accessed 24 November 2017).

20 Friedrich, B., Weinstein, J.D., deCarvalho, R., and Doyle, J.M. (1999). Zeeman spectroscopy of CaH molecules in a magnetic trap. *J. Chem. Phys.* **110**: 2376.

21 Radford, H.E. and Broida, H.P. (1962). Rotational perturbations in CN. Zero-field theory, optical Zeeman effect, and microwave transition probabilities. *Phys. Rev.* **128**: 231.

22 Veseth, L. (1976) Theory of high-precision Zeeman effect in diatomic molecules. *J. Mol. Spectrosc.* **63** (2), 180.

23 Krems, R., Egorov, D., Helton, J.S. et al. (2004). Zeeman effect in CaF($^2\Pi_{3/2}$). *J. Chem. Phys.* **121**: 11639.

24 Wallis, A.O.G. and Krems, R.V. (2014). Magnetic Feshbach resonances in collisions of nonmagnetic closed-shell $^1\Sigma$ molecules. *Phys. Rev. A* **89**: 032716.

25 Aldegunde, J., Rivington, B.A., Żuchowski, P.S., and Hutson, J.M. (2008). Hyperfine energy levels of alkali-metal dimers: ground-state polar molecules in electric and magnetic fields. *Phys. Rev. A* **78**: 033434.

26 Yanovsky, V., Chvykov, V., Kalinchenko, G. et al. (2008). Ultra-high intensity- 300-TW laser at 0.1 Hz repetition rate. *Opt. Express* **16**: 2109.

27 Amann, H. (1990). *Ordinary Differential Equations: An Introduction to Nonlinear Analysis*. Walter de Gruyter.

28 Shirley, J.H. (1965). Solution of the schrödinger equation with a hamiltonian periodic in time. *Phys. Rev.* **138**: B979.

29 Ho, T.S., Chu, S.I., and Tietz, J.V. (1983). Semiclassical many-mode Floquet theory. *Chem. Phys. Lett.* **96**: 464.

30 Ho, T.S. and Chu, S.I. (1984). Semiclassical many-mode Floquet theory. II. Non-linear multiphoton dynamics of a 2-level system in a strong bichromatic field. *J. Phys. B: At. Mol. Opt. Phys.* **17**: 2101.

31 Bonin, K.D. and Kresin, V.V. (1997). *Electric-Dipole Polarizabilities of Atoms, Molecules and Clusters.* Singapore: World Scientific.
32 Happer, W. and Mathur, B.S. (1967). Effective operator formalism in optical pumping. *Phys. Rev.* **163**: 12.
33 Brieger, M. (1984). Stark effect, polarizabilities and the electric dipole moment of heteronuclear diatomic molecules in $^1\Sigma$ states. *Chem. Phys.* **89**: 275.
34 Bishop, D.M., Lam, B., and Epstein, S.T. (1988). The Stark effect and polarizabilities for a diatomic molecule. *J. Chem. Phys.* **88**: 337.
35 Bishop, D.M. (1990). Molecular vibrational and rotational motion in static and dynamic electric fields. *Rev. Mod. Phys.* **62**: 343.
36 Friedrich, B. and Herschbach, D. (1995). Alignment and trapping of molecules in intense laser fields. *Phys. Rev. Lett.* **74**: 4623.
37 DeMille, D., Glenn, D., and Petricka, J. (2004). Microwave traps for cold polar molecules. *Eur. Phys. J. D* **31**: 375.
38 Alyabyshev, S.V., Tscherbul, T.V., and Krems, R.V. (2009). Microwave-laser-field modification of molecular collisions at low temperatures. *Phys. Rev. A* **79**: 060703.
39 Grynberg, G., Aspect, A., and Fabre, C. (2010). *Introduction to Quantum Optics.* Cambridge: Cambridge University Press.
40 Weiner, J. and Ho, P.-H. (2003). *Light-Matter Interaction.* Hoboken, NJ: Wiley.
41 Jackson, J.D. (1998). *Classical Electrodynamics.* New York: Wiley.
42 Arfken, G. and Weber, H.J. (2005). *Mathematical Methods for Physicists.* USA: Elsevier Academic Press.
43 Meyenn, K. (1970). Rotation von zweiatomigen Dipolmolekülen in Starken elektrischen Feldern. *Z. Phys.* **231**: 154.
44 Bernstein, R.B., Herschbach, D., and Levine, R.D. (1987). Dynamical aspects of stereochemistry. *J. Phys. Chem.* **91**: 5365.
45 Brooks, P.R. (1976). Reactions of oriented molecules. *Science* **193**: 11.
46 Loesch, H.J. and Remscheid, A. (1990). Brute force in molecular reaction dynamics: a novel technique for measuring steric effects. *J. Chem. Phys.* **93**: 4779.
47 Friedrich, B. and Herschbach, D. (1991). Spatial orientation of molecules in strong electric fields and evidence for pendular states. *Nature* **353**: 412.
48 Friedrich, B. and Herschbach, D. (1991). On the possibility of orienting rotationally cooled polar molecules in an electric field. *Z. Phys. D* **18**: 153.
49 Rost, J., Griffin, J., Friedrich, B., and Herschbach, D. (1992). Pendular states and spectra of oriented linear molecules. *Phys. Rev. Lett.* **68**: 1299.
50 Friedrich, B., Rubahn, H., and Sathyamurthy, N. (1992). State-resolved scattering of molecules in pendular states: ICl + Ar. *Phys. Rev. Lett.* **69**: 2487.

51 Friedrich, B. and Herschbach, D.R. (1993). Thermodynamic functions of pendular molecules. *Collect. Czech. Chem. Commun.* **58**: 2458.
52 Friedrich, B., Herschbach, D., Rost, J., and Rubahn, H. (1993). Optical spectra of spatially oriented molecules: ICl in a strong electric field. *J. Chem. Soc., Faraday Trans.* **89**: 1539.
53 Friedrich, B., Slenczka, A., and Herschbach, D. (1994). Spectroscopy of pendular molecules in strong parallel electric and magnetic fields. *Can. J. Phys.* **72**: 897.
54 Slenczka, A., Friedrich, B., and Herschbach, D. (1994). Determination of the electric dipole moment of IC1(B $^3\Pi_0$) from pendular spectra. *Chem. Phys. Lett.* **224**: 238.
55 Friedrich, B. and Herschbach, D. (1996). Statistical mechanics of pendular molecules. *Int. Rev. Phys. Chem.* **15**: 325.
56 Aoiz, F.J., Friedrich, B., Herrero, V.J. et al. (1998). Effect of pendular orientation on the reactivity of H + DCl: a quasiclassical trajectory study. *Chem. Phys. Lett.* **289**: 132.
57 Friedrich, B. (2006). Net polarization of a molecular beam by strong electrostatic or radiative fields. *Eur. Phys. J. D* **38**: 209.
58 Coisson, R., Vernizzi, G., and Yang, X. (2009). Mathieu functions and numerical solutions of the Mathieu equation. 2009 IEEE International Workshop on Open-Source Software for Scientific Computation (OSSC), p. 3.
59 Friedrich, B. and Herschbach, D. (1992). On the possibility of aligning paramagnetic molecules or ions in a magnetic field. *Z. Phys. D* **24**: 25.
60 Slenczka, A. (1998). Electric linear dichroism with a simple interpretation in terms of molecular pendular states. *Phys. Rev. Lett.* **80**: 2566.
61 Kim, W. and Felker, P.M. (1996). Spectroscopy of pendular states in optical-field-aligned species. *J. Chem. Phys.* **104**: 1147.
62 Kim, W. and Felker, P.M. (1997). Ground-state intermolecular spectroscopy and pendular states in benzene–argon. *J. Chem. Phys.* **107**: 2193.
63 Kim, W. and Felker, P.M. (1998). Optical-field-induced pendular states and pendular band contours in symmetric tops. *J. Chem. Phys.* **108**: 6763.
64 Kumar, G.R., Gross, P., Safvan, C.P. et al. (1996). Pendular motion of linear in intense laser fields. *J. Phys. B* **29**: L95.
65 Kumar, G., Gross, P., Safvan, C. et al. (1996). Molecular pendular states in intense laser fields. *Phys. Rev. A* **53**: 3098.
66 Safvan, C.P., Vijayalakshmi, K., Rajgara, F.A. et al. (1999). Dissociation dynamics of in intense laser fields: directional specificity of and fragments. *J. Phys. B* **29**: L481.
67 Bhardwaj, V., Vijayalakshmi, K., and Mathur, D. (1997). Spatial alignment of gas-phase polyatomic molecules by an intense laser field. *Phys. Rev. A* **56**: 2455.

68 Bhardwaj, V.R., Safvan, C.P., Vijayalakshmi, K., and Mathur, D. (1999). On the spatial alignment of bent triatomic molecules by intense, picosecond laser fields. *J. Phys. B* **30**: 3821.
69 Sakai, H., Safvan, C., Larsen, J.J. et al. (1999). Controlling the alignment of neutral molecules by a strong laser field. *J. Chem. Phys.* **110**: 10235.
70 Larsen, J., Wendt-Larsen, I., and Stapelfeldt, H. (1999). Controlling the branching ratio of photodissociation using aligned molecules. *Phys. Rev. Lett.* **83**: 1123.
71 Larsen, J.J., Sakai, H., Safvan, C.P. et al. (1999). Aligning molecules with intense nonresonant laser fields. *J. Chem. Phys.* **111**: 7774.
72 Larsen, J.J., Hald, K., Bjerre, N. et al. (2000). Three dimensional alignment of molecules using elliptically polarized laser fields. *Phys. Rev. Lett.* **85**: 2470.
73 Sugita, A., Mashino, M., Kawasaki, M. et al. (2000). Control of photofragment velocity anisotropy by optical alignment of CH_3I. *J. Chem. Phys.* **112**: 2164.
74 Kumarappan, V., Bisgaard, C.Z., Viftrup, S.S. et al. (2006). Role of rotational temperature in adiabatic molecular alignment. *J. Chem. Phys.* **125**: 194309.
75 Pentlehner, D., Nielsen, J.H., Christiansen, L. et al. (2013). Laser-induced adiabatic alignment of molecules dissolved in helium nanodroplets. *Phys. Rev. A* **87**: 063401.
76 Madsen, C.B., Madsen, L.B., Viftrup, S.S. et al. (2009). Manipulating the torsion of molecules by strong laser pulses. *Phys. Rev. Lett.* **102**: 073007.
77 Madsen, C.B., Madsen, L.B., Viftrup, S. et al. (2009). A combined experimental and theoretical study on realizing and using laser controlled torsion of molecules. *J. Chem. Phys.* **130**: 234310.
78 Hansen, J.L., Nielsen, J.H., Madsen, C.B. et al. (2012). Control and femtosecond time-resolved imaging of torsion in a chiral molecule. *J. Chem. Phys.* **136**: 204310.
79 Karczmarek, J., Wright, J., Corkum, P., and Ivanov, M. (1999). Optical centrifuge for molecules. *Phys. Rev. Lett.* **82**: 3420.
80 Korobenko, A., Milner, A.A., and Milner, V. (2014). Direct observation, study, and control of molecular superrotors. *Phys. Rev. Lett.* **112**: 113004.
81 Korobenko, A., Milner, A.A., Hepburn, J.W., and Milner, V. (2014). Rotational spectroscopy with an optical centrifuge. *Phys. Chem. Chem. Phys.* **16**: 4071.
82 Korobenko, A., Hepburn, J.W., and Milner, V. (2015). Observation of nondispersing classical-like molecular rotation. *Phys. Chem. Chem. Phys.* **17**: 951.
83 Korobenko, A. and Milner, V. (2015). Dynamics of molecular superrotors in an external magnetic field. *J. Phys. B* **48**: 164004.

84 Korobenko, A. and Milner, V. (2016). Adiabatic field-free alignment of asymmetric top molecules with an optical centrifuge. *Phys. Rev. Lett.* **116**: 183001.
85 Milner, A.A., Korobenko, A., Floss, J. et al. (2015). Magneto-optical properties of paramagnetic superrotors. *Phys. Rev. Lett.* **115**: 033005.
86 Milner, A.A., Korobenko, A., Hepburn, J.W., and Milner, V. (2014). Effects of ultrafast molecular rotation on collisional decoherence. *Phys. Rev. Lett.* **113**: 043005.
87 Milner, A.A., Korobenko, A., and Milner, V. (2015). Sound emission from the gas of molecular superrotors. *Opt. Express* **23**: 8603.
88 Milner, A.A., Korobenko, A., and Milner, V. (2014). Coherent spinâ rotational dynamics of oxygen superrotors. *New J. Phys.* **16**: 093038.
89 Milner, A.A., Korobenko, A., and Milner, V. (2016). Field-free long-lived alignment of molecules with a two-dimensional optical centrifuge. *Phys. Rev. A* **93**: 053408.
90 Parker, D. and Bernstein, R.B. (1989). Oriented molecule beams via the electrostatic hexapole: preparation, characterization, and reactive scattering. *Annu. Rev. Phys. Chem.* **40**: 561.
91 Cho, V.A. and Bernstein, R.B. (1991). Tight focusing of beams of polar polyatomic molecules via the electrostatic hexapole lens. *J. Phys. Chem.* **95**: 8129.
92 Brooks, P.R. (1995). Orientation effects in electron transfer collisions. *Int. Rev. Phys. Chem.* **14**: 327.
93 Brooks, P.R., McKillop, J.S., and Pippin, H.G. (1979). Molecular beam reaction of K atoms with sideways oriented CF_3I. *Chem. Phys. Lett.* **66**: 144.
94 Brooks, P.R. and Jones, E.M. (1966). Reactive scattering of K atoms from oriented CH3I molecules. *J. Chem. Phys.* **45**: 3449.
95 Beuhler, R.J. Jr., Bernstein, R.B., and Kramer, K.H. (1966). Observation of the reactive asymmetry of methyl iodide. Crossed beam study of the reaction of rubidium with oriented methyl iodide molecules. *J. Am. Chem. Soc.* **88**: 5331.
96 Parker, D.H., Chakravorty, K.K., and Bernstein, R.B. (1982). Crossed beam reaction of oriented methyl iodide with rubidium: steric factor and reactive asymmetry versus scattering angle. *Chem. Phys. Lett.* **82**: 113.
97 Brooks, P.R. (1969). Molecular beam reaction of K with oriented CF_3I. Evidence for harpooning? *J. Chem. Phys.* **50**: 5031.
98 van Beek, M., ter Meulen, J., and Alexander, M. (2000). Rotationally inelastic collisions of OH ($X^2\Pi$) + Ar. II. The effect of molecular orientation. *J. Chem. Phys.* **113**: 637.
99 van Beek, M., Berden, G., Bethlem, H., and ter Meulen, J. (2001). Molecular reorientation in collisions of OH + Ar. *Phys. Rev. Lett.* **86**: 4001.
100 Van Leuken, J., Bulthuis, J., Stolte, S., and Loesch, H. (1995). KBr angular and velocity distributions from a crossed molecular beam study between K

and brute force oriented and nonoriented CH_3Br molecules. *J. Phys. Chem.* **99**: 13582.

101 Loesch, H. and Möller, J. (1993). Brute force in reactive scattering: steric effects in the reaction K + ICl → KI + Cl, KCl + I at E_{tr} = 3.03 eV. *J. Phys. Chem.* **97**: 2158.

102 Loesch, H.J. and Stienkemeier, F. (1993). Steric effects in the state specific reaction Li + HF($v = 1, j = 1, m = 0$) → LiF + H. *J. Chem. Phys.* **98**: 9570.

103 Loesch, H.J. and Moller, J. (1997). Reactive scattering from brute force oriented asymmetric top molecules: K + C_6H_5I → KI + C_6H_5. *J. Phys. Chem. A* **101**: 7534.

104 Loesch, H.J. (1995). Orientation and alignment in reactive beam collisions: recent progress. *Annu. Rev. Phys. Chem.* **46**: 555.

105 Parker, D.H., Jalink, H., and Stolte, S. (1987). Dynamics of molecular stereochemistry via oriented molecule scattering. *J. Phys. Chem.* **91**: 5427.

106 Franks, K.J., Li, H., and Kong, W. (1999). Orientation of pyrimidine in the gas phase using a strong electric field: spectroscopy and relaxation dynamics. *J. Chem. Phys.* **110**: 11779.

107 Kong, W., Pei, L., and Zhang, J. (2009). Linear dichroism spectroscopy of gas phase biological molecules embedded in superfluid helium droplets. *Int. Rev. Phys. Chem.* **28**: 33.

108 Gijsbertsen, A., Siu, W., Kling, M.F. et al. (2007). Direct determination of the sign of the NO dipole moment. *Phys. Rev. Lett.* **99**: 213003.

109 Paul, W. (1990). Electromagnetic traps for charged and neutral particles. *Rev. Mod. Phys.* **62**: 531.

110 Gerlach, W. and Stern, O. (1922). Das magnetische Moment des Silberatoms. *Z. Phys. A* **9**: 353.

111 Kallmann, H. and Reiche, F. (1921). Über den Durchgang bewegter Moleküle durch inhomogene Kraftfelder. *Z. Phys. A* **6**: 352.

112 Küpper, G.M.J., Filsinger, F., and Stapelfeldt, H. (2012). Manipulating the motion of complex molecules: deflection, focusing, and deceleration of molecular beams for quantum-state and conformer selection. In: *Methods in Physical Chemistry* (ed. R. Schafer and P.C. Schmidt), Wiley.

113 Wrede, E. (1927). Über die Ablenkung von Molekularstrahlen elektrischer Dipolmoleküle im inhomogenen elektrischen Feld. *Z. Phys. A* **44**: 261.

114 Rabi, I.I., Millman, S., Kusch, P., and Zacharias, J.R. (1939). The molecular beam resonance method for measuring nuclear magnetic moments. The magnetic moments of $_3Li^6$, $_3Li^7$ and $_9F^{19}$. *Phys. Rev.* **55**: 526.

115 Aquilanti, V., Ascenzi, D., Cappelletti, D., and Pirani, F. (1995). Magnetic analysis of nearly effusive and moderately supersonic beams of oxygen molecules. *Int. J. Mass Spectrom. Ion Proc.* **149–150**: 355.

116 Gordon, J., Zeiger, H., and Townes, C. (1954). Molecular microwave oscillator and new hyperfine structure in the microwave spectrum of NH_3. *Phys. Rev.* **95**: 282.

117 Gordon, J., Zeiger, H., and Townes, C. (1955). The maser—new type of microwave amplifier, frequency standard, and spectrometer. *Phys. Rev.* **99**: 1264.

118 Auerbach, D., Bromberg, E.E.A., and Wharton, L. (1966). Alternate-gradient focusing of molecular beams. *J. Chem. Phys.* **45**: 2160.

119 Rangwala, S., glen, T., Rieger, T. et al. (2003). Continuous source of translationally cold dipolar molecules. *Phys. Rev. A* **67**: 043406.

120 Junglen, T., Rieger, T., Rangwala, S. et al. (2004). Two-dimensional trapping of dipolar molecules in time-varying electric fields. *Phys. Rev. Lett.* **92**: 223001.

121 Rieger, T., glen, T., Rangwala, S. et al. (2005). Continuous loading of an electrostatic trap for polar molecules. *Phys. Rev. Lett.* **95**: 173002.

122 Tsuji, H., Sekiguchi, T., Mori, T. et al. (2010). Stark velocity filter for nonlinear polar molecules. *J. Phys. B* **43**: 095202.

123 Chervenkov, S., Wu, X., Bayerl, J. et al. (2014). Continuous centrifuge decelerator for polar molecules. *Phys. Rev. Lett.* **112**: 013001.

124 Bethlem, H., van Roij, A., Jongma, R., and Meijer, G. (2002). Alternate gradient focusing and deceleration of a molecular beam. *Phys. Rev. Lett.* **88**: 133003.

125 Tarbutt, M., Bethlem, H., Hudson, J. et al. (2004). Slowing heavy, ground-state molecules using an alternating gradient decelerator. *Phys. Rev. Lett.* **92**: 173002.

126 Bethlem, H.L., Tarbutt, M.R., Küpper, J. et al. (2006). Alternating gradient focusing and deceleration of polar molecules. *J. Phys. B* **39**: R263.

127 Wohlfart, K., Filsinger, F., Grätz, F. et al. (2008). Stark deceleration of OH radicals in low-field-seeking and high-field-seeking quantum states. *Phys. Rev. A* **78**: 033421.

128 Küpper, J., Filsinger, F., and Meijer, G. (2009). Manipulating the motion of large neutral molecules. *Faraday Discuss.* **142**: 155.

129 Kalnins, J., Lambertson, G., and Gould, H. (2002). Improved alternating gradient transport and focusing of neutral molecules. *Rev. Sci. Instrum.* **73**: 2557.

130 Wohlfart, K., Grätz, F., Filsinger, F. et al. (2008). Alternating-gradient focusing and deceleration of large molecules. *Phys. Rev. A* **77**: 031404.

131 Putzke, S., Filsinger, F., Küpper, J., and Meijer, G. (2012). Alternating-gradient focusing of the benzonitrile-argon van der Waals complex. *J. Chem. Phys.* **137**: 104310.

132 Putzke, S., Filsinger, F., Haak, H. et al. (2011). Rotational-state-specific guiding of large molecules. *Phys. Chem. Chem. Phys.* **13**: 18962.

133 Filsinger, F., Erlekam, U., von Helden, G. et al. (2008). Selector for structural isomers of neutral molecules. *Phys. Rev. Lett.* **100**: 133003.

134 Filsinger, F., Küpper, J., Meijer, G. et al. (2009). Pure samples of individual conformers: the separation of stereoisomers of complex molecules using electric fields. *Angew. Chem. Int. Ed.* **48**: 6900.

135 Filsinger, F., Putzke, S., Haak, H. et al. (2010). Optimizing the resolution of the alternating-gradient m/μ selector. *Phys. Rev. A* **82**: 052513.

136 Trippel, S., Chang, Y.-P., Stern, S. et al. (2012). Spatial separation of state- and size-selected neutral clusters. *Phys. Rev. A* **86**: 033202.

137 Stapelfeldt, H., Sakai, H., Constant, E., and Corkum, P. (1997). Deflection of neutral molecules using the nonresonant dipole force. *Phys. Rev. Lett.* **79**: 2787.

138 Sakai, H., Tarasevitch, A., Danilov, J. et al. (1998). Optical deflection of molecules. *Phys. Rev. A* **57**: 2794.

139 Zhao, B.S., Sung Chung, H., Cho, K. et al. (2000). Molecular lens of the nonresonant dipole force. *Phys. Rev. Lett.* **85**: 2705.

140 Zhao, B.S., Lee, S.H., Chung, H.S. et al. (2003). Separation of a benzene and nitric oxide mixture by a molecule prism. *J. Chem. Phys.* **119**: 8905.

141 Chung, H.S., Zhao, B.S., Lee, S.H. et al. (2001). Molecular lens applied to benzene and carbon disulfide molecular beams. *J. Chem. Phys.* **114**: 8293.

142 Averbukh, I.S., Vrakking, M.J.J., Villeneuve, D.M., and Stolow, A. (1996). Wave packet isotope separation. *Phys. Rev. Lett.* **77**: 3518.

143 Leibscher, M. and Averbukh, I.S. (2001). Optimal control of wave-packet isotope separation. *Phys. Rev. A* **63**: 043407.

144 Odashima, H., Merz, S., Enomoto, K. et al. (2010). Microwave lens for polar molecules. *Phys. Rev. Lett.* **104**: 253001.

145 Enomoto, K., Djuricanin, P., Gerhardt, I. et al. (2012). Superconducting microwave cavity towards controlling the motion of polar molecules. *Appl. Phys. B* **109**: 149.

146 Gershnabel, E. and Averbukh, I.S. (2011). Deflection of rotating symmetric top molecules by inhomogeneous fields. *J. Chem. Phys.* **135**: 084307.

147 Gershnabel, E. and Averbukh, I.S. (2010). Deflection of field-free aligned molecules. *Phys. Rev. Lett.* **104**: 153001.

148 Arndt, M., Ekers, A., von Klitzing, W., and Ulbricht, H. (2012). Focus on modern frontiers of matter wave optics and interferometry. *New J. Phys.* **14**: 125006.

149 Gershnabel, E. and Averbukh, I. (2010). Controlling molecular scattering by laser-induced field-free alignment. *Phys. Rev. A* **82**: 033401.

150 Floß, J., Gershnabel, E., and Averbukh, I. (2011). Motion of spinning molecules in inhomogeneous fields. *Phys. Rev. A* **83**: 025401.

151 Gershnabel, E. and Averbukh, I.S. (2011). Electric deflection of rotating molecules. *J. Chem. Phys.* **134**: 054304.

152 Purcell, S. and Barker, P. (2010). Controlling the optical dipole force for molecules with field-induced alignment. *Phys. Rev. A* **82**: 033433.

153 Krems, R.V., Stwalley, W.C., and Friedrich, B. ed. (2009). *Cold Molecules: Theory, Experiment and Applications*. Boca Raton, FL: Taylor and Francis.

154 Bethlem, H., Berden, G., and Meijer, G. (1999). Decelerating neutral dipolar molecules. *Phys. Rev. Lett.* **83**: 1558.

155 Maddi, J., Dinneen, T., and Gould, H. (1999). Slowing and cooling molecules and neutral atoms by time-varying electric-field gradients. *Phys. Rev. A* **60**: 3882.

156 Heiner, C.E., Bethlem, H.L., and Meijer, G. (2006). Molecular beams with a tunable velocity. *Phys. Chem. Chem. Phys.* **8**: 2666.

157 Friedrich, B. (2004). A quasi-analytic model of a linear Stark accelerator/decelerator for polar molecules. *Eur. Phys. J. D* **31**: 313.

158 Gubbels, K., Meijer, G., and Friedrich, B. (2006). Analytic wave model of Stark deceleration dynamics. *Phys. Rev. A* **73**: 063406.

159 Bethlem, H., Crompvoets, F., Jongma, R. et al. (2002). Deceleration and trapping of ammonia using time-varying electric fields. *Phys. Rev. A* **65**: 053416.

160 van de Meerakker, S., Vanhaecke, N., Bethlem, H., and Meijer, G. (2005). Higher-order resonances in a Stark decelerator. *Phys. Rev. A* **71**: 053409.

161 van de Meerakker, S., Vanhaecke, N., Bethlem, H., and Meijer, G. (2006). Transverse stability in a Stark decelerator. *Phys. Rev. A* **73**: 023401.

162 van de Meerakker, S.Y.T., Bethlem, H.L., and Meijer, G. (2008). Taming molecular beams. *Nat. Phys.* **4**: 595.

163 Bethlem, H.L., Berden, G., Crompvoets, F.M.H. et al. (2000). Electrostatic trapping of ammonia molecules. *Nature* **406**: 491.

164 Bochinski, J.R., Hudson, E.R., Lewandowski, H.J. et al. (2003). Phase space manipulation of cold free radical OH molecules. *Phys. Rev. Lett.* **91**: 243001.

165 van de Meerakker, S., Smeets, P., Vanhaecke, N. et al. (2005). Deceleration and electrostatic trapping of OH radicals. *Phys. Rev. Lett.* **94**: 023004.

166 Scharfenberg, L., Haak, H., Meijer, G., and van de Meerakker, S. (2009). Operation of a Stark decelerator with optimum acceptance. *Phys. Rev. A* **79**: 023410.

167 Hoekstra, S., Gilijamse, J., Sartakov, B. et al. (2007). Optical pumping of trapped neutral molecules by blackbody radiation. *Phys. Rev. Lett.* **98**: 133001.

168 van de Meerakker, S.Y.T., Labazan, I., Hoekstra, S. et al. (2006). Production and deceleration of a pulsed beam of metastable NH ($a^1\Delta$) radicals. *J. Phys. B* **39**: S1077.

169 Hudson, E., Ticknor, C., Sawyer, B. et al. (2006). Production of cold formaldehyde molecules for study and control of chemical reaction dynamics with hydroxyl radicals. *Phys. Rev. A* **73**: 063404.

170 Jung, S., Tiemann, E., and Lisdat, C. (2006). Cold atoms and molecules from fragmentation of decelerated SO_2. *Phys. Rev. A* **74**: 040701.
171 Tokunaga, S.K., Dyne, J.M., Hinds, E.A., and Tarbutt, M.R. (2009). Stark deceleration of lithium hydride molecules. *New J. Phys.* **11**: 055038.
172 Wall, T.E., Tokunaga, S.K., Hinds, E.A., and Tarbutt, M.R. (2010). Nonadiabatic transitions in a Stark decelerator. *Phys. Rev. A* **81**: 033414.
173 Deng, L.-Z., Fu, G.-B., and Yin, J.-P. (2009). Theoretical study of slowing supersonic CH_3F molecular beams using electrostatic Stark decelerator. *Chin. Phys. B* **18**: 149.
174 Fu, G.-B., Deng, L.-Z., and Yin, J.-P. (2008). A new desirable molecular species for Stark deceleration. *Chin. Phys. Lett.* **25**: 923.
175 Berg, J.E., Turkesteen, S.H., Prinsen, E.B., and Hoekstra, S. (2012). Deceleration and trapping of heavy diatomic molecules using a ring-decelerator. *Eur. Phys. J. D* **66**: 235.
176 Osterwalder, A., Meek, S.A., Hammer, G. et al. (2010). Deceleration of neutral molecules in macroscopic traveling traps. *Phys. Rev. A* **81**: 051401.
177 Bulleid, N., Hendricks, R., Hinds, E. et al. (2012). Traveling-wave deceleration of heavy polar molecules in low-field-seeking states. *Phys. Rev. A* **86**: 021404.
178 Marian, A., Haak, H., Geng, P., and Meijer, G. (2010). Slowing polar molecules using a wire Stark decelerator. *Eur. Phys. J. D* **59**: 179.
179 Meek, S.A., Bethlem, H., Conrad, H., and Meijer, G. (2008). Trapping molecules on a chip in traveling potential wells. *Phys. Rev. Lett.* **100**: 153003.
180 Meek, S.A., Conrad, H., and Meijer, G. (2009). Trapping molecules on a chip. *Science* **324**: 1699.
181 Meek, S.A., Conrad, H., and Meijer, G. (2009). A Stark decelerator on a chip. *New J. Phys.* **11**: 055024.
182 Meek, S.A., Santambrogio, G., Sartakov, B. et al. (2011). Suppression of nonadiabatic losses of molecules from chip-based microtraps. *Phys. Rev. A* **83**: 033413.
183 Schulz, S., Bethlem, H., van Veldhoven, J. et al. (2004). Microstructured switchable mirror for polar molecules. *Phys. Rev. Lett.* **93**: 020406.
184 Isabel González Flórez, A., Meek, S.A., Haak, H. et al. (2011). An electrostatic elliptical mirror for neutral polar molecules. *Phys. Chem. Chem. Phys.* **13**: 18830.
185 Yamakita, Y., Procter, S.R., Goodgame, A.L. et al. (2004). Deflection and deceleration of hydrogen Rydberg molecules in inhomogeneous electric fields. *J. Chem. Phys.* **121**: 1419.
186 Seiler, C., Hogan, S.D., and Merkt, F. (2011). Trapping cold molecular hydrogen. *Phys. Chem. Chem. Phys.* **13**: 19000.
187 Vanhaecke, N., Meier, U., Andrist, M. et al. (2007). Multistage Zeeman deceleration of hydrogen atoms. *Phys. Rev. A* **75**: 031402.

188 Hogan, S., Sprecher, D., Andrist, M. et al. (2007). Zeeman deceleration of H and D. *Phys. Rev. A* **76**: 023412.

189 Hogan, S.D., Wiederkehr, A.W., Andrist, M. et al. (2008). Slow beams of atomic hydrogen by multistage Zeeman deceleration. *J. Phys. B* **41**: 081005.

190 Narevicius, E., Parthey, C.G., Libson, A. et al. (2007). An atomic coilgun: using pulsed magnetic fields to slow a supersonic beam. *New J. Phys.* **9**: 358.

191 Narevicius, E., Libson, A., Parthey, C.G. et al. (2008). Stopping supersonic beams with a series of pulsed electromagnetic coils: an atomic coilgun. *Phys. Rev. Lett.* **100**: 093003.

192 Narevicius, E., Libson, A., Parthey, C. et al. (2008). Stopping supersonic oxygen with a series of pulsed electromagnetic coils: a molecular coilgun. *Phys. Rev. A* **77**: 051401.

193 Wiederkehr, A.W., Schmutz, H., Motsch, M., and Merkt, F. (2012). Velocity-tunable slow beams of cold O2 in a single spin-rovibronic state with full angular-momentum orientation by multistage Zeeman deceleration. *Mol. Phys.* **110**: 1807.

194 Wiederkehr, A., Hogan, S., and Merkt, F. (2010). Phase stability in a multistage Zeeman decelerator. *Phys. Rev. A* **82**: 043428.

195 Narevicius, E., Parthey, C.G., Libson, A. et al. (2007). Towards magnetic slowing of atoms and molecules. *New J. Phys.* **9**: 96.

196 Trimeche, A., Bera, M.N., Cromières, J.P. et al. (2011). Trapping of a supersonic beam in a traveling magnetic wave. *Eur. Phys. J. D* **65**: 263.

197 Lavert-Ofir, E., Gersten, S., Henson, A.B. et al. (2011). A moving magnetic trap decelerator: a new source of cold atoms and molecules. *New J. Phys.* **13**: 103030.

198 Lavert-Ofir, E., David, L., Henson, A.B. et al. (2011). Stopping paramagnetic supersonic beams: the advantage of a co-moving magnetic trap decelerator. *Phys. Chem. Chem. Phys.* **13**: 18948.

199 Friedrich, B. (2000). Slowing of supersonically cooled atoms and molecules by time-varying nonresonant induced dipole forces. *Phys. Rev. A* **61**: 025403.

200 Maher-McWilliams, C., Douglas, P., and Barker, P. (2012). Laser-driven acceleration of neutral particles. *Nat. Photonics* **6**: 386.

201 Fulton, R., Bishop, A.I., Shneider, M.N., and Barker, P.F. (2006). Controlling the motion of cold molecules with deep periodic optical potentials. *Nat. Phys.* **2**: 465.

202 Fulton, R., Bishop, A., and Barker, P. (2004). Optical stark decelerator for molecules. *Phys. Rev. Lett.* **93**: 243004.

203 Barker, P. and Shneider, M. (2002). Slowing molecules by optical microlinear deceleration. *Phys. Rev. A* **66**: 065402.

204 Enomoto, K. and Momose, T. (2005). Microwave Stark decelerator for polar molecules. *Phys. Rev. A* **72**: 061403.

205 Merz, S., Vanhaecke, N., Jäger, W. et al. (2012). Decelerating molecules with microwave fields. *Phys. Rev. A* **85**: 063411.

206 Ilinova, E., Ahmad, M., and Derevianko, A. (2011). Doppler cooling with coherent trains of laser pulses and a tunable velocity comb. *Phys. Rev. A* **84**: 033421.

207 Leibfried, D., Blatt, R., Monroe, C., and Wineland, D. (2003). Quantum dynamics of single trapped ions. *Rev. Mod. Phys.* **75**: 281.

208 Wing, W. (1980). Electrostatic trapping of neutral atomic particles. *Phys. Rev. Lett.* **45**: 631.

209 Peik, E. (1999). Electrodynamic trap for neutral atoms. *Eur. Phys. J. D* **6**: 179.

210 van Veldhoven, J., Bethlem, H., and Meijer, G. (2005). AC electric trap for ground-state molecules. *Phys. Rev. Lett.* **94**: 083001.

211 Jongma, R.T., von Helden, G., Berden, G., and Meijer, G. (1997). Confining CO molecules in stable orbits. *Chem. Phys. Lett.* **270**: 304.

212 Gilijamse, J.J., Hoekstra, S., Meek, S.A. et al. (2007). The radiative lifetime of metastable CO ($a^3\Pi, v = 0$). *J. Chem. Phys.* **127**: 221102.

213 Hoekstra, S., Metsälä, M., Zieger, P. et al. (2007). Electrostatic trapping of metastable NH molecules. *Phys. Rev. A* **76**: 063408.

214 Kleinert, J., Haimberger, C., Zabawa, P., and Bigelow, N. (2007). Trapping of ultracold polar molecules with a thin-wire electrostatic trap. *Phys. Rev. Lett.* **99**: 143002.

215 Sawyer, B.C., Lev, B.L., Hudson, E.R. et al. (2007). Magneto-electrostatic trapping of ground state OH molecules. *Phys. Rev. Lett.* **98**: 253002.

216 Schnell, M., Lützow, P., van Veldhoven, J. et al. (2007). A linear AC trap for polar molecules in their ground state. *J. Phys. Chem. A* **111**: 7411.

217 Lützow, P., Schnell, M., and Meijer, G. (2008). Instabilities of molecule motion in a linear AC trap. *Phys. Rev. A* **77**: 063402.

218 Bethlem, H., Veldhoven, J., Schnell, M., and Meijer, G. (2006). Trapping polar molecules in an AC trap. *Phys. Rev. A* **74**: 063403.

219 Katz, D.P. (1997). A storage ring for polar molecules. *J. Chem. Phys.* **107**: 8491.

220 Crompvoets, F.M.H., Bethlem, H.L., Jongma, R.T., and Meijer, G. (2001). A prototype storage ring for neutral molecules. *Nature* **411**: 174.

221 Heiner, C.E., Carty, D., Meijer, G., and Bethlem, H.L. (2007). A molecular synchrotron. *Nature* **3**: 115.

222 Heiner, C., Meijer, G., and Bethlem, H. (2008). Motional resonances in a molecular synchrotron. *Phys. Rev. A* **78**: 030702.

223 Zieger, P., van de Meerakker, S., Heiner, C. et al. (2010). Multiple packets of neutral molecules revolving for over a mile. *Phys. Rev. Lett.* **105**: 173001.

224 Krems, R.V. (2007). Molecular physics: set for collision course. *Nat. Phys.* **3**: 77.

225 Kügler, K.-J., Paul, W., and Trinks, U. (1978). A magnetic storage ring for neutrons. *Phys. Lett. B* **72**: 422.

226 Migdall, A.L., Prodan, J.V., Phillips, W.D. et al. (1985). First observation of magnetically trapped neutral atoms. *Phys. Rev. Lett.* **54**: 2596.

227 Weinstein, J.D., deCarvalho, R., Guillet, T. et al. (1998). Magnetic trapping of calcium monohydride molecules at millikelvin temperatures. *Nature* **395**: 148.

228 Prodan, J., Migdall, A., Phillips, W.D. et al. (1985). Stopping atoms with laser light. *Phys. Rev. Lett.* **54**: 992.

229 Shuman, E.S., Barry, J.F., and DeMille, D. (2010). Laser cooling of a diatomic molecule. *Nature* **467**: 820.

230 Phillips, D. (1998). Nobel lecture: laser cooling and trapping of neutral atoms. *Rev. Mod. Phys.* **70**: 721.

231 Phillips, W.D. (1997). Nobel prize lecture. http://www.nobelprize.org/nobel_prizes/physics/laureates/1997/phillips-lecture.pdf (accessed 24 November 2017).

232 Doyle, J., Friedrich, B., Kim, J., and Patterson, D. (1995). Buffer-gas loading of atoms and molecules into a magnetic trap. *Phys. Rev. A* **52**: R2515.

233 Friedrich, B., deCarvalho, R., Kim, J. et al. (1998). Towards magnetic trapping of molecules. *Phys. Chem. Chem. Phys.* **94**: 1783.

234 Weinstein, J.D. (2002) Magnetic trapping of atomic chromium and molecular calcium monohydride. PhD thesis. Harvard University.

235 Weinstein, J.D., deCarvalho, R., Amar, K. et al. (1998). Spectroscopy of buffer-gas cooled vanadium monoxide in a magnetic trapping field. *J. Chem. Phys.* **109**: 2656.

236 Posthumus, J.H., Plumridge, J., Frasinski, L.J. et al. (1999). Double-pulse measurements of laser-induced alignment of molecules. *J. Phys. B* **31**: L985.

237 Stoll, M., Bakker, J., Steimle, T. et al. (2008). Cryogenic buffer-gas loading and magnetic trapping of CrH and MnH molecules. *Phys. Rev. A* **78**: 032707.

238 Campbell, W., Tsikata, E., Lu, H.-I. et al. (2007). Magnetic trapping and Zeeman relaxation of NH (X $^3\Sigma^-$). *Phys. Rev. Lett.* **98**: 213001.

239 Tsikata, E., Campbell, W.C., Hummon, M.T. et al. (2010). Magnetic trapping of NH molecules with 20 s lifetimes. *New J. Phys.* **12**: 065028.

240 Vanhaecke, N., de Souza Melo, W., Laburthe Tolra, B. et al. (2002). Accumulation of cold cesium molecules via photoassociation in a mixed atomic and molecular trap. *Phys. Rev. Lett.* **89**: 063001.

241 Hogan, S., Wiederkehr, A., Schmutz, H., and Merkt, F. (2008). Magnetic trapping of hydrogen after multistage Zeeman deceleration. *Phys. Rev. Lett.* **101**: 143001.

242 Riedel, J., Hoekstra, S., Jäger, W. et al. (2011). Accumulation of Stark-decelerated NH molecules in a magnetic trap. *Eur. Phys. J. D* **65**: 161.

243 Takekoshi, T., Patterson, B., and Knize, R. (1998). Observation of optically trapped cold cesium molecules. *Phys. Rev. Lett.* **81**: 5105.

244 Jochim, S., Bartenstein, M., Altmeyer, A. et al. (2003). Pure gas of optically trapped molecules created from fermionic atoms. *Phys. Rev. Lett.* **91**: 240402.

245 Herbig, J. (2003). Preparation of a pure molecular quantum gas. *Science* **301**: 1510.

246 Jochim, S., Bartenstein, M., Altmeyer, A. et al. (2003). Bose-Einstein condensation of molecules. *Science* **302**: 2101.

247 Winkler, K., Thalhammer, G., Theis, M. et al. (2005). Atom-molecule dark states in a Bose-Einstein condensate. *Phys. Rev. Lett.* **95**: 063202.

248 Ni, K.-K., Ospelkaus, S., de Miranda, M. et al. (2008). A high phase-space-density gas of polar molecules. *Science* **322**: 231.

249 Debatin, M., Takekoshi, T., Rameshan, R. et al. (2011). Molecular spectroscopy for ground-state transfer of ultracold RbCs molecules. *Phys. Chem. Chem. Phys.* **13**: 18926.

250 Stellmer, S., Pasquiou, B., Grimm, R., and Schreck, F. (2012). Creation of ultracold Sr_2 molecules in the electronic ground state. *Phys. Rev. Lett.* **109**: 115302.

251 DeMille, D., Glenn, D.R., and Petricka, J. (2004). Microwave traps for cold polar molecules. *Eur. Phys. J. D.* **31**: 375.

252 Grynberg, G. and Robilliard, C. (2001). Cold atoms in dissipative optical lattices. *Phys. Rep.* **355**: 335.

253 Letokhov, V., Minogin, V., and Pavlik, B. (1976). Cooling and trapping of atoms and molecules by a resonant laser field. *Opt. Commun.* **19**: 72.

254 Kirste, M., Sartakov, B., Schnell, M., and Meijer, G. (2009). Nonadiabatic transitions in electrostatically trapped ammonia molecules. *Phys. Rev. A* **79**: 051401.

255 Englert, B., Mielenz, M., Sommer, C. et al. (2011). Storage and adiabatic cooling of polar molecules in a microstructured trap. *Phys. Rev. Lett.* **107**: 263003.

256 Zeppenfeld, M., Englert, B.G.U., Glöckner, R. et al. (2012). Sisyphus cooling of electrically trapped polyatomic molecules. *Nature* **491**: 570.

257 Sawyer, B.C., Stuhl, B.K., Yeo, M. et al. (2011). Cold heteromolecular dipolar collisions. *Phys. Chem. Chem. Phys.* **13**: 19059.

258 Stuhl, B.K., Yeo, M., Hummon, M.T., and Ye, J. (2013). Electric-field-induced inelastic collisions between magnetically trapped hydroxyl radicals. *Mol. Phys.* **111**: 1798.

259 Wang, D., Qi, J., Stone, M. et al. (2004). Photoassociative production and trapping of ultracold KRb molecules. *Phys. Rev. Lett.* **93**: 243005.

260 Zirbel, J.J., Ni, K.-K., Ospelkaus, S. et al. (2008). Collisional stability of fermionic Feshbach molecules. *Phys. Rev. Lett.* **100**: 143201.

261 Ospelkaus, S., Ni, K.-K., Quéméner, G. et al. (2010). Controlling the hyperfine state of rovibronic ground-state polar molecules. *Phys. Rev. Lett.* **104**: 030402.

262 Ni, K.-K., Ospelkaus, S., Wang, D. et al. (2010). Dipolar collisions of polar molecules in the quantum regime. *Nature* **464**: 1324.

263 Ospelkaus, S., Ni, K.-K., Wang, D. et al. (2010). Quantum-state controlled chemical reactions of ultracold potassium-rubidium molecules. *Science* **327**: 853.

264 Chotia, A., Neyenhuis, B., Moses, S.A. et al. (2012). Long-lived dipolar molecules and Feshbach molecules in a 3D optical lattice. *Phys. Rev. Lett.* **108**: 080405.

265 de Miranda, M.H.G., Chotia, A., Neyenhuis, B. et al. (2011). Controlling the quantum stereodynamics of ultracold bimolecular reactions. *Nat. Phys.* **7**: 502.

266 Danzl, J.G., Mark, M.J., Haller, E. et al. (2010). An ultracold high-density sample of rovibronic ground-state molecules in an optical lattice. *Nature* **6**: 265.

267 Doyle, J., Friedrich, B., Krems, R.V., and Masnou-Seeuws, F. (2004). Editorial: Quo vadis, cold molecules? *Eur. Phys. J. D* **31**: 149.

268 Carr, L.D., Demille, D., Krems, R.V., and Ye, J. (2009). Cold and ultracold molecules: science, technology and applications. *New J. Phys.* **11**: 5049.

269 Krems, R.V. (2008). Cold controlled chemistry. *Phys. Chem. Chem. Phys.* **10**: 4079.

270 Dong, F. and Miller, R.E. (2002). Vibrational transition moment angles in isolated biomolecules: a structural tool. *Science* **298**: 1227.

271 Kanya, R. and Ohshima, Y. (2003). Determination of dipole moment change on the electronic excitation of isolated Coumarin 153 by pendular-state spectroscopy. *Chem. Phys. Lett.* **370**: 211.

272 Kanya, R. and Ohshima, Y. (2004). Pendular-state spectroscopy of the S_1–S_0 electronic transition of 9-cyanoanthracene. *J. Chem. Phys.* **121**: 9489.

273 Schnell, M. and Küpper, J. (2011). Tailored molecular samples for precision spectroscopy experiments. *Faraday Discuss.* **150**: 33.

274 Kirste, M., Wang, X., Meijer, G. et al. (2012). Communication: magnetic dipole transitions in the OH A $^2\Sigma^+ \leftarrow$ X $^2\Pi$ system. *J. Chem. Phys.* **137**: 101102.

275 Abel, M.J., Marx, S., Meijer, G., and Santambrogio, G. (2012). Vibrationally exciting molecules trapped on a microchip. *Mol. Phys.* **110**: 1829.

276 Schulz, S., Bethlem, H., van Veldhoven, J. et al. (2011). Driving rotational transitions in molecules on a chip. *ChemPhysChem* **12**: 1799.

277 Veldhoven, J., K pper, J., Bethlem, H.L. et al. (2004). Decelerated molecular beams for high-resolution spectroscopy. *Eur. Phys. J. D* **31**: 337.

278 Hudson, E., Lewandowski, H., Sawyer, B., and Ye, J. (2006). Cold molecule spectroscopy for constraining the evolution of the fine structure constant. *Phys. Rev. Lett.* **96**: 143004.

279 Bethlem, H.L., Kajita, M., Sartakov, B. et al. (2008). Prospects for precision measurements on ammonia molecules in a fountain. *Eur. Phys. J. Spec. Top.* **163**: 55.

280 van de Meerakker, S., Vanhaecke, N., van der Loo, M. et al. (2005). Direct measurement of the radiative lifetime of vibrationally excited OH radicals. *Phys. Rev. Lett.* **95**: 013003.

281 Dzuba, V.A. and Flambaum, V.V. (2012). Parity violation and electric dipole moments in atoms and molecules. *Int. J. Mod. Phys. E* **21**: 1230010.

282 Hudson, J.J., Kara, D.M., Smallman, I.J. et al. (2011). Improved measurement of the shape of the electron. *Nature* **473**: 493.

283 Vutha, A.C., Campbell, W.C., Gurevich, Y.V. et al. (2010). Search for the electric dipole moment of the electron with thorium monoxide. *J. Phys. B* **43**: 074007.

284 Baron, J., Campbell, W.C., DeMille, D. et al. (2014). Order of magnitude smaller limit on the electric dipole moment of the electron. *Science* **343**: 269.

285 Demille, D., Sainis, S., Sage, J. et al. (2008). Enhanced sensitivity to variation of me/mp in molecular spectra. *Phys. Rev. Lett.* **100**: 043202.

286 Zelevinsky, T., Kotochigova, S., and Ye, J. (2008). Precision test of mass-ratio variations with lattice-confined ultracold molecules. *Phys. Rev. Lett.* **100**: 043201.

287 Bethlem, H.L. and Ubachs, W. (2009). Testing the time-invariance of fundamental constants using microwave spectroscopy on cold diatomic radicals. *Faraday Discuss.* **142**: 25.

288 de Nijs, A., Ubachs, W., and Bethlem, H. (2012). Sensitivity of rotational transitions in CH and CD to a possible variation of fundamental constants. *Phys. Rev. A* **86**: 032501.

289 Rios, J.P., Herrera, F., and Krems, R.V. (2010). External field control of collective spin excitations in an optical lattice of $^2\Sigma$ molecules. *New J. Phys.* **12**: 103007.

290 Cahn, S.B., Ammon, J., Kirilov, E. et al. (2014). Zeeman-tuned rotational level-crossing spectroscopy in a diatomic free radical. *Phys. Rev. Lett.* **112**: 163002.

291 Friedrich, B. and Herschbach, D. (2000). Steric proficiency of polar $^2\Sigma$ molecules in congruent electric and magnetic fields. *Phys. Chem. Chem. Phys.* **2**: 419.

292 Alyabyshev, S.V., Lemeshko, M., and Krems, R.V. (2012). Sensitive imaging of electromagnetic fields with paramagnetic polar molecules. *Phys. Rev. A* **86**: 013409.

293 Tscherbul, T.V. and Krems, R.V. (2006). Controlling electronic spin relaxation of cold molecules with electric fields. *Phys. Rev. Lett.* **97**: 083201.

294 Böhi, P., Riedel, M.F., Hänsch, T.W., and Treutlein, P. (2010). Imaging of microwave fields using ultracold atoms. *Appl. Phys. Lett.* **97**: 051101.

295 Cai, L., Marango, J., and Friedrich, B. (2001). Time-dependent alignment and orientation of molecules in combined electrostatic and pulsed nonresonant laser fields. *Phys. Rev. Lett.* **86**: 775.

296 Micheli, A., Brennen, G., and Zoller, P. (2006). A toolbox for lattice-spin models with polar molecules. *Nat. Phys.* **2**: 341.

297 Micheli, A., Pupillo, G., Büchler, H.P., and Zoller, P. (2007). Cold polar molecules in two-dimensional traps: tailoring interactions with external fields for novel quantum phases. *Phys. Rev. A* **76**: 43604.

298 Büchler, H., Micheli, A., and Zoller, P. (2007). Three-body interactions with cold polar molecules. *Nat. Phys.* **3**: 726.

299 Büchler, H.P., Demler, E., Lukin, M. et al. (2007). Strongly correlated 2D quantum phases with cold polar molecules: controlling the shape of the interaction potential. *Phys. Rev. Lett.* **98**: 060404.

300 Micheli, A., Idziaszek, Z., Pupillo, G. et al. (2010). Universal rates for reactive ultracold polar molecules in reduced dimensions. *Phys. Rev. Lett.* **105**: 073202.

301 Gorshkov, A.V., Rabl, P., Pupillo, G. et al. (2008). Suppression of inelastic collisions between polar molecules with a repulsive shield. *Phys. Rev. Lett.* **101**: 73201.

302 Gorshkov, A.V., Manmana, S.R., Chen, G. et al. (2011). Quantum magnetism with polar alkali-metal dimers. *Phys. Rev. A* **84**: 033619.

303 Gorshkov, A.V., Manmana, S.R., Chen, G. et al. (2011). Tunable superfluidity and quantum magnetism with ultracold polar molecules. *Phys. Rev. Lett.* **107**: 115301.

304 Kuns, K.A., Rey, A.M., and Gorshkov, A.V. (2011). d-wave superfluidity in optical lattices of ultracold polar molecules. *Phys. Rev. A* **84**: 063639.

305 Manmana, S.R., Stoudenmire, E.M., Hazzard, K.R.A. et al. (2013). Topological phases in ultracold polar-molecule quantum magnets. *Phys. Rev. B* **87**: 081106.

306 Krems, R.V. (2006). Controlling collisions of ultracold atoms with DC electric fields. *Phys. Rev. Lett.* **96**: 123202.

307 Herrera, F. (2008). Magnetic-field-induced interference of scattering states in ultracold collisions. *Phys. Rev. A* **78**: 054702.

308 Messiah, A. (1999). *Quantum Mechanics*. Dover Publications, Inc.

309 Huo, W.M. and Green, S. (1996) Quantum calculations for rotational energy transfer in nitrogen molecule collisions. *J. Chem. Phys.* **104**: 7572.

310 Mott, N.F. and Massey, H.S.W. (1965). *The Theory of Atomic Collisions*. Oxford Press.

311 Johnson, B. (1973). The multichannel log-derivative method for scattering calculations. *J. Comput. Phys.* **13**: 445.
312 Manolopoulos, D.E. (1986). An improved log derivative method for inelastic scattering. *J. Chem. Phys.* **85**: 6425.
313 Krems, R.V. and Dalgarno, A. (2004). Quantum-mechanical theory of atom-molecule and molecular collisions in a magnetic field: spin depolarization. *J. Chem. Phys.* **120**: 2296.
314 Arthurs, A. and Dalgarno, A. (1960). The theory of scattering by a rigid rotator. *Proc. R. Soc. London, Ser. A* **256**: 540.
315 Tscherbul, T.V. and Dalgarno, A. (2010). Quantum theory of molecular collisions in a magnetic field: efficient calculations based on the total angular momentum representation. *J. Chem. Phys.* **133**: 184104.
316 Suleimanov, Y.V., Tscherbul, T.V., and Krems, R.V. (2012). Efficient method for quantum calculations of molecule-molecule scattering properties in a magnetic field. *J. Chem. Phys.* **137**: 024103.
317 Wigner, E.P. (1948). On the behavior of cross sections near thresholds. *Phys. Rev.* **73**: 1002.
318 Bethe, H.A. and Placzek, G. (1937). Resonance effects in nuclear processes. *Phys. Rev.* **51**: 450.
319 Balakrishnan, N., Forrey, R.C., and Dalgarno, A. (1998). Quenching of H_2 vibrations in ultracold ^{3}He and ^{4}He collisions. *Phys. Rev. Lett.* **80**: 3224.
320 Alexander, M.H. (1982). Rotationally inelastic collisions between a diatomic molecule in a $^2\Pi$ electronic state and a structureless target. *J. Chem. Phys.* **76**: 5974.
321 Krems, R.V., Groenenboom, G.C., and Dalgarno, A. (2004). Electronic interaction anisotropy between atoms in arbitrary angular momentum states. *J. Phys. Chem. A* **108**: 8941.
322 Feshbach, H. (1958). Unified theory of nuclear reactions. *Ann. Phys.* **5**: 357.
323 Feshbach, H. (1967). The unified theory of nuclear reactions. *Ann. Phys.* **43**: 410.
324 Newton, R.G. (1988). *Scattering Theory of Waves and Particles*. New York: Springer.
325 Hemming, C.J. and Krems, R.V. (2008). Multiple-state feshbach resonances mediated by high-order couplings. *Phys. Rev. A* **77**: 022705.
326 Ashton, C.J., Child, M.S., and Hutson, J.M. (1983). Rotational predissociation of the Ar–HCl van der Waals complex: close-coupled scattering calculations. *J. Chem. Phys.* **78**: 4025.
327 Hazi, A.U. (1979). Behavior of the eigenphase sum near a resonance. *Phys. Rev. A* **19**: 920.
328 Hutson, J.M. (2007). Feshbach resonances in ultracold atomic and molecular collisions: threshold behaviour and suppression of poles in scattering lengths. *New J. Phys.* **9**: 152.

329 Gordon, R.G. (1969). New method for constructing wavefunctions for bound states and scattering. *J. Chem. Phys.* **51**: 14.

330 Johnson, B.R. (1977). New numerical methods applied to solving the one-dimensional eigenvalue problem. *J. Chem. Phys.* **67**: 4086.

331 Johnson, B.R. (1978). The renormalized Numerov method applied to calculating bound states of the coupled channel Schroedinger equation. *J. Chem. Phys.* **69**: 4678.

332 Hutson, J.M. (1994). Coupled channel methods for solving the bound-state Schrödinger equation. *Comput. Phys. Commun.* **84**: 1.

333 González-Martínez, M.L. and Hutson, J.M. (2011). Effect of hyperfine interactions on ultracold molecular collisions: NH($^3\Sigma^-$) with Mg(^{1}S) in magnetic fields. *Phys. Rev. A* **84**: 052706.

334 Suleimanov, Y.V. and Krems, R.V. (2011). Efficient numerical method for locating Feshbach resonances of ultracold molecules in external fields. *J. Chem. Phys.* **134**: 014101.

335 Juanes-Marcos, J.C., Althorpe, S.C., and Wrede, E. (2005). Theoretical study of geometric phase effects in the hydrogen-exchange reaction. *Science* **309**: 1227.

336 Kim, J.B., Weichman, M.L., Sjolander, T.F. et al. (2015). Spectroscopic observation of resonances in the F + H_2 reaction. *Science* **349**: 510.

337 Volpi, A. and Bohn, J.L. (2002). Magnetic-field effects in ultracold molecular collisions. *Phys. Rev. A* **65**: 052712.

338 Avdeenkov, A.V. and Bohn, J.L. (2002). Collisional dynamics of ultracold OH molecules in an electrostatic field. *Phys. Rev. A* **66**: 52718.

339 Avdeenkov, A.V., Bortolotti, D.C., and Bohn, J.L. (2004). Field-linked states of ultracold polar molecules. *Phys. Rev. A* **69**: 12710.

340 Cybulski, H., Krems, R., Sadeghpour, H. et al. (2005). Interaction of NH(X $^3\Sigma^-$) with He: potential energy surface, bound states, and collisional Zeeman relaxation. *J. Chem. Phys.* **122**: 094307.

341 Tscherbul, T.V. and Krems, R.V. (2006). Manipulating spin-dependent interactions in rotationally excited cold molecules with electric fields. *J. Chem. Phys.* **125**: 194311.

342 González-Martínez, M.L. and Hutson, J.M. (2007). Ultracold atom-molecule collisions and bound states in magnetic fields: tuning zero-energy Feshbach resonances in He-NH($^3\Sigma^-$). *Phys. Rev. A* **75**: 022702.

343 Abrahamsson, E., Tscherbul, T.V., and Krems, R.V. (2007). Inelastic collisions of cold polar molecules in nonparallel electric and magnetic fields. *J. Chem. Phys.* **127**: 044302.

344 Tscherbul, T.V., Klos, J., Rajchel, L., and Krems, R.V. (2007). Fine and hyperfine interactions in cold YbF - He collisions in electromagnetic fields. *Phys. Rev. A* **75**: 033416.

345 Campbell, W.C., Tscherbul, T.V., Lu, H.-I. et al. (2009). Mechanism of collisional spin relaxation in $^3\Sigma$ molecules. *Phys. Rev. Lett.* **102**: 013003.

346 Tscherbul, T.V., Groenenboom, G.C., Krems, R.V., and Dalgarno, A. (2009). Dynamics of OH($^2\Pi$)–He collisions in combined electric and magnetic fields. *Faraday Discuss.* **142**: 127.

347 Tscherbul, T.V., Suleimanov, Y.V., Aquilanti, V., and Krems, R.V. (2009). Magnetic field modification of ultracold molecule–molecule collisions. *New J. Phys.* **11**: 055021.

348 Hutson, J.M., Beyene, M., and Leonardo Gonzalez-Martinez, M. (2009). Dramatic reductions in inelastic cross sections for ultracold collisions near feshbach resonances. *Phys. Rev. Lett.* **103**: 163201.

349 Żuchowski, P.S. and Hutson, J.M. (2009). Low-energy collisions of NH_3 and ND_3 with ultracold Rb atoms. *Phys. Rev. A* **79**: 062708.

350 Meyer, E.R. and Bohn, J.L. (2010). Product-state control of bi-alkali-metal chemical reactions. *Phys. Rev. A* **82**: 042707.

351 Janssen, L.M.C., Zuchowski, P.S., van der Avoird, A. et al. (2011). Cold and ultracold NH-NH collisions in magnetic fields. *Phys. Rev. A* **83**: 022713.

352 Wallis, A.O.G., Longdon, E.J.J., Zuchowski, P.S., and Hutson, J.M. (2011). The prospects of sympathetic cooling of NH molecules with Li atoms. *Eur. Phys. J. D* **65**: 151.

353 Tscherbul, T.V. and Krems, R.V. (2015). Tuning bimolecular chemical reactions by electric fields. *Phys. Rev. Lett.* **115**: 023201.

354 Krems, R.V. (2005). Molecules near absolute zero and external field control of atomic and molecular dynamics. *Int. Rev. Phys. Chem.* **24**: 99.

355 Tscherbul, T.V. (2008). Differential scattering of cold molecules in superimposed electric and magnetic fields. *J. Chem. Phys.* **128**: 244305.

356 Olshanii, M. (1998). Atomic scattering in the presence of an external confinement and a gas of impenetrable bosons. *Phys. Rev. Lett.* **81**: 938.

357 Petrov, D.S. and Shlyapnikov, G.V. (2001). Interatomic collisions in a tightly confined Bose gas. *Phys. Rev. A* **64**: 012706.

358 Sadeghpour, H., Bohn, J., Cavagnero, M., and Esry, B. (2000). Collisions near threshold in atomic and molecular physics. *J. Phys. B* **33**: R93.

359 Naidon, P., Tiesinga, E., Mitchell, W.F., and Julienne, P.S. (2007). Effective-range description of a Bose gas under strong one- or two-dimensional confinement. *New J. Phys.* **9**: 19.

360 Li, Z., Alyabyshev, S.V., and Krems, R.V. (2008). Ultracold inelastic collisions in two dimensions. *Phys. Rev. Lett.* **100**: 073202.

361 Li, Z. and Krems, R.V. (2009). Inelastic collisions in an ultracold quasi-two-dimensional gas. *Phys. Rev. A* **79**: 050701.

362 Krems, R.V. and Dalgarno, A. (2002). Shape resonances and nonadiabatic dynamics in O(3Pj)+He collisions at cold and ultracold temperatures. *Phys. Rev. A* **66**: 012702.

363 Schwenke, D.W. and Truhlar, D.G. (1985). The effect of wigner singularities on low-temperature vibrational relaxation rates. *J. Chem. Phys.* **83**: 3454.

364 Forrey, R.C., Kharchenko, V., Balakrishnan, N., and Dalgarno, A. (1999). Vibrational relaxation of trapped molecules. *Phys. Rev. A* **59**: 2146.
365 Balakrishnan, N., Dalgarno, A., and Forrey, R.C. (2000). Vibrational relaxation of CO by collisions with ^{4}He at ultracold temperatures. *J. Chem. Phys.* **113**: 621.
366 Forrey, R.C. (2001). Cooling and trapping of molecules in highly excited rotational states. *Phys. Rev. A* **63**: 051403.
367 Zhu, C., Balakrishnan, N., and Dalgarno, A. (2001). Vibrational relaxation of CO in ultracold ^{3}He collisions. *J. Chem. Phys.* **115**: 1335.
368 Forrey, R.C. (2002). Prospects for cooling and trapping rotationally hot molecules. *Phys. Rev. A* **66**: 023411.
369 Flasher, J.C. and Forrey, R.C. (2002). Cold collisions between argon atoms and hydrogen molecules. *Phys. Rev. A* **65**: 032710.
370 Balakrishnan, N., Groenenboom, G.C., Krems, R.V., and Dalgarno, A. (2003). The He-CaH($^2\Sigma^+$) interaction. II. Collisions at cold and ultracold temperatures. *J. Chem. Phys.* **118**: 7386.
371 Volpi, A. and Bohn, J.L. (2003). Fine-structure effects in vibrational relaxation at ultralow temperatures. *J. Chem. Phys.* **119**: 866.
372 Florian, P.M., Hoster, M., and Forrey, R.C. (2004). Rotational relaxation in ultracold CO + He collisions. *Phys. Rev. A* **70**: 032709.
373 Balakrishnan, N. (2004). On the role of van der waals interaction in chemical reactions at low temperatures. *J. Chem. Phys.* **121**: 5563.
374 Uudus, N., Magaki, S., and Balakrishnan, N. (2005). Quantum mechanical investigation of rovibrational relaxation of H_2 and D_2 by collisions with ar atoms. *J. Chem. Phys.* **122**: 024304.
375 Lara, M., Bohn, J.L., Potter, D. et al. (2006). Ultracold Rb-OH collisions and prospects for sympathetic cooling. *Phys. Rev. Lett.* **97**: 183201.
376 Lara, M., Bohn, J.L., Potter, D.E. et al. (2007). Cold collisions between OH and Rb: the field-free case. *Phys. Rev. A* **75**: 012704.
377 al-Qady, W.H., Forrey, R.C., Yang, B.H. et al. (2011). Cold collisions of highly rotationally excited CO_2 with He: the prospects for cold chemistry with super-rotors. *Phys. Rev. A* **84**: 054701.
378 Balakrishnan, N. (2016). Perspective: ultracold molecules and the dawn of cold controlled chemistry. *J. Chem. Phys.* **145**: 150901.
379 Balakrishnan, N. and Dalgarno, A. (2001). Chemistry at ultracold temperatures. *Chem. Phys. Lett.* **341**: 652.
380 Bodo, E., Gianturco, F.A., and Dalgarno, A. (2002). F + D_2 reaction at ultracold temperatures. *J. Chem. Phys.* **116**: 9222.
381 Soldán, P., Cvitaš, M.T., Hutson, J.M. et al. (2002). Quantum dynamics of ultracold Na + Na_2 collisions. *Phys. Rev. Lett.* **89**: 153201.
382 Bodo, E., Gianturco, F.A., Balakrishnan, N., and Dalgarno, A. (2004). Chemical reactions in the limit of zero kinetic energy: virtual states and Ramsauer minima in F + H_2 → HF + H. *J. Phys. B* **37**: 3641.

383 Weck, P.F. and Balakrishnan, N. (2004). Chemical reactivity of ultracold polar molecules: investigation of H + HCl and H + DCl collisions. *Eur. Phys. J. D* **31**: 417.

384 Weck, P.F. and Balakrishnan, N. (2005). Quantum dynamics of the Li + HF → H + LiF reaction at ultralow temperatures. *J. Chem. Phys.* **122**: 154309.

385 Cvitaš, M.T., Soldán, P., Hutson, J.M. et al. (2005). Ultracold Li + Li_2 collisions: bosonic and fermionic cases. *Phys. Rev. Lett.* **94**: 033201.

386 Cvitaš, M.T., Soldán, P., Hutson, J.M. et al. (2005). Ultracold collisions involving heteronuclear alkali metal dimers. *Phys. Rev. Lett.* **94**: 200402.

387 Krems, R., Dalgarno, A., Balakrishnan, N., and Groenenboom, G.C. (2003). Spin-flipping transitions in $^2\Sigma$ molecules induced by collisions with structureless atoms. *Phys. Rev. A* **67**: 060703(R).

388 Krems, R., Sadeghpour, H.R., Dalgarno, A. et al. (2003). Low-temperature collisions of NH (X $^3\Sigma^-$) molecules with He atoms in a magnetic field: An ab initio study. *Phys. Rev. A* **68**: 051401(R).

389 Soldán, P. and Hutson, J.M. (2004). Interaction of NH($X\,^3\Sigma^-$) molecules with rubidium atoms: implications for sympathetic cooling and the formation of extremely polar molecules. *Phys. Rev. Lett.* **92**: 163202.

390 Wallis, A.O.G. and Hutson, J.M. (2009). Production of ultracold NH molecules by sympathetic cooling with Mg. *Phys. Rev. Lett.* **103**: 183201.

391 Zuchowski, P.S. and Hutson, J.M. (2011). Cold collisions of N(^4S) atoms and NH($^3\Sigma$) molecules in magnetic fields. *Phys. Chem. Chem. Phys.* **13**: 3669.

392 Tscherbul, T.V. (2012). Total-angular-momentum representation for atom-molecule collisions in electric fields. *Phys. Rev. A* **85**: 052710.

393 Jin, D.S. and Ye, J. (2011). Polar molecules in the quantum regime. *Phys. Today* **64**: 27.

394 Loesch, H.J. and Remsheid, A. (1990). Brute force in molecular reaction dynamics - a novel technique for measuring steric effects. *J. Chem. Phys.* **93**: 4779.

395 Herschbach, D. (2009). Molecular collisions, from warm to ultracold. *Faraday Discuss.* **142**: 9.

396 de Miranda, M., Chotia, A., Neyenhuis, B. et al.(2011). Controlling the quantum stereodynamics of ultracold bimolecular reactions. *Nat. Phys.* **7**: 502.

397 Wark, S.J. and Opat, G.I. (1992). An electrostatic mirror for neutral polar molecules. *J. Phys. B* **25**: 4229.

398 Opat, G.I., Wark, S.J., and Cimmino, A. (1992). Electric and magnetic mirrors and gratings for slowly moving neutral atoms and molecules. *Appl. Phys. B* **54**: 396.

399 Gilijamse, J.J., Hoekstra, S., van de Meerakker, S.Y.T. et al. (2006). Near-threshold inelastic collisions using molecular beams with a tunable velocity. *Science* **313**: 1617.

400 Von Zastrow, A., Onvlee, J., Vogels, S.N. et al. (2014). State-resolved diffraction oscillations imaged for inelastic collisions of NO radicals with He, Ne and Ar. *Nat. Chem.* **6**: 216.

401 Onvlee, J., Gordon, S.D., Vogels, S.N. et al. (2016) Imaging quantum stereodynamics through fraunhofer scattering of NO radicals with rare-gas atoms. *Nat. Chem.* **9**: 226.

402 Vogels, S.N., Onvlee, J., Chefdeville, S. et al. (2015). Imaging resonances in low-energy NO-He inelastic collisions. *Science* **350**: 787.

403 Vogels, S.N., Onvlee, J., von Zastrow, A. et al. (2014). High-resolution imaging of velocity-controlled molecular collisions using counterpropagating beams. *Phys. Rev. Lett.* **113**: 263202.

404 Kirste, M., Wang, X., Schewe, H.C. et al. (2012). Quantum-state resolved bimolecular collisions of velocity-controlled OH with NO radicals. *Science* **338**: 1060.

405 Patterson, D. (2016). Decelerating and trapping large polar molecules. *ChemPhysChem* **17**: 3790.

406 Kozyryev, I., Baum, L., Matsuda, K., and Doyle, J.M. (2016). Proposal for laser cooling of complex polyatomic molecules. *ChemPhysChem* **17**: 3641.

407 Wu, X., Gantner, T., Zeppenfeld, M. et al. (2016). Thermometry of guided molecular beams from a cryogenic buffer-gas cell. *ChemPhysChem* **17**: 3631.

408 Patterson, D., Tsikata, E., and Doyle, J.M. (2010). Cooling and collisions of large gas phase molecules. *Phys. Chem. Chem. Phys.* **12**: 9736.

409 Piskorski, J., Patterson, D., Eibenberger, S., and Doyle, J.M. (2014). Cooling, spectroscopy and non-sticking of trans-stilbene and nile red. *ChemPhysChem* **15**: 3800.

410 Griesmaier, A., Werner, J., Hensler, S. et al. (2005). Bose-Einstein condensation of chromium. *Phys. Rev. Lett.* **94**: 160401.

411 Lu, M., Burdick, N.Q., Youn, S.H., and Lev, B.L. (2011). Strongly dipolar Bose-Einstein condensate of dysprosium. *Phys. Rev. Lett.* **107**: 190401.

412 Lu, M., Burdick, N.Q., and Lev, B.L. (2012). Quantum degenerate dipolar fermi gas. *Phys. Rev. Lett.* **108**: 215301.

413 Frisch, A., Aikawa, K., Mark, M. et al. (2012). Narrow-line magneto-optical trap for erbium. *Phys. Rev. A* **85**: 051401.

414 Aikawa, K., Frisch, A., Mark, M. et al. (2012). Bose-Einstein condensation of Erbium. *Phys. Rev. Lett.* **108**: 210401.

415 McGuyer, B., McDonald, M., Iwata, G. et al. (2015). Precise study of asymptotic physics with subradiant ultracold molecules. *Nat. Phys.* **11**: 32.

416 Kendrick, B., Hazra, J., and Balakrishnan, N. (2015) The geometric phase controls ultracold chemistry. *Nat. Commun.* **6**: 7918.

417 Wallis, A.O., Gardiner, S., and Hutson, J.M. (2009). Conical intersections in laboratory coordinates with ultracold molecules. *Phys. Rev. Lett.* **103**: 083201.

418 Moore, M. and Vardi, A. (2002). Bose-enhanced chemistry: amplification of selectivity in the dissociation of molecular Bose-Einstein condensates. *Phys. Rev. Lett.* **88**: 160402.

419 Cui, J. and Krems, R.V. (2015). Gaussian process model for collision dynamics of complex molecules. *Phys. Rev. Lett.* **115**: 073202.

420 Cui, J., Li, Z., and Krems, R.V. (2015). Gaussian process model for extrapolation of scattering observables for complex molecules: from benzene to benzonitrile. *J. Chem. Phys.* **143**: 154101.

421 Leitner, D.M. and Wolynes, P.G. (1996). Many-dimensional quantum energy flow at low energy. *Phys. Rev. Lett.* **76**: 216.

422 Bloch, I., Dalibard, J., and Nascimbene, S. (2012). Quantum simulations with ultracold quantum gases. *Nat. Phys.* **8**: 267.

423 Bandrauk, A.D., Fillion-Gourdeau, F., and Lorin, E. (2013). Atoms and molecules in intense laser fields: gauge invariance of theory and models. *J. Phys. B* **46**: 153001.

424 http://www.wolfram.com/mathematica.

425 Devanathan, V. (2002). *Angular Momentum Techniques in Quantum Mechanics*. New York: Kluwer Academic Publishers.

Index